APS
Advances in Pharmacological Sciences

Temperature Regulation

**Recent Physiological and
Pharmacological Advances**

Edited by
A.S. Milton

Birkhäuser Verlag
Basel · Boston · Berlin

Editor:

Professor A.S. Milton
Division of Pharmacology
University of Aberdeen
Marischal College
Aberdeen AB9 1AS
Scotland

A CIP catalogue record for this book is available from the Library of Congress,
Washington D.C., USA

Deutsche Bibliothek Cataloging-in-Publication Data
Temperature regulation : recent physiological and
pharmacological advances / ed. by A.S. Milton. – Basel ;
Boston ; Berlin : Birkhäuser, 1994
    (Advances in pharmacological sciences)
    ISBN 3-7643-2992-0 (Basel ...)
    ISBN 0-8176-2992-0 (Boston)
NE: Milton, Anthony S. [Hrsg.]

Wilhelm Siegmund Feldberg
19th November 1900 to 23rd October 1993

'Perhaps it was that I had brought with me a key that would open the doors'

ipse dixit

Contents

Preface XIII

Effect of peripheral corticotrophin-releasing factors on febrile responses
N. G. N. Milton 01

Fever induction by a cytokine network in guinea pigs: The roles of tumour necrosis factor
and interleukin-6
J. Roth, M. Bock, J. L. McClellan, M. J. Kluger and E. Zeisberger 11

Cytokines in endotoxin tolerance
E. Zeisberger, K. E. Cooper, M. J. Kluger, J. L. McClellan and J. Roth 17

Effects of immunomodulators and protease inhibitors on fever
P. P. Murzenok and V. N. Gourine 23

Endogenous steroids limit the magnitude of febrile responses
J. Davidson 29

The role of heat shock proteins (HSPs) and interleukin-1 interaction in suppression of fever
M. Kosaka, J. M. Lee, G. J. Yang, T. Matsumoto, K. Tsuchiya, N. Ohwatari
and M. Shimazu 35

Body temperature elevation per se induces the late phase syndrome
A. A. Romanovsky and C. M. Blatteis 41

The absence of fever in rat malaria is associated with increased turnover of
5-hydroxytryptamine in the brain
M. J. Dascombe and J. Y. Sidara 47

Prostaglandins and the acute phase immune response
D. Rotondo 53

Pryogenic immunomodulator stimulation of a novel thymic corticotrophin releasing factor
in vitro
N. G. N. Milton, E. Swanton and E. W. Hillhouse 59

Apparent dissociation between lipopolysaccharide-induced intrapreoptic release of
prostaglandin E_2 and fever in guinea pigs
M. Szekély, E. Sehic, V. Menon and C. M. Blatteis 65

Thermoregulatory responses of rabbits to combined heat exposure, pyrogen
and dehydration
C. M. Blatteis, A. L. Ungar and R. B. Howell 71

Fever and the Organum vasculosum laminae terminalis: Another look
W. S. Hunter, E. Sehic and C. M. Blatteis 75

Lipopolysaccharide (LPS)-induced *fos* expression in the brains of febrile rats
A. Oladehin, J. A. Barriga-Briceno and C. M. Blatteis *81*

Analysis of body temperature and blood protein in hypothermic syrian hamsters and rats
N. Ohwatari, M. Yamauchi, M. Shimazu, T. Matsumoto, K. Pleschka and M. Kosaka . . . *87*

Neurophysiology of thermoregulation: Role of hypothalamic neuronal networks
J. A. Boulant *93*

Effect of 5-HT receptor agonists or antagonists on hypothalamic 5-HT release or colonic temperature in rats
M. T. Lin and H. J. Liu *103*

Temperature sensitivity of rat spinal cord neurons recorded *in vitro*
H. A. Schmid, U. Pehl and E. Simon *109*

The effect of ambient temperature on the modulation of thermoregulatory mechanisms by selective opioid peptides
C. M. Handler, T. C. Piliero, E. B. Geller and M. W. Adler *115*

Regulation of body temperature: Involvement of opioid and hypothalamic GABA
S. Ghosh and M. K. Poddar *121*

Bilateral difference in tympanic temperatures reflects that in brain temperature
T. Ogawa, J. Sugenoya, N. Ohnishi, K. Imai, T. Umeyama, M. Nishida, Y. Kandori and A. Ishizuka *127*

Temperature interhemispheric brain asymmetry as a sign of functional activity
I. K. Yaitchnikov and V. S. Gurevitch *133*

Cerebral and related temperatures in normothermic subjects
Z. Mariak, J. Lewko, M. Jadeszko and H. Dudek *139*

Non-thermometric means of assessing changes of brainstem temperature: The question of selective brain cooling in humans
T. Langer, B. Nielsen and C. Jessen *145*

Effects of selective α-adrenergic blockade on control of human skin blood flow during exercise
W. L. Kenney *151*

Acclimation to 3 different climates with the same wet bulb globe temperature
B. Griefahn and P. Schwarzenau *159*

Sensible heat loss after systemic anticholinergic treatment
M. A. Kolka, L. A. Stephenson and R. R. Gonzalez *165*

Neuropeptide-Y (NPY) reduces cutaneous microcirculatory blood flow and increases total blood flow in the rat tail
M. E. Heath and J. R. Thomas *171*

Effect of pretreatment with delta-9-tetrahydrocannabinol on the ability of certain cannabimimetic agents to induce hypothermia in mice
R. G. Pertwee and L. A. Stevenson 177

Central motor command affects the sweating activity during exercise
N. Ohnishi, T. Ogawa, J. Sugenoya, K. Natsume, Y. Yamashita, R. Imamura and K. Imai . 183

Selective brain cooling in the horse during exercise
F. F. McConaghy, J. R. S. Hales and D. R. Hodgson 189

Skin blood flow during severe heat stress: Regional variations and failure to maintain maximal levels
J. R. S. Hales, B. Nielsen and M. Yanase 195

Heat transfer via the blood
J. Werner and H. Brinck 201

Fall in body core temperature during the previous heat exposure time in rats after subjection to heat loads at a fixed time daily
O. Shido, S. Sakurada and T. Nagasaka 207

Treatment of immersion hypothermia by forced-air warming
G. G. Giesbrecht, M. Schroeder and G. K. Bristow 213

Factorial effects on contact cooling
F. Chen, H. Nilsson and I. Holmér 219

Convective and metabolic heat in human fingers
M. B. Ducharme and P. Tikuisis 223

From foetus to neonate: Implications for the ontogeny of thermoregulation
H. Laburn, D. Mitchell and K. Goelst 229

Effects of several factors on the enlargement of brown adipose tissue
H. Yamashita, T. Ookawara, T. Kizaki, M. Yamamoto, Y. Ohira, T. Wakatsuki, Y. Sato and H. Ohno 241

Thermal preference behaviour following β_3-agonist stimulation
H. J. Carlisle, S. Rothberg and M. J. Stock 247

Sympathetic tone and noradrenaline responsiveness of brown adipocytes from rats with high levels of sexual steroids
M. Puerta 253

The tail of the rat in temperature regulation: effect of angiotensin II
M. J. Fregly, N. E. Rowland and J. R. Cade 261

Selective vulnerability of rat hippocampus in heat stress
H. S. Sharma, J. Westman, F. Nyberg, C. Zimmer, J. Cervós-Navarro and P. K. Dey 267

Development of temperature regulation in precocial chicks: Patterns in shorebirds and ducks
G. H. Visser and R. E. Ricklefs *273*

Central venous pressure and cardiovascular responses to hyperthermia
T. Morimoto, A. Takamata and H. Nose *279*

Effects of bright and dim light intensities during daytime upon circadian rhythm of core temperature in man
H. Tokura, M. Yutani, T. Morita and M. Murakami *285*

Long term heat acclimation: Acquired peripheral cardiovascular adaptations and their stability under multifactorial stressors
M. Horowitz, W. Haddad, M. Shochina and U. Meiri *291*

Effects of solar radiation and feed quality on heart rate and heat balance parameters in cattle
A. Brosh, S. Fennell, D. Wright, G. Beneke and B. Young *297*

Cytological changes in brown adipose tissue of lean and obese mice: Acclimation to mild cold with and without a warm refuge
D. Challoner, S. McBennett, J. F. Andrews and M. E. Jakobson *303*

Manipulation of brown adipose tissue development in neonatal and postnatal lambs
M. E. Symonds, J. A. Bird, L. Clarke, C. J. Darby, J. J. Gate and M. A. Lomax *309*

Resting muscle: A source of thermogenesis controlled by vasomodulators
M. G. Clark, E. Q. Colquhoun, K. A. Dora, S. Rattigan, T. P. D. Eldershaw, J. L. Hall, A. Matthias and J.-M. Ye *315*

Thyroid status modulates hypothalamic thermosensitivity, vasopressin and corticosteroid secretion in rabbits
R. Keil, W. Riedel and E. Simon *321*

Role of prolactin in brown adipose tissue thermogenic activity
T. Yahata and A. Kuroshima *327*

Biology of adaptive heat production: Studies on brown adipose tissue
P. Trayhurn *333*

Brown adipose tissue: Receptors and recruitment
J. Nedergaard and B. Cannon *345*

High-energy food supplement, energy substrate mobilization and heat balance in coldexposed humans
A. L. Vallerand and I. Jacobs *351*

Emerging themes in thermoregulation and fever
K. E. Cooper and W. L. Veale *357*

Author Index *369*

Keyword Index *371*

Preface

Many advances have been made in the field of thermoregulation in the past few years. These include our understanding of Fever, which is now considered not simply a rise in deep body temperature following infection, but just one aspect, though perhaps the most easily measured, of the Acute Phase of the Immune Response. Classification and identification of the Cytokines and the availability of recombinant material has greatly aided this research. Similarly, our understanding of the Hypothalamo-Pituitary Adrenal Axis has altered our way of thinking about temperature regulation. Of importance are the problems associated with adverse climatic conditions and survival, and the problems encountered by the neonate and the hibernator. At the biochemical level, our knowledge of the control of heat production and the role of brown adipose tissue is rapidly advancing. All these issues and many others were discussed at a Symposium 'Thermal Physiology 1993' held in Aberdeen, Scotland in August 1993 under the auspices of the Thermal Physiology Commission of the International Union of Physiological Sciences.

Six main aspects of the subject of temperature regulation are included in this book, namely, Fever (including the Acute Phase of the Immune Response and Thermoregulatory Peptides), Neurophysiology of Thermoregulation, Neonatal Thermoregulation, Mechanisms of Heat Production, Ecological and Behavioural Thermoregulation, and Emerging Themes in Thermoregulation.

This monograph consists of five comprehensive review chapters written by the principal contributors, namely N.G.N. Milton, Newcastle, England; J.A. Boulant, Columbus Ohio, USA; H.P. Laburn, Johannesburg, South Africa; P. Trayhurn, Aberdeen, Scotland, and K.E. Cooper, Calgary, Canada, together with fifty-two selected short contributions, covering all aspects of temperature regulation.

I hope that this Monograph will interest all those working in the field of Body Temperature Regulation, will provide them with a comprehensive survey of the progression in this area over the past few years and will stimulate them to delve deeper.

I am extremely grateful to all the authors for preparing their manuscripts with great care and sending them to me in such good time. I am indebted to my secretary, Miss Nicky Portz, for typing the introduction, the contents and indices and also for retyping some of the manuscripts.

A.S. Milton
Aberdeen, Scotland
October 1993

Postscript

On 23rd October 1993 Professor Wilhelm Feldberg died at his home in London, at the age of ninety-two. His research into the Inner Surfaces of the Brain, in particular into the actions of Drugs on Body Temperature Regulation, is known to all who worked in this field. He was a colleague and friend to many of us and to his memory this book is dedicated.

EFFECT OF PERIPHERAL CORTICOTROPHIN-RELEASING FACTORS ON FEBRILE RESPONSES.

N.G.N. Milton

Department of Clinical Biochemistry,
University of Newcastle-upon-Tyne, UK.

SUMMARY:

The hypothalamo-pituitary-adrenocortical (HPA) axis is regulated by a corticotrophin-releasing factor (CRF) complex which includes corticotrophin-releasing factor-41 (CRF-41) and arginine vasopressin (AVP). The release of both these neuropeptides is activated by pyrogens and when administered centrally both have antipyretic actions during fever. Endogenous CRF-41 and AVP have also recently been shown to have peripheral pro-pyretic actions, mediated via actions on prostanoid generation. This suggests that the CRFs have multiple actions which contribute to the pathophysiology of the febrile response.

FEVER AND ACTIVATION OF THE HYPOTHALAMO-PITUITARY-ADRENOCORTICAL AXIS:

The changes in body temperature during fever are caused by changes in both heat gain and heat loss mechanisms and appear to be regulated via actions at hypothalamic thermoregulatory centres. Fever is associated with the activation of endogenous pyrogens such as interleukin-1 (IL-1), tumour necrosis factor (TNF) and interferons (INF). The febrile response is dependent on prostaglandin (PG) production and central administration of prostanoids will cause fever[1]. Pyrogens will stimulate increases in both circulating blood and cerebrospinal fluid PG levels. Since pyrogens such as lipopolysaccharide (LPS) and IL-1 appear unable to cross the blood-brain barrier it is thought that peripheral PGs play a major role in the febrile response to peripheral pyrogens[2].

The febrile response can be activated by both exogenous and endogenous pyrogens and constitutes a stressful stimulus. The hypothalamo-pituitary-adrenocortical (HPA) axis is activated by stress to maintain homeostasis and regulated at the hypothalamic level by a corticotrophin-releasing factor (CRF) complex. Two of the major hormonal components of this CRF complex are corticotrophin-releasing factor-41 (CRF-41) and arginine vasopressin (AVP). The release of both CRF-41 and AVP from the hypothalamus into the hypophysial-portal vasculature is under neurotransmitter control[3]. Activation of the pro-opiomelanocortin (POMC) gene and release of adrenocorticotrophin (ACTH) plus other POMC-gene products are regulated, in the anterior-pituitary gland, by a synergistic interaction between CRF-41 and AVP[4]. The end products of HPA-axis activation, the adrenal glucocorticoid hormones and anterior pituitary hormones, are well known immunosuppressive agents which can also inhibit febrile responses.

Evidence for HPA-axis activation by pyrogens includes the activation of ACTH and corticosterone release in rats by both LPS and IL-1. The pyrogenic INF inducer polyinosinic:polycytidylic acid [poly (I):poly (C)] has also been shown to activate the rabbit HPA-axis *in vivo*[5] and this response has a CRF-41-dependent component, an AVP-independent component and a PG-independent component. Changes in hypothalamic immunoreactive (ir)-CRF-41 levels have also been observed during febrile challenge with leukocyte pyrogen in rabbits[6]. The site and mechanism of HPA-axis activation by pyrogens remains a controversial area of research. The failure of some pyrogens to cross the blood brain barrier suggests that an induced messenger may be involved, both in activation of hypothalamic thermoregulatory centres and the hypothalamic components of the HPA-axis.

Both irCRF-41 and irAVP have been found to be widely distributed throughout the central nervous system (CNS) and peripheral tissues as have specific receptor binding sites for these neuropeptides. The ACTH releasing activity of CRF-41 and AVP are not their sole physiological actions and both these neuropeptides have been shown to have cardiovascular and immunoregulatory actions. CRF-41 has been shown to peripherally mediate stimulation of lymphocytes[7] and AVP can regulate γ-INF production by immune cells[8]. These actions of CRF-41 and AVP on the immune system suggest that both neuropeptides have peripheral actions on the immune system which could, in turn, influence febrile responses. Combined with there abilities to induce other hormones and their cardiovascular activities this suggests that CRF-41 and AVP may either directly or indirectly play multiple roles in the febrile responses to pyrogens.

ACTIONS OF CRF'S DURING THE FEBRILE RESPONSE:
(1) CORTICOTROPHIN-RELEASING FACTOR-41:

The role of CRF-41 in fever has proved a controversial area due to the generation of apparently conflicting sets of data by a number of groups. Much of the apparent confusion, however, has arisen due to methodological and species differences. The multiple actions of CRF-41 *in vivo* have also undoubtedly contributed to the problem. The methodology employed to study the actions of endogenous CRF-41 has used antibodies or a CRF-41 receptor antagonist (α-helical CRF 9-41). In the case of antibody experiments the interpretation is complicated by actions of the foreign immunoglobulin in the test animal. The CRF-41 receptor antagonist has also been shown to be a partial agonist in some systems and the frequent use of high doses to block endogenous actions of CRF-41 may therefore lead to misinterpretation of observations. These problems notwithstanding there have been a number of interesting studies suggesting roles for CRF-41 in fever.

The hypothalamic origin of CRF-41 and the low circulating levels of the hormone have led to a number of studies to determine the central actions of this neuropeptide. In rabbits central administration of CRF-41 has been shown to have antipyretic actions[9]. This observation has been repeated[10] and the current suggestion is that central CRF-41 may also have a second body temperature raising action. The failure of peripherally administered CRF-41 to effect pyrogen elevated body temperature provides further evidence for a central antipyretic action in rabbits. In the rat, studies have suggested that central CRF-41 has a pro-pyretic action[11], however, unlike the experiments in rabbits the pyrogen was administered centrally and the observed fever may therefore be controlled via different regulatory mechanisms. The comparison of anti-CRF-41 antibody with a saline control[11] may not be appropriate, since control antibodies can have effects, and adds doubt to the interpretation of these results in rats. Observations of non-specific fevers and actions due to foreign immunoglobulin are numerous and clearly antipyretic actions of the antibody may not result from its ability to neutralise CRF-41. The use of high doses of CRF-41 receptor antagonist also complicates the interpretation of the actions of endogenous central CRF-41 in fever. The observation that CRF-41 administered centrally can increase body temperature in rabbits but not potentiate a pyrogen induced fever[10] is supportive of a pro-pyretic role for central CRF-41, however, in the same study central CRF-41 was shown to be antipyretic. Thus central actions of CRF-41 remain controversial and in reality the peptide appears to have multiple actions.

Central CRF-41 also has cardiovascular and immunoregulatory actions and these may be responsible for some of the *in vivo* actions attributed to this peptide during fever. Similarly CRF-41 can stimulate the release of other peptides from CNS tissue. Central ACTH, β-endorphin and α-MSH have both been suggested as thermoregulatory peptides and their secretion in response to CRF-41 could also play a role in the observed responses. The ability of CRF-41 to activate pituitary release of PGE_2 suggests that the antipyresis caused by antagonism of central CRF-41[11] could be the result of interference in PG generation. It has been suggested that such CRF-41 antagonism inhibits the effects of $PGF_{2\alpha}$, but not PGE_2, on body temperature[12]. The observation that $PGF_{2\alpha}$, but not PGE_2, stimulates the release of CRF-41 combined with the observations of CRF-41 induced PGE_2 release[13] raises the possibility that $PGF_{2\alpha}$ acts via CRF-41 to activate PGE_2 release and that the consequent increases in body temperature are due to the

PGE$_2$ end product. However, PGF$_{2\alpha}$ is pyrogenic in the presence of cyclo-oxegenase inhibitors, suggesting that it does not require another prostanoid to elicit its effects *in vivo*.

The hormonal nature of CRF-41, with its release into the hypophysial portal vasculature, suggests that peripheral CRF-41 could also play a role in fever. The immunoregulatory actions of peripheral CRF-41 and the identification of peripheral CRF-41 receptors further strengthen this hypothesis. It has recently been shown that peripheral administration of the pyrogenic INF inducer poly (I):poly (C) activates the HPA-axis, febrile responses and circulating PGs in a CRF-41 dependent manner in rabbits. A peripheral site of action for endogenous CRF-41 during the febrile response is suggested by the following observations[14]:

(1) Peripherally administered anti-CRF-41 antibodies or low doses of the CRF-41 receptor antagonist are antipyretic.

(2) Centrally administered CRF-41 receptor antagonist has no effect on the febrile response.

(3) Peripheral administration of anti-CRF-41 antibody after the onset of temperature rises causes an immediate defervesence.

The failure of the centrally administered CRF-41 antagonist to inhibit fever contrasts with observations of others for fevers induced by central IL-1β in rats[11]. The doses of CRF-41 antagonist administered both peripherally and centrally were, however, significantly lower than that used by other groups. The low dose of CRF-41 antagonist administered peripherally was able to reduce an basal circulating cortisol levels, but unable to significantly effect poly (I):poly (C) stimulated cortisol. The high dose of antagonist administered centrally by other research groups may be sufficient to reach peripheral receptors and their results could therefore be a result of a peripheral action of CRF-41 rather than the suggested central action. The use of different pyrogens, routes of administration and species differences will also contribute to the observations and may explain any apparent contradictions.

Peripherally administered pyrogens have been suggested to cause fever via the stimulation of peripheral PG release[15]. CRF-41 can stimulate the release of PGs *in vitro*[13] and the proposed mechanism of action of endogenous peripheral CRF-41 in maintaining a febrile response is via an action on the pyrogen induced PG generation[16]. We have observed that peripheral antagonism of CRF-41 causes a reduction in the circulating levels of both PGE$_2$ and PGF$_{2\alpha}$ stimulated by poly (I):poly (C). The antagonism of PG release can be observed in the presence of either the CRF-41 receptor antagonist or anti-CRF-41 antibodies, with the effects much more apparent on PGF$_{2\alpha}$ than PGE$_2$. The failure of peripherally administered CRF-41 to cause a fever itself indicates that the mode of action may require the presence of another pyrogenic substance.

Peripheral CRF-41 also has other actions which can contribute to modification of the febrile response[17]. Peripheral administration of high doses of CRF-41 will stimulate increases in circulating PGE$_2$ and PGF$_{2\alpha}$. However, the high dose of CRF-41 used in these experiments does not itself raise body temperature or potentiate a poly (I):poly (C) induced fever. Instead this high dose of CRF-41 itself causes a drop in body temperature and also a decrease in the pyrogen stimulated fever. The mechanism for this action of CRF-41 appears to be cardiovascular, since a marked vasodilation is caused by the CRF-41. These hypothermic effects and vasodilation are only caused by high doses of CRF-41, outwith the normal physiological range, suggesting that this action of CRF-41 may have a different physiological basis. Alternatively other hormones activated by CRF-41 could play a role. Two hormones activated by CRF-41, ACTH and α-MSH, have been suggested as endogenous antipyretics. However, both peptides have also been shown to effect normal thermoregulatory processes. The observed hypothermia caused by CRF-41 is similar to that seen with ACTH 1-24 and α-MSH[18]. The apparent antipyresis observed when high doses of CRF-41 are administered peripherally is also accompanied by vasodilation and may therefore be a direct result of the well known cardiovascular actions of CRF-41.

The site of production of centrally acting CRF-41 is undoubtedly contained within the CNS where significant concentrations of the peptide have been found and shown to be released. Fever specific changes in hypothalamic CRF-41 content have been observed and further strengthen this suggestion. However, the circulating blood levels of CRF-41 are extremely low and although marked changes have been observed in the hypophysial portal vasculature following stress, little hypothalamic CRF-41 appears to enter the general circulation. Peripheral sites of immunoreactive CRF-41 production have been suggested but few studies have been able to accurately confirm that material is authentic CRF-41. Recently a thymic CRF has been observed[19]. Characterization of thymic CRF has revealed a high molecular weight form of CRF which is chromatographically distinct from hypothalamic CRF-41 but shares immunological and biological characteristics with CRF-41. Of particular interest in the field of thermoregulation, and in particular fever, is the observation that the synthesis and release of thymic CRF can be activated by pyrogenic INF inducers or the endogenous pyrogen INF.

These observations suggest that both central and peripheral CRF-41 may have multiple actions during fever. Peripheral CRF-41 has both pro-pyretic and apparently antipyretic activities with the former mediated via stimulation of prostanoids and the later via cardiovascular actions. In the case of the apparent antipyretic actions of peripheral CRF-41 the mechanisms of action suggest that this is not so much an antipyresis as an interaction with the general heat loss mechanisms.

(2) ARGININE VASOPRESSIN:

Studies have suggested roles for AVP as an endogenous antipyretic[20]. This action appears to be mediated centrally since in sheep peripheral administration of AVP had no effect on LPS induced fever whilst central administration in the septal region had profound antipyretic actions[21].

The marked secretion of AVP into this septal region during fever[21] further suggests that endogenous AVP may have such an action during fever. The enhancement of LPS fever by perfusion of the rabbit septum with AVP antiserum provides a further substantiation[22] and confirms the role of endogenous AVP. The ability of centrally injected AVP, but not oxytocin or a behaviourally active AVP fragment, to suppress LPS fever in rats[23] further strengthens the suggested role for AVP. The failure of intra-hypothalamic and septal injections of AVP to reduce fever in febrile rabbits[24] points to the controversy regarding the role of AVP in fever. The observations, however, may be due to differences in the precise site of administration, doses of pyrogen or doses of AVP. The precise role is also complicated by the cardiovascular, anti-diuretic and ACTH inducing activities of this substance, all of which potentially could influence the febrile response. The ability of an AVP V_1-receptor antagonist to significantly potentiate an IL-1 fever in rats points to the antipyretic action of AVP requiring central receptors of this subtype[25]. A recent report showed that antagonism of central AVP V_1-receptors in the ventral septal area of rats prevented the antipyretic action of indomethacin[26], an observation which led the authors to question the ability of PG synthesis inhibition by this drug as the mechanism for its antipyretic activity. In the same study a V_2-receptor antagonist had no effect. The studies showing salicylate blockade of PGE_1 induced fevers[27] also call into question the mechanism of antipyresis caused by this substance and have been suggested to also involve AVP. Thus, AVP has been proposed as both an antipyretic substance and the mediator of antipyresis induced by some cyclo-oxygenase inhibitors, results which clearly await further characterization to substantiate the precise mechanisms of AVP action in fever.

Peripheral levels of AVP are increased by pyrogens such as LPS[28]. Interestingly dehydration, which also activates peripheral AVP release, has been shown to have an antipyretic effect during LPS fever[29]. This action is, however, thought to be mediated via activation of central AVP since centrally administered anti-AVP antiserum enhances the fever in dehydrated guinea pigs[29]. A recent report has shown that a peripherally administered AVP V_1-receptor antagonist has antipyretic effects in rabbits[30]. This is accompanied by a change in the pattern of a poly (I):poly (C) fever from a biphasic rise in body temperature to a monophasic rise in body temperature in the presence of the V_1 antagonist. These observations raise a number of questions as to the actions of peripheral AVP during fever.

The apparent drop in the poly (I):poly (C) induced core body temperature which occurs after 2.5 h to give a biphasic temperature rise pattern could be due to stimulation of AVP having a transient antipyretic effect. The consistency of this observation and the failure of the V_1 antagonist to effect body temperature itself, suggests that the antagonist is interfering with a poly (I):poly (C) activated response and not some thermoregulatory mechanism. The large number of previous observations of AVP induced antipyresis, combined with the observation that this may be mediated by an AVP V_1-receptor[25] are consistent with the suggestion that the biphasic temperature change pattern is due to transient AVP induced antipyresis. However, the lag phase and speed of the initial rises in body temperature are altered by the V_1 antagonist. The increased lag phase and reduced rate of temperature increase suggests that the antagonist may in fact be inhibiting the onset of fever. This in turn suggests that AVP in this situation has a pro-pyretic role. The peripheral route of administration, whilst not guaranteed to have peripheral effects, is less likely to have central actions and it could be suggested that this pro-pyretic action of AVP has a peripheral site of action. This is in agreement with the observations that peripheral AVP can effect cytokine production[8], which in turn may modify febrile responses regulated by pyrogenic endogenous cytokines.

The mechanism of action of peripheral AVP in maintaining a febrile response appears, like that of peripheral CRF-41, to be mediated via actions on pyrogen activated PG generation. In the case of AVP V_1-receptor antagonism the main effect appears to be a reduction in the poly (I):poly (C) stimulated PGE_2 rises to virtually those levels seen in saline treated control experiments. The rises in $PGF_{2\alpha}$ are reduced but not so dramatically. Observations of AVP induced PGE_2 release from the anterior pituitary gland *in vitro*[13] and stimulatory effects of PGE_2 on AVP release *in vitro* raise the possibility of positive feedback loops involving AVP and PGE_2. The failure of the observed PGE_2 antagonism to significantly antagonise the febrile response brings into question the suggestions that poly (I):poly (C) stimulated peripheral PGE_2 is responsible for the activation of the hypothalamic thermoregulatory centres and the resultant fever[15]. However, in the rabbit $PGF_{2\alpha}$ is also pyrogenic[1] and, whilst this prostanoid has a much lower potency[1] (1/27th the activity of PGE_2), the circulating levels of $PGF_{2\alpha}$ are sufficiently high to suggest that this prostanoid could fulfil the role of a peripheral fever inducing prostanoid during these responses.

The *in vivo* observations reflect the net result of multiple effects on both the responses studied and other related and unrelated responses. As a result the conclusions drawn from such studies, whilst limited to the actual observations, must take into account these other effects. The role of AVP V_1-receptors in the stimulation of blood vessel vasoconstriction and antipyresis could produce both potentiating and inhibitory effects on the poly (I):poly (C) stimulated body temperature and/or PGE_2 responses causing the net changes observed.

CONCLUSIONS:

The studies of the roles of CRFs during fever suggest that both CRF-41 and AVP have multiple actions during the febrile response. These neuropeptides also appear to have different effects which are dependent on their site of action. The observations that, in rabbits, both peptides have pro-pyretic actions in the periphery, mediated via actions on pyrogen activated prostanoid generation, suggests that there may be a CRF-Prostanoid axis (Fig 1). The *in vitro* observations that both CRF-41 and AVP can stimulate release of PGs, from a number of tissues in a range of species, suggest that this axis may mediate the regulation of host responses during stress and immune system activation as well as the febrile response.

Fig 1 - THE CRF-PROSTANOID AXIS

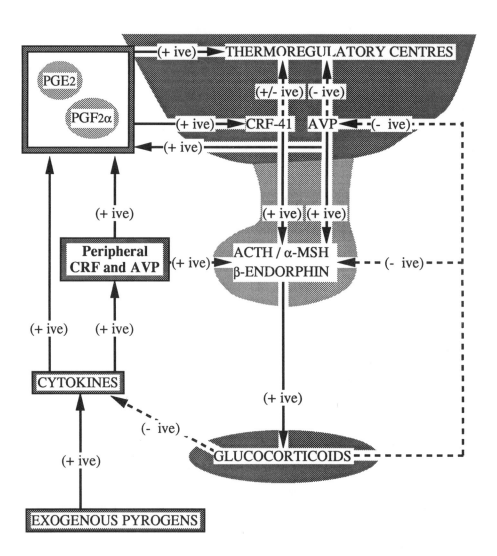

REFERENCES:

[1]   Milton AS, Wendlandt S. Effects on body temperature of prostaglandins of the A, E and F series on injection into the third ventricle of unanaesthetized cats and rabbits. J Physiol (Lond) 1971; 218: 325-36.

[2]   Milton AS. Prostaglandins in fever and the mode of action of antipyretic drugs. In Milton AS (Ed) Pyretics and Antipyretics, Handbook of Experimental Pharmacology , Vol 60, Springer Verlag, Berlin. pp 257-303, 1982.

[3]   Hillhouse EW, Milton NGN. Neurotransmitter regulation of CRF-41 secretion *in vitro*. In Rose FC (Ed) The control of the hypothalamo-pituitary adrenocortical axis, Int Univers Press, Madison, CT, USA. 1989, pp 243-60.

[4]   Gillies GE, Linton EA, Lowry PJ. Corticotropin releasing activity of the new CRF is potentiated several times by vasopressin. Nature 1982; 299: 355-7.

[5]   Milton NGN, Hillhouse EW, Milton AS. Activation of the hypothalamo-pituitary-adrenocortical axis in the conscious rabbit by the pyrogen polyinosinic:polycytidylic acid is dependent on corticotrophin releasing factor-41. J Endocrinol 1992; 135: 69-75.

[6]   Holdeman M, Khorram O, Samson WK, Lipton JM. Fever specific changes in central MSH and CRF concentrations. Am J Physiol 1985; 242: R125-9.

[7]   McGillies JP, Park A, Ruben-Fletter P, Turck C, Dallman MF, Payan DG. Stimulation of rat B-lymphocyte proliferation by corticotropin-releasing factor. J Neurosci Res 1989; 23: 346-52.

[8]   Johnson HM, Torres BA. Regulation of lymphokine production by arginine vasopressin and oxytocin: modulation of lymphocyte function by neurohypophyseal hormones. J Immunol 1985; 135: 743s-45s.

[9]   Bernardini GL, Richards DB, Lipton JM. Antipyretic effect of centrally administered CRF. Peptides 1984; 5: 57-9.

[10]   Opp M, Obal FJr, Krueger JM. Corticotropin releasing factor attenuates interleukin-1-induced sleep and fever in rabbits. Am J Physiol 1989; 257: R528-35.

[11]   Rothwell NJ. CRF is involved in the pyrogenic and thermogenic effects of interleukin 1β in the rat. Am J Physiol 1989; 256: E111-5.

[12]   Rothwell NJ. Central activation of thermogenesis by prostaglandins: dependence on CRF. Horm Metab Res 1990; 22: 616-8.

[13]   Vlaskovska MA, Hertting G, Knepel W. Adrenocorticotropin and β-endorphin release from rat adenohypophysis *in vitro*: inhibition by prostaglandin E2 formed locally in response to vasopressin and corticotrophin-releasing factor. Endocrinology 1984; 115: 895-903.

[14]   Milton NGN, Hillhouse EW, Milton AS. A possible role for peripheral corticotrophin-releasing factor-41 in the febrile response of conscious rabbits. J Physiol (Lond) 1993; 465: 419-30.

[15]   Rotondo D, Abul HT, Milton AS, Davidson J. Pyrogenic immunomodulators increase the level of prostaglandin E2 in the blood simultaneously with the onset of fever. Eur J Phamacol 1988; 154: 145-52.

[16]   Milton NGN, Hillhouse EW, Milton AS. Modulation of the prostaglandin responses of conscious rabbits to the pyrogen polyinosinic:polycytidylic acid by corticotrophin-releasing factor-41. J Endocrinol 1993; 138: 7-11.

[17]   Milton AS, Milton NGN. The modulatory role of peripheral corticotrophin-releasing factor (CRF) in the febrile response to polyinosinic:polycytidylic acid (poly-I:C). In Lomax P, Schonbaum E (Eds) Thermoregulation: The Pathophysiological Basis of Clinical Disorders, Karger, Basel. 1992, pp 25-27.

[18]   Lipton JM, Glyn-Ballinger JR, Murphy MT, Zimmer JA, Bernardini G, Samson WK. The central neuropeptides ACTH and α-MSH in fever control. J Therm Biol 1984; 9: 139-43.

[19]   Milton NGN, Swanton E, Hillhouse E. Pyrogenic immunomodulator stimulation of a novel thymic corticotrophin-releasing factor in vitro. (see this volume pp 59-64).

[20]   Veale WL, Cooper KE, Ruwe WD. Vasopressin: its role in antipyresis and febrile convulsion. Brain Res Bull 1984; 12: 161-5.

[21]   Cooper KE, Kasting NW, Lederis K, Veale WL. Evidence supporting a role for endogenous vasopressin in natural suppression of fever in the sheep. J Physiol (Lond) 1979; 295: 33-45.

[22]   Malkinson TJ, Bridges TE, Lederis K, Veale WL. Perfusion of the septum of the rabbit with vasopressin antiserum enhances endotoxin fever. Peptides 1987; 8: 385-9.

[23]   Kovacs GL, De Weid D. Hormonally active vasopressin suppresses endotoxin-induced fever in rats: lack of effect of oxytocin and a behaviourally active vasopressin fragment. Neuroendocrinology 1983; 37: 258-61.

[24]   Bernardini GL, Lipton JM, Clark WG. Intracerebroventricular and septal injections of arginine vasopressin are not antipyretic in the rabbit. Peptides 1983; 4: 195-8.

[25]   Cooper KE, Naylor AM, Veale WL. Evidence supporting a role for endogenous vasopressin in fever suppression in the rat. J Physiol (Lond) 1987; 387: 163-72.

[26]   Wilkinson MF, Kasting NW. Central vasopressin V1-receptors mediate indomethacin-induced antipyresis in rats. Am J Physiol 1989; 256: R1164-8.

[27]   Alexander SJ, Cooper KE, Veale WL. Blockade of prostaglandin E1 hyperthermia by sodium salicylate given into the ventral septal area of the rat brain. J Physiol (Lond) 1987; 384: 223-31.

[28]   Kasting NW, Mazurek MF, Martin JB. Endotoxin increases vasopressin release independently of known physiological stimuli. Am J Physiol 1985; 248: E420-4.

[29]   Roth J, Schulze K, Simon E, Zeisberger E. Alteration of endotoxin fever and release of arginine vasopressin by dehydration in the guinea pig. Neuroendocrinol 1992; 53: 680-6.

[30]   Milton NGN, Hillhouse EW, Milton AS. Does peripheral arginine vasopressin have a role in the febrile response of conscious rabbits? J Physiol (Lond) 1993; 469: 525-534.

Temperature Regulation
Advances in Pharmacological Sciences
© 1994 Birkhäuser Verlag Basel

FEVER INDUCTION BY A CYTOKINE NETWORK IN GUINEA PIGS: THE ROLES OF TUMOR NECROSIS FACTOR AND INTERLEUKIN-6

J. Roth, M. Bock, J.L. McClellan*, M.J. Kluger* and E. Zeisberger
Physiologisches Institut, Klinikum der Justus-Liebig-Universität, Aulweg 129, D-35392
Giessen, Germany and
*Department of Physiology, University of Michigan, Medical School, Ann Arbor, Michigan
48109, U.S.A.

SUMMARY: To investigate the roles of systemic and hypothalamic tumor necrosis factor (TNF) and interleukin-6 (IL-6) in fever, which can be monitored after intramuscular injections of bacterial lipopolysaccharide (LPS) in blood plasma and hypothalamic push-pull perfusates, we tried to simulate the activities of of these cytokines during fever by systemic or intrahypothalamic infusions. Intraarterial infusions of TNF resulted in a 15-fold increase in systemic release of endogenous IL-6 and in a biphasic elevation of body temperature. TNF and IL-6 also proved to be pyrogenic, when microinfused into the anterior hypothalamus. Systemic as well as hypo-thalamic TNF and IL-6 seem thus to participate in the complex process of fever generation.

INTRODUCTION

The febrile increase in body temperature can be regarded as a result of complex interactions between the immune system and that part of the central nervous system where body temperature is regulated, i.e. the anterior hypothalamus. The cascade of events that occur to induce fever after infection with an exogenous pyrogen, for example bacterial LPS, takes place in two compartments, first in the peripheral circulation and then subsequently in the brain. In the circulation LPS acts on membranes of immune competent cells [1], which respond to this stimulus with release of cytokines like IL-1, IL-6 or TNF [2, 3]. The transfer of these febrile signals into the thermoregulatory centers of the brain is not yet understood completely and partly still speculative. It has been proposed that the signal transduction into the brain is mediated by the peripherally released cytokines and occurs within the circumventricular organs, where the blood-brain-barrier is incomplete [4, 5]. Neuronal pathways from the circumventricular organs to the anterior hypothalamus may activate an intrahypothalamic release of substances being finally responsible for the febrile shift of body temperature to a higher level. In a previous investigation in guinea pigs [6] we found that TNF and IL-6 can be detected in increased amounts during LPS induced fever not only in the peripheral circulation but also in hypothalamic push-pull perfusates. Injections of LPS caused a dramatic increase of IL-6-like activity in plasma. The logarithmic values of plasma IL-6 activities showed a linear correlation to the febrile change of body temperature ($r = 0.898$, $p < 0.0001$) during the whole time course of fever. IL-6-like activity in hypothalamic perfusates increased 12-fold in the first hour after pyrogen application, and declined slowly despite the further increase in body temperature. TNF activity in plasma, not detectable before LPS injection, had its peak (680 units/ml) in the first hour after pyrogen application and rapidly declined again to an undetectable value within the next few hours. In the hypothalamus TNF was not detectable before endotoxin application, but it could be monitored in most animals after injection of LPS. To investigate possible roles of systemic and hypothalamic TNF and IL-6 during LPS-induced fever we performed the following two experimental series. In a first series of experiments we tried to simulate the initial rise of TNF in plasma during endotoxin fever and studied the effects on body temperature and systemic levels of TNF, IL-1 and IL-6. In a second series of experiments we simulated the increases of TNF and IL-6 detected in hypothalamic perfusates during LPS fever by slow microinfusions of these cytokines into the anterior hypothalamus and studied the effects on body temperature.

MATERIALS AND METHODS

Animals. The experiments were performed in several groups of guinea pigs. The animals were equipped either with intra-arterial catheters for repeated blood sampling or stereotactically implanted guide cannulae, through which microinfusion cannulae could be introduced into the hypothalamus (for details c.f. [6]).

Measurement of body temperature. Abdominal body temperature was measured by use of battery operated biotelemetry transmitters (VM-FH-discs, Mini-Mitter Co., Sunriver, OR, U.S.A.) implanted intraperitoneally. Output (frequency in Hz) was monitored by a mounted antenna placed under each animal's cage (RA 1000 radioreceivers, Mini-Mitter Co.) and multi-plexed by means of a BCM 100 consolidation matrix to an IBM personal computer system. A Dataquest IV data aquisition system (Data Sciences Inc., St. Paul, MN, U.S.A.) was used for automatic control of data collection and analysis. Body temperatures were monitored and recorded at 5 min intervals.

Experimental procedure. In the first series of experiments 2 µg TNF (specific activity 20000 units/µg) were administered either as intra-arterial bolus injections or slowly infused within 60 min. Plasma samples were collected at selected stages of the experiments. In the second series of experiments microinfusions of TNF or IL-6 into the anterior hypothalamus were performed for 6 h with an infusion speed of 0.1 µl/min. In this time 36 µl of TNF or IL-6 solutions with total specific activities of 1400 units or of sterile pyrogen free 0.9 % NaCl solution were microinfused into the animals' hypothalamus. Endotoxin fever was induced by intramuscular injection of LPS from E. coli (0111:B4, 20µg/kg, Sigma Co., St. Louis, U.S.A.). The integrated areas between the thermal responses of normothermic and febrile animals, the fever index, was expressed in °C·h for 6 h.

Assays. Determination of TNF was based on the cytotoxic effect of TNF on the mouse fibrosarcoma cell line WEHI 164 subclone 13. The percentage of viable cells incubated either with different concentrations of TNF standard or with biological samples were measured by use of the dimethylthiazol-diphenyltetrazolium bromide (MTT) colorimetric assay. Determination of IL-6 was based on the dose dependent growth stimulation of IL-6 on the B 9 hybridoma cell line. After incubation with different concentrations of IL-6 standard or with biological samples, the MTT assay was used to measure the number of living cells (for literature on details c.f. [6]).

RESULTS AND DISCUSSION

Intra-arterial bolus injections as well as infusions within 1 h of 2μg TNF (specific activity: 20000 units/μl) resulted in biphasic elevations of body temperature lasting 6 h, as shown on fig.1.

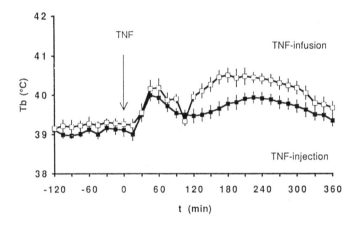

Fig.1: Thermal responses in 2 groups of guinea pigs (N = 8) to intra-arterial bolus injections of 2 μg TNF (black symbols) or intra-arterial infusions of the same amount of TNF within 1h (open symbols). The arrow indicates the time of bolus injections and start of infusions.

At selected stages of the experiment we measured the plasma levels of TNF and IL-6. 1 min after the bolus injection of 2 μg TNF a mean activity of 2976 ± 788 TNF units/ml was detectable in plasma. This value rapidly declined to 255 ± 122 TNF units/ml 60 min after and to 43 ± 15 units/ml 180 min after injection of TNF. At later stages of the experiment no TNF-like activity was detectable in plasma. At corresponding stages of the experiment, IL-6-like activity was 199 ± 45 units/ml 1 min after TNF injection, which significantly increased to 946 ± 147 units/ml 1 h after injection and 959 ± 175 units/ml 3 h after injection of TNF. When 2μg TNF were intra-arterially infused within 1 h, TNF-like activity rose from an undetectable value before the start of the infusion to 339 ± 61 units/ml 15 min after the start of the infusion and further to 438 ± 83 units/ml at the end of the infusion and declined to 56 ± 7 units/ml within the next 2 h. The infusion of TNF caused a much stronger increase in plasma activity of IL-6, which rose from a baseline value of 305 ± 107 units/ml to 4382 ± 696 units/ml at the end of the TNF infusion and 3321 ± 792 units/ml 2 h later (180 min after start of the infusion). 6 h after the start of TNF infusion, almost at the end of the febrile response, IL-6-like activity in plasma was 774 ± 143 units/ml. It seemed, thus, that systemic activity of IL-6 was increased as long as body temperature stayed elevated and that the higher body temperature in TNF-infused than in TNF- injected animals corresponded to the observed higher activity of IL-6 in plasma. The results of these experiments therefore confirm the role of IL-6 as a circulating endogenous pyrogen [3] and the potential of TNF to induce release of endogenous IL-6 in vivo, which has already been observed in vitro [7]. The potential to induce release of IL-6 has also been ascribed to IL-1 [7] and the increase in IL-6-like activity during LPS fever seems to be partly mediated by IL-1 [8]. IL-1-like activity after injection or infusion of TNF could not be detected in our experiments in guinea pigs.

In the second series of experiments we simulated the increased amounts of TNF and IL-6 detectable in hypothalamic push-pull perfusates during LPS-induced fever by intrahypothalamic microinfusions of these cytokines. During the duration of the microinfusions the mean fever index was calculated and compared to the fever index during LPS fevers in the same group of animals. Endotoxin fevers, which lasted about 6 h, had a mean fever index of 6.2 °C·h. During the time of IL-6 microinfusions the mean fever index was 3.6°C·h, in TNF infused animals 5.8 °C·h. TNF fevers developed as quickly as endotoxin fevers and reached the fever maximum at almost the same time (after 3 h). Intrahypothalamic infusions of IL-6 elevated body temperature with a much smaller slope, so that the fever maximum was not reached before the end of the infusion. The results of this series of experiments are summarized on fig.2. Taken these results together, IL-6 and TNF proved to be pyrogenic when microinfused directly into the guinea pigs' hypothalamus. Since during LPS fever highest intrahypothalamic activity of IL-6 ranging from several 100 to several 1000 units/ml perfusate were measured in the first hour after injection of endotoxin, the slow microinfusion of about 5 units IL-6/min was possibly not able to cause the same febrile effect as endogenously released IL-6. How might hypothalamic cytokines like IL-6 and TNF be capable to modify the thermoregulatory set point ? On the one hand hypothalamic cytokines might induce an intrahypothalamic release of prostaglandin E in a similar way as it occurs in the peripheral circulation [9]. Indeed prostaglandin E is regarded by some investigators as the neural mediator of the febrile response [10] and increased levels of

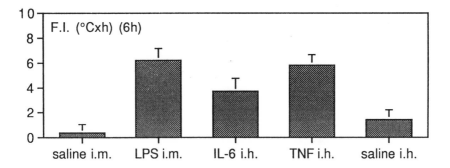

Fig. 2: Thermal responses of a group of guinea pigs (N = 6) to intramuscular injections of saline, 20µg/kg LPS or to intrahypothalamic microinfusions of 1400 units of IL-6 or TNF or sterile pyrogen free saline (duration 6 h, infusion speed 0.1 µl/min). The temperature increase is expressed as the mean fever index (c.f. methods).

prostaglandins have been detected in hypothalamic push-pull perfusates of febrile rats [11]. On the other hand cytokines themselves have been shown to alter the activity of thermosensitive and other neurons. Thus, IL-6 is able to depress the activity of preoptic warm-sensitive neurons and to excite cold sensitive neurons [12], a mechanism which might contribute to the febrile shift of body temperature. Another well documented action of hypothalamic cytokines like IL-1, IL-6 and TNF is the activation of the hypothalamic-pituitary-adrenal axis due to the stimulation of neurons producing corticotropin releasing hormone (CRH) in the hypothalamic paraventricular nucleus. This kind of stimulation results in increased releases of ACTH, other derivatives of the pro-opiomelanocortin, like melanocyte-stimulating hormones, and finally hormones of the adrenal cortex, like cortisol or corticosterone. The activation of this axis during fever has been described and reviewed recently [13] and all the substances released under the influence of CRH have been shown to contribute to a limitation of height and duration of the febrile response. It seems thus, that hypothalamic cytokines activate at the same time mechanism that cause fever and mechanisms that control and finally suppress the febrile response by a negative feedback mechanism aiming to reestablish a normal state in the immune system. As it becomes obvious from the presented results it is not easy to ascribe one special function in this complex process to a certain cytokine, since one cytokine is able to induce the release of another or more cytokines in the peripheral circulation and perhaps also in the central nervous system. Therefore, in the title of this paper, we called it a "cytokine network" that induces fever in our experimental animal model.

ACKNOWLEDGEMENTS

This studies were supported by the DFG (project Ze 183/4-1) and by the NIH grant AI 27556. The excellent technical assistance of Birgit Störr is gratefully acknowledged.

REFERENCES

1. Raetz CR, Ulevitch RJ, Wright, SD, Sibley CH, Ding A, Nathan CF. Gram-negative endotoxin: an extraordinary lipid with profound effects on eukaryotic signal transduction. FASEB J 1991; 5: 2652-2660.

2. Dinarello CA, Cannon JG, Wolff SM. New concepts on the pathogenesis of fever. Rev Infect Dis 1988; 10: 168-189.

3. Kluger MJ. Fever: Role of pyrogens and cryogens. Physiol Rev 1991; 71: 93-127.

4. Blatteis CM, Bealer SL, Hunter WS, Llanos QJ, Ahokas RA, Mashburn TA. Suppression of fever after lesions of the anteroventral third ventricle in guinea pigs. Brain Res Bull 1983; 11: 519-526.

5. Stitt JT. Evidence for the involvement of the organum vasculosum laminae terminalis in the febrile response of rabbits and rats. J Physiol 1985; 368: 501-511.

6. Roth J, Conn CA, Kluger MJ, Zeisberger E. Kinetics of systemic and intrahypothalamic interleukin 6 and tumor necrosis factor during endotoxin fever in the guinea pig. Am J Physiol 1993; in press.

7. Dinarello CA. The proinflammatory cytokines interleukin-1 and tumor necrosis factor and treatment of the septic shock syndrome. J Infect Dis 1991; 163: 1177-1184.

8. LeMay LG, Otterness I, Vander AJ, Kluger MJ. In vivo evidence that the rise in plasma IL 6 following injection of fever-inducing dose of LPS is mediated by IL 1β. Cytokine 1990; 2: 199-204.

9. Rotondo D, Abul HT, Milton AS, Davidson J. Pyrogenic immunomodulators increase the level of prostaglandin E in the blood simultaneously with the onset of fever. Eur J Pharmac 1988; 154: 145-152.

10. Stitt JT. Prostaglandin E as the neural mediator of the febrile response. Yale J Biol Med 1986; 59: 137-149.

11. Sirko S, Bishai I, Coceani F. Prostaglandin formation in the hypothalamus in vivo: effect of pyrogens. Am J Physiol 1989;256: R616-R624.

12. Xin L, Blatteis CM. Hypothalamic neuronal responses to interleukin-6 in tissue slices: effects of indomethacin and naloxone. Brain Res Bull 1992; 29: 27-35.

13. Zeisberger E, Roth J, Kluger MJ. Interactions between the immune system and the hypothalamic neuroendocrine system during fever and endogenous antipyresis. In: Pleschka K, Gerstberger R (eds.). Integrative and cellular aspects of autonomic functions. London• Paris, Libbey Eurotext Ltd, 1993: in press.

Temperature Regulation
Advances in Pharmacological Sciences
© 1994 Birkhäuser Verlag Basel

CYTOKINES IN ENDOTOXIN TOLERANCE

E. Zeisberger, K.E. Cooper[+], M.J. Kluger*, J.L. McClellan*, and J. Roth

Physiologisches Institut, Klinikum der Justus-Liebig-Universität, Aulweg 129, D-35392 Giessen, Germany, [+]Department of Medical Physiology, The University of Calgary, Calgary, Alberta, Canada T2N 4N1, *Department of Physiology, University of Michigan, Medical School, Ann Arbor, Michigan 48109, USA

SUMMARY: The effects of different modes of lipopolysaccharide (LPS) administration, the variation of the time schedule of repeated LPS injections, and the kinetics of the release of cytokines were studied in guinea pigs in order to understand the mechanisms of endotoxin tolerance. Although the down-regulation of the cytokine response seems to be the main mechanism, the reduced febrile rise in tolerant animals is probably caused by several mechanisms.

INTRODUCTION

It is long known that repeated administrations of LPS from membranes of gram-negative bacteria result in progressive attenuation of the febrile response, a phenomenon, which is called endotoxin tolerance [1]. The mechanisms underlying this tolerance are not well understood. Whereas some authors describe the existence of inhibitor substances responsible for reduced sensitivity to LPS [2], others reported an increased neutralization of LPS [3], and some suggested a contribution of endogenous antipyretic mechanisms to the endotoxin tolerance [4, 5]. In the present study we analyze different modes of LPS administration, modify the time pattern of repeated LPS injections, and study the kinetics of the release of cytokines in guinea pigs in order to understand better the hyporesponsiveness to LPS.

MATERIALS AND METHODS

The experimental animals were raised in the animal house of the Physiological Institute in Giessen. For induction of fever, the guinea pigs were either equipped with mini-osmotic pumps (Alzet) delivering LPS continuously for 14 days at a dose of 0.33 µg/kg • min., or were injected with LPS from E. coli (0111: B4, Sigma Co.) intramuscularly at a dose of 20 µg/kg. All experiments were performed at a room temperature of 23-25°C. The LPS was injected at 9.00 a.m. into the thigh muscle of unrestrained, freely moving animals. In the first series the body temperature was measured at 15 minutes intervals by means of a thin thermocouple inserted 6 cm deep into the colon. In the second and third series, the abdominal temperature and motor activity were measured continuously by means of implanted radiotelemetric transmitters (VM-FH Discs, Mini-Mitter Co.) A Dataquest IV data acquisition system (Data Sciences Inc.) was used for the automatic control of data collection and analysis. In the third series blood samples were taken by means of chronically implanted arterial catheters at times indicated in the text. The blood plasma samples were analyzed in the Dept. of Physiology in Ann Arbor for activities of two cytokines, tumor necrosis factor (TNF), and interleukin-6 (IL-6) by specific bioassays described by Roth et al. in this volume. For more details on methods cf. [6].

RESULTS AND DISCUSSION

We are reporting on three experimental series in guinea pigs. In the first series we attempted to evoke a long-term febrile response by means of a continuous infusion of the lipopolysaccharide (LPS) from E. coli applied by implanted mini-osmotic pumps (Alzet). The response was compared

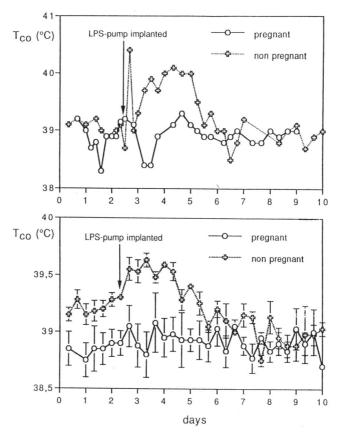

Fig. 1: Reactions to prolonged administration of LPS in single animals (above) and in the two investigated groups (below).

in a group of 4 nonpregnant female guinea pigs (mean body weight 866g) and a group of 4 pregnant guinea pigs (mean body weight 990g) near term exhibiting an endogenous antipyresis (Fig. 1).

Although the pumps administered LPS for 2 weeks at a daily dose 20 times higher than a single effective dose for a fever response when injected intramuscularly, the nonpregnant control group exhibited only a slight increase in colonic temperature (0.5°C) during the first three days after implantation, and the pregnant group showed no changes at all in colonic temperature. From these experiments we concluded that the continuous infusion of LPS increases the body temperature only temporarily because of a partial development of LPS tolerance. The pregnant animals were tolerant to LPS from the very beginning because they had activated their endogenous antipyretic systems for other reasons [4]. However, there was a large difference in the dynamic pattern of the pyrogen increase between a single application, in which 20 µg/kg of LPS were injected within a minute, and the continuous infusion, during which only 0.33 µg/kg were applied per minute amounting to a daily dose of 480 µg/kg. Therefore we used the same pattern of pyrogen increase to further study the development of LPS tolerance.

In the second series in a group of 6 guinea pigs we determined the febrile responses to single injections of LPS (20 µg/kg) repeated 4 times in intervals of 7 days (Fig.2).

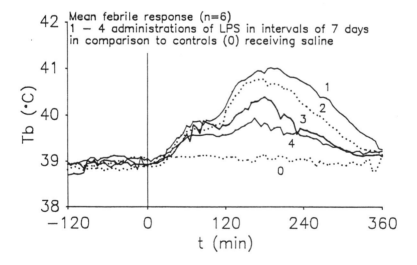

Fig.2: Mean febrile responses in a group of 6 guinea pigs during 4 consecutive administrations of LPS in comparison to control administration of saline (0).

The animals were allowed to recover for 3 weeks before the injections were repeated another 4 times in 3-day intervals. The first two tests resulted in normal febrile responses characterized by an increase in body temperature lasting 6 h and reaching its maximum ($\Delta t = 2\ °C$) 3 h after the LPS injection. The magnitude of this febrile response described by the fever index (the increase in temperature integrated over 6 h) surpassed 6 °C • h (Fig. 3).

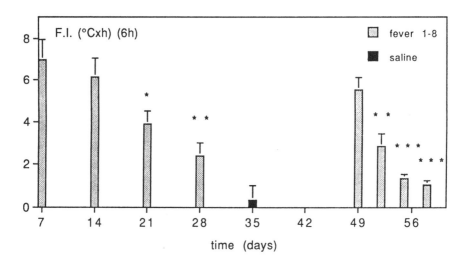

Fig. 3: Comparison of magnitudes of febrile responses, expressed as fever index (F.I.) for 6 hours, during the first series of consecutive administrations of LPS in intervals of 7 days and during the second series of consecutive administrations of LPS in intervals of 3 days repeated in the same group of 6 guinea pigs 3 weeks after the end of the first series. A control response to administration of saline is shown for week 5.

With the third test the febrile response started to decline towards a significantly lower level below 4 °C • h and reached levels below 3 °C • h in the fourth test. After the three weeks of recovery the first febrile response was almost normal, about 5.5 °C • h and decreased significantly in the second test three days later to levels below 3 °C • h, and even further to highly significantly lower levels of about 1 °C • h and below 1 °C • h in the third and fourth tests, respectively. It seemed that the earlier experience with the LPS pyrogen had accelerated the development of tolerance.

To investigate this we undertook the third experimental series (6 animals) in which LPS pyrogen was applied 5 times, again in 3-day intervals, but this time to animals that had no previous experience with LPS. In these animals blood samples were taken by means of chronically implanted arterial catheters 1 h before, 1 h after, and 3 h after each LPS injection. These time intervals corresponded to the nonfebrile stage, the stage of fever rise and the stage of fever maximum. In the blood plasma the activities of two cytokines, tumor necrosis factor (TNF) and interleukin-6 (IL-6) were determined in specific bioassays. Indeed, in this series the tolerance developed more slowly from the normal response above 6 °C • h in the first test to about 5 °C • h in the second, above 4 °C • h in the third, 3.5 °C • h in the fourth, and about 2 °C • h in the fifth test (Fig. 4).

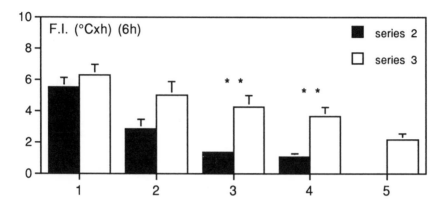

Fig. 4: Comparison of the development of tolerance in the series 2 and 3. The tolerance developed more rapidly in series-2 animals having a previous experience with LPS. In series-3 animals, without previous LPS experience, the magnitude of febrile response declined more slowly. The difference between the series 2 and 3 became statistically highly significant ($p < 0.01$) after the 2nd test.

The analyses of cytokine activities in the blood plasma revealed a characteristic sequence. Plasma activity of TNF, not detectable before pyrogen application, had its peak (above 500 U/ml) in the first hour after LPS application and rapidly disappeared from circulation within the next few hours. The activity of IL-6 strongly increased after pyrogen application parallel to the changes in body temperature, reaching its maximum of nearly 250000 U/ml in the third hour after LPS application. During the development of tolerance plasma activities of TNF as well as of IL-6 decreased parallel to the febrile responses, reaching in the fourth test maximum levels of only 10% of the levels found in the first test. For more details, methods and literature cf. [7]. Similar down-regulation of the cytokine response was described in rats [8], rabbits [9], and cancer patients [10].

CONCLUSION

From these investigations we conclude that the reduced febrile temperature rise in tolerant animals is probably caused by several mechanisms. Part of the LPS is inactivated more rapidly as suggested by [2], the cytokine response is to large part down-regulated as evidenced by the present study, and also central antipyretic mechanisms may contribute to some extent to this phenomenon, as proposed previously [4, 5].

ACKNOWLEDGEMENTS

The studies were supported by the DFG (projects Ze 183/2-6 and Ze 183/4-1). The excellent technical assistance of Birgit Störr is gratefully acknowledged.

REFERENCES

1. Atkins E. Pathogenesis of fever. Physiol Rev 1960; 40:580-646.

2. Höche A, Dahlbokum B, Iriki M, Riedel W. Contribution of elevated levels of cortisol and tumor necrosis factor a to bacterial endotoxin tolerance in rabbits. Am J Physiol 1993; (in press)

3. Warren HS, Knights CV, Siber GR. Neutralization and lipoprotein binding of lipopolysaccharides in tolerant rabbit serum. J Infect Dis 1986; 154:784-791.

4. Cooper KE, Blähser S, Malkinson TJ, Merker G, Roth J, Zeisberger E. Changes in body temperature and vasopressin content of brain neurons, in pregnant and non-pregnant guinea pigs, during fevers produced by Poly I:Poly C. Pflügers Arch 1988; 412:292-296.

5. Wilkinson MF, Kasting NW. Centrally acting vasopressin contributes to endotoxin tolerance. Am J Physiol 1990; 258:R443-R449.

6. Roth J, Conn CA, Kluger MJ, Zeisberger E. Kinetics of systemic and intrahypothalamic interleukin-6 and tumor necrosis factor during endotoxin fever in the guinea pig. Am J Physiol 1993; (in press)

7. Roth J, McClellan JL, Kluger MJ, Zeisberger E. Attenuation of fever and systemic release of cytokines after repeated injections of bacterial lipopolysaccharide in guinea pigs. J Physiol (Lond) 1993; (in press)

8. He W, Fong Y, Marano MA, et al. Tolerance to endotoxin prevents mortality in infected thermal injury: Association with attenuated cytokine responses. J Infect Dis 1992; 165:859-864.

9. Mathison JC, Virca GD, Wolfson E, Tobias PS, Glaser K, Ulevitch RJ. Adaptation to bacterial lipopolysaccharide controls lipopolysaccharide- induced tumor necrosis factor production in rabbit macrophages. J Clin Invest 1990; 85:1108-1118.

10. Mackensen A, Galanos C, Engelhardt R. Modulating activity of interferon-gamma on endotoxin-induced cytokine production in cancer patients. Blood 1991; 78:3254-3258.

Temperature Regulation
Advances in Pharmacological Sciences
© 1994 Birkhäuser Verlag Basel

EFFECTS OF IMMUNOMODULATORS AND PROTEASE INHIBITORS ON FEVER

Murzenok, P.P. & Gourine V.N.
Institute of Physiology, Academy of Sciences, Minsk, Republic of Belarus

Summary

Correlations between the value of the febrile response to LPS or endogenous pyrogens (EPs) and some immunologic functions were shown. It was established that these were adjuvant effect of T cell mitogen on IL-1β-induced fever in mice. It was found that oxytocin decreases and protease inhibitors prevent a diminution of EP production in culture of human peripheral blood mononuclear cells (PBMC). Oxytocin attenuates and protease inhibitor, contrical, potentiates the febrile response to LPS in rabbits.

Introduction

Development of fever depends on EPs such as IL-1α, IL-1β, IL-6, IL-8, TNF-α, IFN-γ, PG_0 (1-7). The most part of EPs is produced by immunocompetent cells. That's why the great interest is the evaluation of the febrile response to pyrogens in relation to initial state of the immune system and searching for a new approaches to modulation of the EPs production and their activity aimed at controlling cooperative thermoregulatory and immune processes during fever.

Materials and Methods

Experiments were performed on Mongrel rabbits weighing 2-2.5 kg, and mice C57B1/6 - 18-22 g. Rabbits were kept at 18-22°C, and mice in a special thermochamber at 23-25°C (lights on 08.00 - 20.00) during 2 week before experimental influences. Patients with chronic bronchitis and volunteers were also under investigation.

Human IL-1β was from Dainippon Corp. and had an activity of 10^7 units per ml. Culture medium RPMI-1640, concanavalin A (con A), met-enkephalin, oxytocin, phenylmethylsulfonylfluoride (PMSF) were obtained from Serva Corp., lipopolysaccharide (LPS) (E. coli, 0.111:B4) was from Sigma Chemical Corp., Germany, protease inhibitor, contrical, was from VEB Artzneimittelwerk, Germany.

Pyrogenic activity of IL-1β was evaluated in mice in thermochamber at 33°C (8). The methods of investigation of immunologic functions and registration of body temperature were previously described in detail (9, 10).

The statistical significance of the differences between the groups were evaluated by the Student-t test.

Results and Discussion

We studied the value of temperature response induced by pyrogens in animals with activated or suppressed immune functions.

It was shown that intraperitoneal injection of T lymphocyte mitogen - con A (50 μg) to C57B1/6 mice two days before testing resulted in increased proliferative capacity of thymocytes and splenocytes (table 1). Large doses of the stimulator (1 mg) does not enhance DNA synthesis in these cells (may be due to its *in vivo* toxic effects and changes in qualitative cellular composition). Temperature response to IL-1β (25 U/mouse) was changed in accordance with a degree of immunocompetent cells activation. Significant elevation in body temperature was found only in animals with activated immune system.

Table 1. Adjuvant effect of T cell mitogen on febrile response to IL-1β in mice

Index	control	50 μg con A	1 mg con A
DNA-synthesis, cpm thymocytes	7161±172(6)	18983±671(6)*	13268±897(6)*
splenocytes	12288±1169(6)	27950±1664(6)	10925±683(6)
rectal temperature, ΔT°C			
20 min after	0.49±0.10(8)	0.93±0.10(11)*	0.77±0.13(g)
IL-1β injection after 40 min	0.15±0.08(8)	0.43±0.10(11)*	0.30±0.09(g)

Number of investigated animals in parenthesis; asterisk - $p < 0.01$ (as compared with control value).

Suppression of immune response was induced by cyclophosphamide (CP). This immunodepressant (2 and 50 mg/kg) injected 24 hours before fever induction in rabbits decreases febrile response to LPS (2 μg/kg), especially its second phase, and proliferative response of splenocytes to mitogens (11). Probably, the decrease in febrile response is related with suppression of EPs production. It is shown that the TNF production under the CP influence is inhibited (12).

Similar diminution of cellular proliferative activity is observed in LPS-tolerant rabbits which as is known have a weak febrile response to repeated injection of LPS (13). There is also a decrease in secretion of IL-1-like activity by spleen and blood of mononuclear cells in such animals while intracellular content of EP is increased (14). The latter may be explained by corticosteroid action, the high level of which was registered in LPS tolerant animals (15).

It is necessary to note that CP had no effect on temperature response in LPS-tolerant animals, but it partially reconstitutes febrile response to EP in such animals (11). We believe that decrease in temperature response to EP in LPS-tolerant rabbits resulted from enhanced consumption of EP by immunocompetent cells.

So, animals with activated immune system respond to pyrogens more significantly but with suppressed immune system their response is much lower than in control animals. These results were used for development of a new approach to preliminary evaluation of the human immune system state.

The correlative analysis of the date showed a direct relation between the value of the temperature response to LPS and the capacity of PBMC to produce EP, and respond to mitogens, with the release of leukocytes and neutrophils into blood stream after intramuscular injection of exogenous pyrogen in chronical bronchitis patients (16). Particularly, the relations between changes in body temperature, quantity of EP and neutrophils may be depicted by regression equations:

$$y = 11018 \pm 101x \quad (p<0.02) \quad (1)$$
$$y = 2083 \pm 34x \quad (0<0.05) \quad (2)$$

where x - is the value of temperature index, and y - extra- and intracellular immunostimulating activity of EP (1) and number of neutrophils (2). We believe that the value of induced fever may depict to a certain extent the immune system state.

There are various mechanisms of fever regulation mediated by hormones, neuromediators etc. (1, 2, 4, 6, 17). In our experiments new evidence was obtained about the participation of some peripheral immune and nonimmune mechanisms in modulation of EP production, and in such a way in determination of the initial stages of fever development.

Such neuropeptides as oxytocin and met-enke·,halin inhibit the EP production by human PBMC in culture. By *in vivo* studies we have evidence of antipyrogenic activity of oxytocin (9). Constantly infused oxytocin (0.4 and 4 µg/kg/h) decreases the temperature response to LPS in rabbits. Antipyretic effects of oxytocin are related with the inhibition of EP production because oxytocin does not block the hyperthermia induced by EP.

It was shown that on excess of potassium ions in culture of stimulated PBMC amplifies the IL-1 synthesis because we observe elevation of extra- and intracellular content of IL-1 which was registrated by immunostimulating and pyrogenic activity.

One of the cause of weak temperature response to exogenous pyrogens is the high level of EP inhibitor in serum of non-febrile patients with chronic bronchitis (16).

By *in vitro* investigation it was shown that the decrease of immunostimulating and pyrogenic activity of EPs, produced in culture PBMC, is related with action of proteases (10). the addition of protease inhibitors, PMSF or contrical, to culture saves EP content as in human as in rabbit blood. It was possible to enhance the febrile response to LPS in rabbits by protease inhibitor, contrical (table 2). We believe that in such conditions the protease inhibitor prevents inactivation of EP.

Table 2. Potentiating effect of protease inhibitor on febrile response to LPS in rabbits

---

Time	LPS 0.15 µg/kg ΔT°C, (n=6)	LPS, 0.15 µg/kg + protease inhibitor (contrical), 150 U/kg ΔT°C, (n=4)
60 min	0.67 ± 0.11	1.70 ± 0.26 *
120 min	0.75 ± 0.11	1.03 ± 0.13
180 min	0.83 ± 0.08	1.43 ± 0.17*
240 min	0.58 ± 0.09	1.15 ± 0.32
300 min	0.32 ± 0.09	0.78 ± 0.27

Footnotes: as in table 1

The results indicate a close coupling between the immune and thermoregulatory systems during fever and allow us to use peripheral immune and nonimmune mechanisms for increasing or decreasing the febrile response to pyrogens in humans and experimental animals.

References

1.   ATKINS, E. (1984). Fever: The old and the new. *J. Infect. Diseases* **149**, 339-48.
2.   BLATTEIS, C.M. (1987). Neural mechanisms in the pyrogenic and acute phase responses in interleukin-1. *J. Neuroscience* **35**.
3.   KLUGER, M.J. (1991). Fever: role of pyrogens and cryogens. *Physiol. Rev.* **71**, 93-127.
4.   MILTON, A. (1982). Prostaglandins in fever and mode of action of antipyretic drugs. In: *Handbook of Exp Pharm* **60**, 257-303.
5.   MORIMOTO, A., MURAKAMI, N., TAKADO, M. ET AL. (1987). Fever and acute phase response induced in rabbits by human recombinant interferon-γ. *J. Physiol.* **391**, 209-18.
6.   OPPENHEIM, J.J., KOVACS, E.J., MATSUSHIMA, K. ET AL. (1986). There is more than one interleukin 1. *Immunol. Today* **7**, 45-56.
7.   SRIJBOS, P.J., HARDWICK, A.J., RELTON, J.K., CAREY, F., ROTHWELL, N.J.

(1992). Inhibition of central actions of cytokines on fever and thermogenesis by lipocortin-1 involves CRF. *Amer. J. Physiol.* **263**, (Endocrinol. Metab. 26): E 632-6.

8.   BODEL, P., MILLER, H. (1976). Pyrogen from mouse macrophages causes fever in mice. *Proc. Soc. Exp. Biol. Med.* **151**, 93-6.

9.   MURZENOK, P.P., KUTSAEVA, L.F., ZHYTKEVICH, T.I., GOURINE, V.N. (1989). Antipyrogenic properties of oxytocin. *Bull. exper. biol. med.* 426-8 (in Russian).

10.   MURZENOK, P.P. (1991). Formation of interleukin-1 under conditions of proteolytic enlymes inhibition. *Voprosy med chimiji* **37**, 38-40 (in Russian).

11.   GOURINE, MURZENOK, P.P. (1991). Properties of fever development induced by exo- or endogenous pyrogen in conditions of immunosuppression or immunostimulation. *Doklady AS BSSR* **35**, 185-7 (in Russian).

12.   MORI, H., MIHARA, M., TESHIMA, K. ET AL. (1987). Effects of immunistimulants and antitumour agents on tumour necrosis factor (TNF) production. *Int. J. Immunopharmacol.* **9**, 881-92.

13.   ATKINS, E., DINARELLO, C.A. (1985). Reflections on the mechanism of the biphasic febrile response and febrile tolerance. In: *Physiological, metabolic and immunologic actions of interleukin-1. Proceedings of Symposium; 1985 June 4-6; Ann Arbor, Mich.: N.V.* 97-106.

14.   MURZENOK, P.P. (1992). Functional state of immunocompetent cells in partially tolerant to LPS rabbits. *Pathol. fiziologiya and experim therapiya* **N1**, 37-40 (in Russian).

15.   BURKOVA, N.P. (1979). Changes in lymphoid tissue of rabbit thymus after pyrogenal or hydrocortisone injection. *Archi. anat. hystol. embryol.* 39-46 (in Russian).

16.   MURZENOK, P.P. MAKAREVITCH, A.E., DANILOV, J.P., GOURINE, V.N. (1991). The evaluation of immune reactivity of the organism based on intersystem relations analysis. *Fiziologiya cheloveka* **17**, 168-74 (in Russian).

17.   MURZENOK, P.P. (1992). Mechanisms of coupling of febrile and immune responses. *Uspechi fiziologitcheskich nauk* **N3**, 3-23 (in Russian).

Temperature Regulation
Advances in Pharmacological Sciences
© 1994 Birkhäuser Verlag Basel

ENDOGENOUS STEROIDS LIMIT THE MAGNITUDE OF FEBRILE RESPONSES.

J. Davidson

Division of Pharmacology, Department of Biomedical Sciences, Marischal College,
University of Aberdeen, Aberdeen. AB9 1AS, Scotland, U.K.

SUMMARY: The involvement of endogenous steroids in pyrogen-induced fever was studied. Pretreatment of rabbits with dexamethasone reduced the febrile response to polyinosinic:polycytidylic acid (poly I:C). This effect was reversed by administration of the type 2 glucocorticoid receptor antagonist RU 38486. RU 38486 enhanced the response to poly I:C but had no effect on normal body temperature. It is suggested that the increase in levels of endogenous glucocorticoids observed during fever may constitute a negative feedback loop which acts to modulate the magnitude of the febrile response.

INTRODUCTION: It is generally accepted that fever occurs in response to a number of exogenous pyrogens, for example bacteria, viruses or components of these organisms. These agents stimulate the host to synthesise and release a range of endogenous pyrogenic mediators or cytokines, such as interleukin-1 (IL-1), interleukin-6 (IL-6) and tumour necrosis factor-α (TNF-α). Endogenous mediators then act on a wide variety of cells such as macrophages and fibroblasts to induce the production of prostaglandins which are thought to be the final mediators of fever. Fever is thought to occur as a result of the actions of prostaglandins, particularly prostaglandins of the E series, on the pre-optic anterior hypothalamus [1].

Results from a number of experiments using either, adrenalectomized animals in which the ability to produce steroids is removed or administering exogenous steroid to intact animals have indicated that corticosteroids are able to modulate the febrile response. For example a recent study by Coelho et al.[2] has shown that intraperitoneal injection of lipopolysaccharide (LPS) produces fevers of significantly greater magnitude in adrenalectomized rats than in sham operated controls, indicating that corticosteroids can modulate the febrile response in rats. It is well established that non-steroidal anti-inflammatory drugs reduce prostanoid synthesis and are antipyretic when given systemically and that the ability to suppress eicosanoid biosynthesis underlies the anti-inflammatory effect of glucocorticoids [1;3]. This suggested that glucocorticoids may also have antipyretic actions. Early studies, however, on the effects of glucocorticoids on fever proved disappointing. In 1980 Willies and Woolf [4] showed that neither bolus injection nor constant infusion of either hydrocortisone or methyl prednisolone simultaneously with i.v. injection of either endotoxin or endogenous pyrogen (EP) had any effect on the resulting fever in rabbits. A study by Nakamura *et al*. in 1985 [5] also failed to show that dexamethasone, which has a 25-fold greater anti-inflammatory activity than hydrocortisone, had any effect on LPS-induced fever in rabbits when given 4 hours before pyrogen. The "antipyretic" effect of a glucocorticoid was not reported until 1987 when Abul *et al*. [6] demonstrated that i.v administration of dexamethasone to rabbits attenuated polyinosinic:polycytidylic acid (poly I:C), LPS- and EP/IL-1-induced fever in a dose-dependent manner when given between 0.5 h and 2 h prior to pyrogen. Further work by this group showed that dexamethasone pretreatment also attenuated the increase in plasma levels of PGE_2 observed during fever induced by these pyrogens [7]. The fact that attenuation of fever was observed only when dexamethasone was given before pyrogen suggested that "antipyresis" required the induction of a mediator. The molecular mechanism by which glucocorticoids inhibit eicosanoid synthesis appears to correlate with the induction of lipocortins, in particular lipocortin-1 (LC-1). A subsequent study by Davidson *et al*. [8] showed that i.v. administration of LC-1 immediately after poly I:C significantly attenuated the resulting fever providing indirect evidence that the "antipyretic" action of dexamethasone could be mediated by LC-1.

Intravenous administration of the polyribonucleotide, poly I:C, a synthetic double stranded polyribonucleotide, structurally related to viral RNA, to rabbits produces a biphasic increase in body temperature. With increasing dose, fevers of longer duration are produced, however, the rise in temperature reaches a ceiling [3]. This suggests that an endogenous mechanism may prevent the change in body temperature reaching lethal levels. Since, adrenalectomy enhances experimentally-induced fever, administration of exogenous glucocorticoid or LC-1 attenuates pyrogen-induced fever and circulating levels of endogenous glucocorticoid increase during fever [9], these molecules may represent endogenous antipyretic mediators which act to limit the magnitude of the febrile response. One approach used to investigate the function of endogenous compounds is the use of antagonists. In the study presented here the effect of the glucocorticoid receptor antagonist, RU 38486 [10], on poly I:C-induced fever was studied in the rabbit. A submaximal dose of poly I:C was chosen to induce fever [3] thus enabling either inhibition or potentiation of the response to be observed.

MATERIALS AND METHODS: Experiments were carried out in male Dutch rabbits (1.6 - 2.6 kg) at an ambient temperature of 22 ± 1 °C between 9.00 h and 17.00 h. Body temperature was measured continuously using rectal thermistor probes. The exogenous pyrogen poly I:C (2.5 μg/kg), the synthetic glucocorticoid dexamethasone (1 or 3 mg/kg) or vehicle (sterile saline) were given intravenously via the marginal ear vein. RU 38486 (20 mg/kg) or its vehicle, which consisted of 1 % carboxymethyl cellulose w/v and 0.5 % Tween v/v in sterile water, were administered orally. Results are expressed as TRI$_5$ values or thermal response indexes calculated by integrating the area under the temperature response curve over a period of 5 h following poly I:C administration. A single TRI unit being equivalent to a 1°C change in temperature lasting for a period of 1 h. Treatments were randomized, each rabbit acting as its own control and data analysed using a paired Students t-test.

RESULTS: Poly I:C (2.5 µg/kg) produced a biphasic increase in temperature, the first peak occurring 1.5 h and the second peak occurring 3.5 h after pyrogen administration with ΔT values of 1.00 ± 0.07 °C and 1.30 ± 0.11 °C respectively. Dexamethasone had no effect on normal body temperature but attenuated poly I:C fever in a dose dependent manner when administered 1 h prior to pyrogen (Fig. 1).

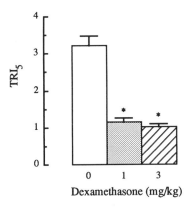

Fig. 1. Effect of dexamethasone on poly I:C fever. Rabbits were injected i.v. with either 1 mg/kg or 3 mg/kg dexamethasone or an equivalent volume of sterile saline. Poly I:C (2.5 µg/kg) was given i.v. 1 h later and and body temperature monitored for a further 5 h. Values represent means ± s.e. mean for n = 4. *P < 0.05 versus the response to administration of sterile saline then poly I:C.

RU 38486 (20 mg/kg) had no effect on normal body temperature, but when administered 2 h prior to poly I:C significantly increased the magnitude of the febrile response. Giving dexamethasone (1 mg/kg) 1 h after RU 38486 and 1 h prior to poly I:C antagonised the enhancement of the febrile response produced by RU 38486. The response was not significantly different from that to poly I:C alone (Fig. 2).

DISCUSSION: Data from this study clearly indicate that endogenous steroids play a role in modulating the febrile response to pyrogen since administration of the glucocorticoid receptor antagonist RU 38486 enhanced poly I:C fever and this was antagonized by giving exogenous glucocorticoid. Neither dexamethasone or RU 38486 had any effect on normal body temperature indicating that they had no effect on the afferent control of thermoregulation. It is unlikely that poly I:C is directly pyrogenic but acts by inducing the production of cytokines such as IL-1. Feedback regulation of IL-1 synthesis appears to involve the brain-pituitary-adrenal axis (Fig. 3).

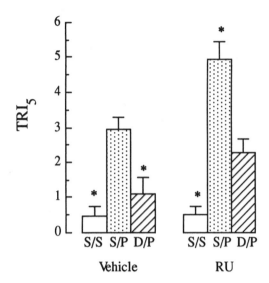

Fig. 2. Effect of dexamethasone and RU 38486 on poly I:C fever. Rabbits were given either vehicle or RU 38486 (20 mg/kg) orally 2 h before pyrogen. Dexamethasone (1 mg/kg) was administered i.v. 1 h after RU 38486 and 1 h prior to poly I:C (cross-hatched columns). Poly I:C (2.5 µg/kg) was injected i.v. at time zero (shaded and cross-hatched columns) and body temperature monitored for a further 5 h. Values represent means ± s.e. mean for n = 4. *P < 0.05 versus the response to administration of vehicle followed by sterile saline then poly I:C.

Fig. 3. A schematic overview of the stimuli for release and potential sites of action of endogenous corticosteroids during immunoactivation.

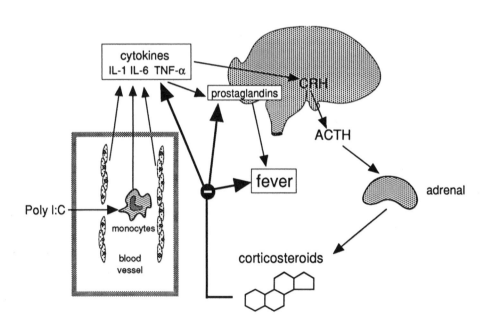

IL-1 is a pleiotropic, hormone-like cytokine with numerous diverse actions and in order to explain this it has been proposed that the various bioactivities of IL-1 may reside on different parts of the molecule. Naito *et al.* [11] have investigated the ACTH-releasing and pyrogenic properties of human IL-1β and its analogues when injected into the rat. These workers found that authentic human IL-1β induced both fever and ACTH release. Modification of the C-terminus resulted in the loss of pyrogenicity but the analogue retained the ability to stimulate ACTH release whereas modification of the N-terminus resulted in the loss of both pyrogenicity and the ability to stimulate ACTH release. These observations clearly indicate that the ACTH-releasing activity of IL-1β presumably via the stimulation of CRH is separate from its pyrogenic activity and IL-1 may therefore play a dual role during fever, both promoting fever by stimulating prostaglandin synthesis and stimulating ACTH release with its subsequent action on the adrenal cortex to promote the release of glucocorticoids. The function of glucocorticoids being to modulate the febrile response in an attempt to restore homeostasis. A 3-fold increase in circulating levels of cortisol in rabbits was measured following i.v. administration of 50 ng/kg of IL-1β (unpublished observations). Glucocorticoids have been found to reduce pyrogen-stimulated increases in both circulating IL-1 [12] and PGE_2 [7]. This suggests that the "antipyretic" actions of glucocorticoids may involve two components, 1) reduction of prostanoid synthesis via the induction of a lipocortin which reduces the availability of substrate arachidonic acid and 2) attenuation of the synthesis of IL-1 (Fig 3).

ACKNOWLEDGEMENTS: The author wishes to thank Roussel-Uclaf, France, for the gift of RU 38486 and Nigel C. Eastmond for his help in the preparation of Fig. 3.

REFERENCES:

[1]   MILTON, A.S. (1982). Prostaglandins in fever and the mode of action of antipyretic drugs. In Pyretics and Antipyretics. Handb. Exp. Pharm. 60, pp257 - 304, Milton, A.S. (Ed.), Springer-Verlag, Berlin, Heidelberg, New York.

[2]   COELHO, M.M., SOUZA, G.E.P. and PELA, I.R. (1992). Endotoxin-induced fever is modulated by endogenous glucocorticoids in rats. *Am. J. Physiol.* **263**, R423 - R427.

[3]   ROTONDO, D., ABUL, H.T., MILTON, A.S. and DAVIDSON, J. (1987). The pyrogenic actions of the interferon-inducer, polyinosinic:polycytidylic acid are antagonised by ketoprofen. *Eur. J. Pharmacol.* **137**, 257 - 260.

[4]   WILLIES, G.H. and WOOLF, C.J. (1980). The site of action of corticosteroid antipyresis in the rabbit. *J. Physiol.* **33**, 1 - 6.

[5] NAKAMURA, H., MIZUSHIMA, Y., SETO, Y., MOTOYOSHI, S. and KADOKAWA, T. (1985). Dexamethasone fails to produce antipyretic and analgesic actions in experimental animals. *Agents Actions*, **16**, 542 - 547.

[6] ABUL, H.T., DAVIDSON, J., MILTON, A.S. and ROTONDO, D. (1987). Dexamethasone pretreatment is antipyretic toward polyinosinic:polycytidylic acid, lipopolysaccharide and interleukin 1/endogenous pyrogen. *Naunyn-Schmiedebergs Arch. Pharmacol.* **335**, 305 - 309.

[7] MILTON, A.S., ABUL, H.T., DAVIDSON, J. and ROTONDO, D. (1989). Antipyretic mechanism of dexamethasone. In: *Thermoregulation: Research and Clinical Applications*, pp 74 - 77, Lomax, P. and Schönbaum, E. (Eds.), Karger, Basel.

[8] DAVIDSON, J., FLOWER, R.J., MILTON, A.S., PEERS, S.H. and ROTONDO, D. (1991). Antipyretic actions of recombinant lipocortin-1. *Br. J. Pharmacol.* **102**, 7 - 9.

[9] MILTON, N.G.N., HILLHOUSE, E.W. and MILTON, A.S (1992). Activation of the hypothalamo-pituitary-adrenocortical axis in the conscious rabbit by the pyrogen polyinosinic:polycytidylic acid is dependent on corticotrphin-releasing factor-41. *J. Endocrinology*, **135**, 69 - 75.

[10] MAO, J., REGELSON, W. and KALIMI, M. (1992). Molecular mechanism of RU 486 action: a review. *Molec. Cell. Biochem.* **109**, 1 - 8.

[11] NAITO, Y., FUKATA, J., MASUI, Y., HIRAI, Y., MURAKAMI, N., TOMINAGA, T., NAKAI, Y., TAMAI, S., MORI, K. and IMURA, H. (1990). Interleukin-1β analogues with markedly reduced pyrogenic activity can stimulate secretion of adrenocorticotropic hormone in rats. *Biochem. Biophys. Res. Commun.* **167**, 103 - 109.

[12] STARUCH, M.J. and WOOD, D.D. (1985). Reduction of serum interleukin-1-like activity after treatment with dexamethasone. *J. Leuk. Biol.* **37**, 193 - 207.

Temperature Regulation
Advances in Pharmacological Sciences
© 1994 Birkhäuser Verlag Basel

THE ROLE OF HEAT SHOCK PROTEINS (HSPs) AND INTERLEUKIN-1 INTERACTION IN SUPPRESSION OF FEVER

Kosaka, M., Lee, J.M., Yang, G.J., Matsumoto, T., Tsuchiya, K.,
Ohwatari, N. & Shimazu, M.
Department of Environmental Physiology, Institute of Tropical Medicine, Nagasaki
University, Nagasaki, 852, Japan

Summary

Introduction of HSP 70 by heat shock was identified in macrophages. Suppression of activities of macrophages including proliferation and LPS-induced IL-1 secretion were also observed during heat shock treatment. This indicates that IL-1 secretion from the macrophage which effects the induction of fever, has a close relationship to the induction of HSP 70. Further, difficulty of induction of HSP by heat stress in a weak heat tolerant pika rabbit was discussed from the view point of thermal physiology.

Introduction

The secretion of interleukin-1 (IL-1) is an initial response that intervenes and promotes many immune responses. The reaction as follows is important during bacterial infection: LPS - macrophage - IL-1 - PGE_2 - fever. One might wonder why even extremely high fevers seldom exceed 41°C or 42°C. It is postulated that there is a self-limiting mechanisms, such as a negative feedback loop. However, the existence of a negative feedback loop has not yet been proved. It would thus be quite important to investigate whether or not fever is in charge of this role. Heat shock proteins (HSP) can be induced by heat stress and the relationship between heat tolerance and HSP 70 was discussed in many papers. However, studies of whether HSP 70 relates to a febrile response or not has seldom been done. On the other hand, IL-1 is the principal substance among endogenous pyrogens which cause varient physiological regulations including a fever response. In the present paper, therefore, influences of heat shock on proliferation and lipopolysaccharide (LPS)-induced IL-1 secretion of macrophage as well as induction of HSP 70 by heat in macrophage were studied and the correlation between them was examined.

Furthermore, today, the pika (whistle rabbit) is thought to be an interesting animal which has a certain character of weak heat tolerance (Table 1), therefore, the second aim of this paper is to identify the induction of HSPs and CSPs (cold shock proteins) in this weak heat- and strong cold-tolerant pika in order to clarify the mechanism of its character of weak heat-tolerance and to examine the physiological correlation between thermo-tolerance and HSPs or CSPs.

Table 1. Comparison of the morphological and thermophysiological characteristics among pika, rat and rabbit.

	Afghan pika	Wistar rat	Albino rabbit
Rectal temp. ℃	39.6	38.4	39.2
Oxygen consump. ml/kg·min	20-25	25-30	9-10
Respiratory rate. min⁻¹	120-125	100-110	80-90
Autonomic thermoregulation			
Thermal panting	±	+	+ + +
Salivation	−	+ + +	+
Ear/body surface ratio. %	7.2		17.0
Behavioral thermoregulation	+ + +	+	+
Sensitivity to pyrogens	+	+	+ +
Sensitivity to anesthetics	+ +	+	+
Heat shock proteins (HSPs)	±	+	+
RBC. ×10⁴/μl	771-781	693-702	631-670
MCV. μ³	50.5-51.6	52.6-56.3	59.9-60.9
MCD. μ	anisocytosis	6.0	7.0
Hematocrit. %	38-39	45-46	40-41
R/L ventricular weight ratio	0.25	0.30	0.35

Materials and Methods

Experiment I RAW 246.7 cells (mouse macrophage cell line) obtained from Dr. Unkeless (Rockefeller University) and U-937 cells (human monocyte-like histiocytic lymphoma) obtained from Japanese Cancer Research Resource Bank were used in this study. Polyclonal anti-HSP 70 antibody kindly supplied from Dr. Ohtsuka (Aichi Cancer Center Research Institute) and human HSP 70 cDNA (Oncogene Science Inc.) were used for Western blotting and Northern blotting, respectively. Cell proliferation was assessed by methyl-[^{3}H]-thymidine incorporation. IL-1 assay was performed by enzyme-linked immunosorbent assay (ELISA) using human IL-1 ELISA assay kit (Otsuka Pharm. Co.). Cells were treated with heat-shock stimuli at various temperature from 37°C to 44°C for various time durations.

Experiment II Wistar rats (n=16, 8 weeks old), albino rabbits (n=8, 9 weeks old) and Afghan pikes (n=5, 12 months old) were subjected in this study. For heat shock on whole body, the anesthetized animal was immersed into a water bath at 45°C and the rectal temperature was kept at 42±0.2°C for 15 min. For cold shock experiment, the rectal temperature was kept at 17-25°C for 30-120 min with ice pad for hypothermia. Animals were killed 20 hours after heat or cold shock and the tissue proteins in the liver, kidney, adrenal gland, spleen, skeletal muscle and brain were analyzed by 10% SDS-PAGE and Western blotting with anti-HSP 70 antibody.

Results

Experiment I Induction of HSP 70 by heat in macrophage was demonstrated in this study. After the cells were treated with heat shock at 42°C and 44°C for 30 min, induction of HSP 70 was identified in RAW as well as U-937 cells by Western blotting with anti-HSP 70 antibody (Fig. 1). Increased level of HSP 70 mRNA was also confirmed by Northern blotting with human HSP 70 cDNA probe in RAW cells. Cell proliferation after exposure to heat was assessed by methyl-$[^3H]$-thymidine incorporation. The proliferation of macrophage (RAW cells) was suppressed by heat. Thymidine uptake after heat treatment at 39°C and 41°C for 2 to 3 hours were significantly suppressed (P<0.01) compared to the control value of the cell cultured at 37°C.

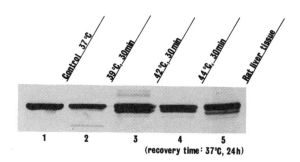

Fig. 1. The induction of HSP 70 by heat shock in RAW 246.7 cells. In Lanes 1 and 2, only a band of 72 kD protein (constitutional type of HSP 70) was shown. In Lanes 3, 4 and 5, a new band of 70 kD protein (inducible type of HSP 70) as well as a band of constitutional HSP 70 (72 kD) was were observed.

Effect of heat treatment on LPS-induced IL-1 secretion was investigated in U-937 cells. LPD was added into the medium at a final concentration of 500 μg/ml. Cells, $6X10^5$, were cultured at 37°C with LPS for the entire 24 hours period. After 15 hours of incubation at 37°C, cells were subjected to heat at 39°C, 41°C and 43°C for 30 min. After that cells were returned to the incubator, still at 37°C. IL-1 was not detected in the medium of control cells cultured without LPS. In the cells cultured with LPS at 37°C for 24 hours, 30 pg/ml of concentration of IL-1 was detected in the medium. LPS-induced IL-1 secretion from U-937 cells was suppressed by heat treatments in temperature dependent manner; 92%, 82% and 74% of control at 39°C, 41°C and 43°C, respectively,

Experiment II Induction of HSP 70 after whole body heat shock was detected in all organs examined in rats as well as rabbit liver by means of 10% SDS-PAGE. Immunoprecipitation of this protein with polyclonal anti-HSP 70 antibody was also confirmed in rat liver and in rabbit liver. In pika, contrarily, HSP 70 was not detected in any determined organs by 10% SDS-PAGE. Furthermore, induction of HSP 70 was not shown even by Western blotting with anti-HSP 70 antibody but only one pika liver tissue.

In the experiment of whole body hypothermia, induction of new protein, "cold shock protein", was not identified in rat, rabbit or pika by using 10% SDS-PAGE. But vanishment of a band of 32 kD protein in cytosol fraction after cold shock treatment was observed in one rat liver.

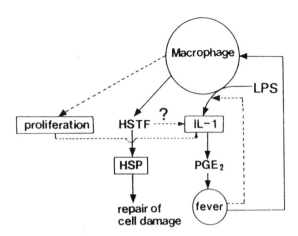

Fig. 2. Hypothetical schema of the negative feedback loop of fever response obtained from the present results. Solid and dashed arrows indicate excitatory and inhibitory effects, respectively.

Discussion

Suppression of IL-1 secretion and proliferation of macrophage by heat shock was demonstrated in this study. Simultaneously, induction of HSP 70 in macrophage by heat shock was identified and its regulation mechanism in mRNA level was shown. These results indicate the LPS-induced fever may act as a negative feedback control on the macrophage's activity. It is assumed that HSP 70 may play a role correlated to this self-limiting mechanism of fever, as shown in Fig. 2.

The release of IL-1 from macrophage is an important step in a series of LPS-induced fever response. A variety of inducing signals suggests that negative regulatory mechanisms must exist to limit IL-1 bio-synthesis but as yet, these are poorly understood. One possibility is that negative feedback loops exist where the IL-2 bioactivity act back on production. Such a role has been postulated for PGE_2, which is in response to IL-1. Schmidt and Abdulla (1988) reported that the synthesis of LPS-induced IL-1 precursor protein, p35, and its mRNA were down regulated by heat shock compared with control cell at 37°C. Inhibition of the release of mature form IL-1 by heat shock was shown in the present study. Heat-shock transcription factor (HSTF) which binds to heat shock element (HSE) during heat shock can bind to many additional chromosomal sites, and may repress normal gene activity (Westwood et al., 1991). Therefore, it might be possible that HSTF binds to IL-1 gene and repress its gene expression during heat shock (see Fig. 2.). This may be a possible reason why IL-1 secretion of macrophage is suppressed by heat shock. The more experimental evidences should be accumulated in order to confirm this hypothesis (Fig. 2.).

Induction of HSP after whole body hyperthermia was determined by 10% SDS-PAGE in pikas and rats. In rats, HSP 70 was clearly induced after heat shock. In pikas, however, changes in liver protein were difficult to observe at least by 10% SDS-PAGE after heat shock. A lack of induction of HSPs might be a disadvantage to survival in hot environment. Behavioural thermoregulation such as lying posture on all fours may compensate for lack of HSPs induction. Regarding CSPs, several reports indicate that four kinds of plasma proteins, 55kD, 27kD, 25kD and 20kD, vanished in chipmunk, a hibernator, during and after the period of hibernation. The present findings showed the similarity to the previous report of chipmunk suggests that not only hibernation but also acute cold-shock may lead to the similar result.

References

1.   KOSAKA, M., YANG, G-J., LEE, J-M., MATSUMOTO, T., TSUCHIYA, K. & OHWATARI, N. (1992). Physiological characteristics of pika (*Ochotona rufescens rufescens*) as a weak heat tolerant animal. In: *High Altitude Medicine, ed. by. G. Ueda et al., Shinshu University Press, Matsumoto, Japan.* pp. 487-491.

2.   LEE, J-M (1991). Effects of heat shock on macrophage. *Trop. Med.,* **33,** 135-143.

3.   SCHMIDT, J.A. & ABDULLA, E. (1988). Down-regulation of IL-1 biosynthesis by inducers of the heat-shock response. *J. Immunol.* **141,** 2027-2034.

4.   WESTWOOD, J.T., CLOS, J. & WU, C. (1991). Stress-induced oligomerization and chromosomal relocalization of heat-shock factor. *Nature,* **353,** 822-827.

Temperature Regulation
Advances in Pharmacological Sciences
© 1994 Birkhäuser Verlag Basel

BODY TEMPERATURE ELEVATION *PER SE* INDUCES THE LATE PHASE SYNDROME

A. A. Romanovsky and C. M. Blatteis
University of Tennessee, Memphis, Tennessee 38163, U.S.A.

SUMMARY

In conscious guinea pigs, the rise in core temperature (T_c) induced by intraperitoneal heating was followed by a decrease to below its initial level when the heating was ended. The magnitude and duration of this decrease depended on the extent and duration of the hyperthermia. When a substantial T_c decrease occurred, it was preceded by the reversal of the hyperthermia-associated ear skin vasodilation to a vasoconstriction, despite high values of T_c and continued heating. "Hyperthermia-induced hypothermia" was accompanied by sleep-like behavior, enhanced lability of T_c, and decreased responsiveness of ear skin vasomotion to changes in T_c. These features resemble the "late phase syndrome", *i.e.*, the thermoregulatory changes characteristic of the late phase of fever.

INTRODUCTION

Various exogenous and endogenous pyrogens administered intravenously induce monophasic fevers at low doses and biphasic fevers at higher doses. The second (late) phase of fever is characterized not only by an elevated core temperature (T_c), but also by certain changes in the character of T_c regulation (1-3). Thus, during the late phase of lipopolysaccharide (LPS) fever, an increased lability of T_c and a greater dependence of T_c on ambient temperature (T_a) have been reported (1, 2). The latter dependence, which is particularly evident in newborn, malnourished, and small-sized animals, often results in a fall rather than a rise in T_c during the late phase of fever if the subjects are exposed to a cool environment (2). The mechanism of this "pyrogen-induced hypothermia" is thought to involve regulation with a wide dead band; this type of regulation is characterized by a "threshold dissociation", *i.e.*, a separation between the threshold T_c for activation of cold defense mechanisms and that for activation of heat defense mechanisms (3). This complex of thermoregulatory changes characteristic of the late phase of fever is often coupled with certain behavioral alterations, such as sleep (4). For the present purposes, we have termed this complex the "late phase syndrome".

The purpose of this study was to investigate whether an elevated level of T_c might induce by itself the changes in T_c regulation resembling the late phase syndrome.

MATERIALS AND METHODS

The experiments were conducted using adult guinea pigs previously implanted with an intraperitoneal thermode and an intrahypothalamic thermocouple. Intraperitoneal heating (IPH) was accomplished by perfusing water through the thermode (50 ml/min). IHP was applied to conscious, lightly restrained animals to raise their T_c (measured as their hypothalamic temperature [T_h]) and thereby mimic its elevation during the first phase of LPS fever. Four experimental conditions were compared; water temperatures and perfusion durations were, respectively: 45°C and 14 min (group A; n = 3), 45°C and 40 min (B; n = 6), 46°C and 40 min (C; n = 5), and 45°C and 80 min (D; n = 3). The experiments were conducted at T_a = 24°C, which is below the thermoneutral zone of guinea pigs. Ear skin temperature (T_{sk}) was recorded, and the heat loss index (HLI) was calculated as HLI = (T_{sk} - T_a)/(T_h - T_a), according to the formula of Székely (5).

RESULTS

In all the experimental groups, the pre-IPH period was characterized by a stable T_h (mean ± SE: 38.5 ± 0.1°C). The thermal effects of IPH were distinct in the different experimental groups (Fig. 1). Depending on perfusion duration and water temperature, IPH increased T_h by 0.8 ± 0.2°C (group A), 1.4 ± 0.1°C (B), 1.5 ± 0.1°C (C), and 1.8 ± 0.2°C (D). These hyperthermic responses were followed by either return of T_h to its pre-IPH level (A) or by hypothermia, viz., slight (-0.7 ± 0.1°C; B), moderate (-1.9 ± 0.7°C; C), or severe (up to -7°C; D). One animal in group D died 3.5 h after IPH.

During the pre-IPH period, the values of HLI (0.32-0.43) indicated a substantial ear skin vasoconstriction. IPH caused a rapid rise in T_{sk} practically to the level of T_h (Fig. 2), which resulted in all the experimental groups in an increase in HLI up to 0.93-0.96, indicating maximal vasodilation. When the IPH-induced hyperthermia was followed by the return of T_h to its initial level (group A) or by a slight fall of T_h (B), the extant ear skin vasodilation was changed to vasoconstriction simultaneously with T_h restoration. In contrast, when "hyperthermia-induced hypothermia" occurred, it was preceded by the development of a paradoxical ear skin vasoconstriction, in spite of a high T_h and the continued IPH (Fig. 2). The development of this "hyperthermia-induced vasoconstriction" resulted in the appearance of a wide hysteresis in the relationship between HLI and T_h (Fig. 3). Another prognostic sign of "hyperthermia-induced hypothermia" was the appearance of coordinated oscillations of T_h (amplitude 0.2-0.5°C) and T_{sk} (2-3°C) at 0.2-0.4 min^{-1} (Figs. 1D, 2).

During IPH-induced hyperthermia, we determined in all the animals the minimal elevation of T_h (the threshold increase in T_h, tΔT_h) required to induce ear skin vasodilation. For all the experimental groups, the mean tΔT_h was 0.4 ± 0.1°C. In some animals, IPH was applied a second time to determine changes in tΔT_h that might have been induced by the first IPH. This second IPH was performed by perfusing water (45°C, 50 ml/min) through the thermode until com-

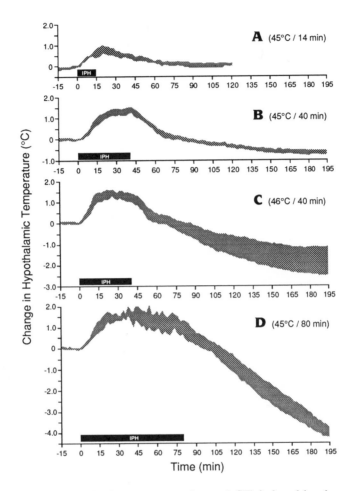

Fig. 1. Changes in hypothalamic temperature (mean ± SE) induced by intraperitoneal heating (IPH). Perfusion temperatures and durations, respectively: A - 45°C and 14 min, B- 45°C and 40 min, C - 46°C and 40 min, D - 45°C and 80 min.

plete vasodilation was observed (usually 10-20 min). In three animals of group A (return of T_h to the initial level), the values of $t\Delta T_h$ during the first and second IPH were identical (0.4 ± 0.1°C). However, when the second IPH was applied to one animal of group B and one animal of group C during "hyperthermia-induced hypothermia" (-0.9°C and -2.2°C, respectively), substantial increases in $t\Delta T_h$ were observed (0.9°C from 0.3°C and 1.2°C from 0.4°C, respectively).

During "hyperthermia-induced hypothermia", the animals exhibited behavioral changes typical of sleep, *i.e.*, postural extension, relaxation of neck muscles, closed eyes, and decreased responsiveness to external stimuli.

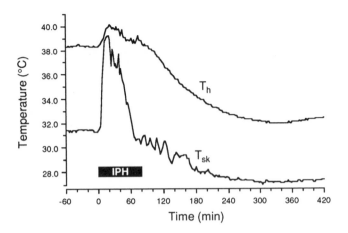

Fig. 2.   Changes in the hypothalamic (T_h) and ear skin (T_{sk}) temperatures during a typical experiment of group D. Intraperitoneal heating (IPH) was performed by perfusing water (45°C, 50 ml/min) through the thermode for 80 min.

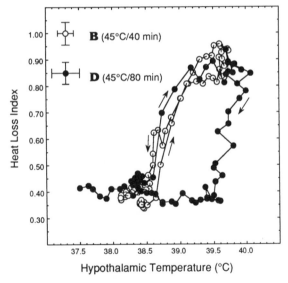

Fig. 3.   The effect of prolongation of intraperitoneal heating (IPH) on the relationship between the mean heat loss index and the mean hypothalamic temperature. Perfusion temperatures and durations were, respectively: group B - 45°C and 40 min; D - 45°C and 80 min. The data are presented from 30 min before to 120 min after the beginning of IPH; time direction is indicated by arrows. Average SEs for each group are shown in the upper left corner.

DISCUSSION

These data show that hyperthermia induced by IPH is followed in some cases by the development of hypothermia after IPH is ended. Similar responses have been reported earlier, e.g., hypothermia after whole-body heat exposure in guinea pigs (6), rats (7), mice (8, 9), and cats (6), and T_h fall after IPH in rats (10). The extent of the T_h decrease in our experiments depended on both the magnitude and duration of the preceeding hyperthermia; the same dependence was reported by Wilkinson et al. (9). It is noteworthy that a small difference in perfusing water temperatures (45 vs. 46°C) and, correspondingly, in IPH-induced T_c rises resulted in a substantial difference in the subsequent T_c falls (Fig. 1B vs. 1C).

"Hyperthermia-induced hypothermia" has been interpreted as a sign of a poikilothermia (6-9). It also was shown to develop at low but not high T_a (9). The latter observation is consistent with the decreased precision of T_c regulation noted in newborn rabbits maintained for 1 h in the hot end of a thermogradient (Székely, personal communication). The present data support the notion that a decreased precision of thermoregulation may exist during "hyperthermia-induced hypothermia", as indicated by the described increase in tΔT_h for vasodilation.

"Hyperthermia-induced hypothermia" was preceded by the occurrence of "hyperthermia-induced vasoconstriction" (Figs. 2, 3) and the appearance of oscillations of T_h and T_{sk} (Figs. 1D, 2). The nature of the "hyperthermia-induced vasoconstriction", its biological significance, and its relationship to the so-called "local heat-induced vasoconstriction" (11) and to the cutaneous vasoconstriction that may be induced by decreased central blood volume and venous pressure during heat stroke (12) await further investigation. The observed oscillations of T_h and T_{sk} resemble those reported during the defervescence phase of prostaglandin E_1-induced hyperthermia in rats (13). These oscillations may be viewed as increased T_c instability, which constitutes a characteristic sign of regulation with a wide dead band.

"Hyperthermia-induced hypothermia" was accompanied by sleep-like behavior. This observation is in line with the fact that T_c regulation with a wide dead band develops during sleep, particularly desynchronized (paradoxical) sleep (14). Coupling between this type of thermoregulation and sleep has also been proposed for the late phase of fever (4).

In summary, the present data show that, in conscious guinea pigs, an IPH-induced T_c rise, designed to mimic the first phase of fever, caused thermoregulatory changes (an increase in tΔT_h for ear skin vasoditation, the appearance of hysteresis in the relationship between HLI and T_h, the development of hypothermia in a cool environment, and instability of T_c) and sleep-like behavior, thus resembling the late phase syndrome. In contrast to the fact that the second T_c rise during LPS fever is independent of the first rise and, consequently, is not induced by an elevation of T_c per se (15), the changes in the character of T_c regulation (the late phase syndrome) seem, according to the present data, to be triggered by an elevated T_c. It is proposed, therefore, that the T_c rise during the first phase of fever may by itself be responsible for the thermoregulatory changes (but not for the changes in T_c) during the second febrile phase.

ACKNOWLEDGMENTS

Supported by NIH grants NS 22716 and HL 47650 to CMB. Presentation of this paper at the Symposium was partly supported by a Mini Mitter Co. (Sunriver, Oregon) travel grant to AAR. AAR is Senior Researcher of the Institute of Physiology, Minsk, Belarus, and the recipient of a postdoctoral fellowship from the University of Tennessee Neuroscience Center of Excellence.

REFERENCES

1. Székely M. Changed set-point and changed sensitivity: either or both? In: Hales JRS, editor. Thermal Physiology. New York: Raven Press, 1984: 525-30.

2. Székely M, Szelényi Z. The pathophysiology of fever in the neonate. In: Milton AS, editor. Handbook of Experimental Pharmacology, Vol. 60. Pyretics and Antipyretics. Berlin: Springer-Verlag, 1982: 479-528.

3. Vybíral S, Székely M, Janský L, Černý L. Thermoregulation of the rabbit during the late phase of endotoxin fever. Pflügers Arch 1987;410:220-2.

4. Romanovsky AA, Székely M, Pastukhov YuF. Sleep and thermoregulation in adaptive responses: a reappraisal [abstract]. Sleep Res 1993;22:470.

5. Székely M. Skin temperature - skin blood flow: assessment of thermoregulatory changes [abstract]. Acta Physiol Acad Sci Hung 1986;68:284.

6. Adolph EF. Tolerance to heat and dehydration in several species of mammals. Am J Physiol 1947;151:564-75.

7. Lord PF, Kapp DS, Hayes T, Weshler Z. Production of systemic hyperthermia in the rat. Eur J Cancer Clin Oncol 1984;20:1079-85.

8. Wright CL. Critical thermal maximum in mice. J Appl Physiol 1976;40:683-7.

9. Wilkinson DA, Burholt DR, Shrivastava PN. Hypothermia following whole-body heating of mice: effect of heating time and temperature. Int J Hyperthermia 1988;4:171-82.

10. Shido O, Yoneda Y, Nagasaka T. Shifts in the hypothalamic temperature of rats acclimated to direct internal heat load with different schedules. J Therm Biol 1991;16:267-71.

11. Sakurada S, Shido O, Nagasaka T. Mechanism of vasoconstriction in the rat's tail when warmed locally. J Appl Physiol 1991;71:1758-63.

12. Khogali M. Heat stroke: an overview with particular reference to the Makkah pilgrimage. In: Hales JRS, Richards DAB, editors. Heat Stress: Physical Exertion and Environment. Amsterdam: Excerpta Medica, 1987: 21-36.

13. Szelényi Z, Székely M, Romanovsky AA. Central thermoregulatory effects of cholecystokinin-octapeptide and prostaglandin E_1. Sechenov Physiol J 1992;78:94-101.

14. Parmeggiani PL. Thermoregulation furing sleep from the viewpoint of homeostasis. In: Lydic R, Biebuyck JF, editors. Clinical Physiology of Sleep. Bethesda: American Physiological Society, 1988: 159-69.

15. Romanovsky AA, Blatteis CM. Different thermoregulatory responses to intraperitoneal cooling (IPC) during the early and the late phases of lipopolysaccharide (LPS) fever [abstract]. FASEB J 1993;7:A17.

Temperature Regulation
Advances in Pharmacological Sciences
© 1994 Birkhäuser Verlag Basel

THE ABSENCE OF FEVER IN RAT MALARIA IS ASSOCIATED WITH INCREASED TURNOVER OF 5-HYDROXYTRYPTAMINE IN THE BRAIN

M.J.Dascombe and J.Y.Sidara, Neurosciences Research Group, School of Biological Sciences, Stopford Building, University of Manchester, Oxford Road, Manchester M13 9PT England

Summary: Infection in mice and rats by the rodent specific malaria parasite, *Plasmodium berghei* is associated with hypothermia, not fever. The mechanism(s) underlying the fall in temperature in this common laboratory model of human malaria may include a general debilitating effect of the infection, such as anaemia, on heat production and/or heat conservation in small animals. But in this study, the hypothermia in rats infected with *Plasmodium berghei* was also associated with an increased turnover in the brain of 5-hydroxytryptamine (5-HT, serotonin), a putative neurotransmitter reducing both food intake and body temperature in this species.

Introduction

Malaria, principally *Plasmodium falciparum*, is estimated to cause between 1 and 2 million deaths each year, the majority being children under five years of age (1), cerebral malaria being the commonest presentation of severe *P. falciparum* malaria in humans, with mortality reported at 22% despite treatment (2). Greater knowledge of the pathophysiological response to the malaria parasite could lead to much needed advances in the treatment of malaria. Laboratory investigations of malaria and antimalarial drugs commonly employ mice or rats inoculated with the rodent specific parasite, *P. berghei*. In these species, however, malaria does not cause pyrexia (fever) as in humans, instead a fall in core temperature occurs (3,4). The mechanisms underlying this temperature response are not known, but it is conceivable that both peripheral and central homeostatic mechanisms are disturbed by the disease.

5-Hydroxytryptamine (5-HT, serotonin) is a putative neurotransmitter reducing both food intake (5) and body temperature (6,7) in the rat. In this study, we have investigated the possible involvement of central 5-HT in the hypothermia associated with rodent malaria by determining the effect of *P. berghei* infection on 5-HT in the rat brain.

Materials & Methods

Male Sprague-Dawley rats weighing 70-90g at the start of the experiment were used, experiments being conducted at an environmental temperature of 21-24°C. Rats were housed in groups, except for food intake studies when animals were housed individually, on a 12h dark/light (08:00 to 20:00) cycle with free access to food and water. *P. berghei* (N/13/1A/4/203), maintained by serial blood passage in MF1 mice, was injected i.v. (2×10^7 parasitised erythrocytes/animal). Control rats were injected with 0.2ml of normal mouse blood diluted to the same degree with 0.85% saline. All animals were treated topically with monosulfiram (2.5%) in alcoholic solution to prevent infection by *Eperythrozoon coccoides*.

Following inoculation, malarial parasitaemia in rats was determined using light microscopy and expressed as the percentage of erythrocytes containing Leishman positive granules. Colonic temperature was measured by a rectal thermistor, inserted under minimal manual restraint, and displayed on a Digitron Model 4701 thermometer. At the times indicated in Results, food intake, body weight and exploratory locomotor activity (LKB Animex Activity Meter) were measured. Rats were assessed visually for behavioural and physiological changes including the adaptation of a huddled posture, vasoconstriction and the development of piloerection. For 5-HT studies, rats were humanely killed seven days after inoculation (Day 7), when parasitaemia was maximal or near maximal, and the concentration of 5-HT in the whole brain measured by fluorometric assay (8). 5-HT turnover in the brain was assessed by the accumulation of 5-HT (9) in other rats after injection i.p. of pargyline hydrochloride (Sigma Chemical Company) 75 mg/kg and L-tryptophan (Sigma Chemical Company) 100 mg/kg. Turnover rates of 5-HT (ng 5-HT/g wet weight/h) were calculated as the differences in brain 5-HT concentration in drug-treated animals compared with saline-treated rats after 1h.

Results are expressed as mean ± s.e.m. The probabilty (P) of the difference between means was evaluated by a 2-tailed Student's t-test and MANOVA.

Results

Seven days after inoculation with *P. berghei* (2×10^7 parasitised erythrocytes/rat i.v.), malarial parasitaemia in rats was $19 \pm 1\%$ (Figure 1).

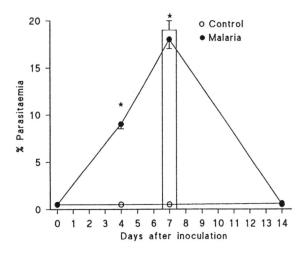

Figure 1 Time course of *P. berghei* parasitaemia, measured as the percentage of erythrocytes staining Leishman positive (% Parasitaemia), in rats after inoculation i.v. on Day 0. Points are the mean and vertical lines the s.e.m. for n > 6. The bar shows the mean parasitaemia for rats (n = 20) used on Day 7 in this study to determine the effect of malaria on brain 5-HT. * P < 0.05 compared with the number of Leishman positive erythrocytes in control rats.

Colonic temperatures in infected rats on Day 7 ($37.2 \pm 0.1°C$) were lower (P < 0.05) than both values for control animals on Day 7 ($37.8 \pm 0.2°C$) and paired values on Day 0 (Figure 2). There were no visible differences in body posture, piloerection or vasomotor tone between groups on Day 7, and locomotor activity was not significantly different (Figure 2). Malarial rats appeared anaemic with loss of blood colouration from the eyes and skin. Food intake, measured between 10:00 and 12:00, was unaffected by *P. berghei* infection on Days 1 to 4 inclusive, but was reduced (P < 0.05) on Day 5 (control 20.9 ± 0.5 g/day, malaria 18.7 ± 0.4 g/day), Day 6 (control 21.5 ± 0.4 g/day, malaria 19.0 ± 0.5 g/day) and Day 7 (control 22.9 ± 0.4 g/day, malaria 16.8 ± 0.6 g/day). Body weights on Day 7 (control 140 ± 4 g, malaria 135 ± 2 g) and weight gains from Day 0 (Figure 2) were similar.

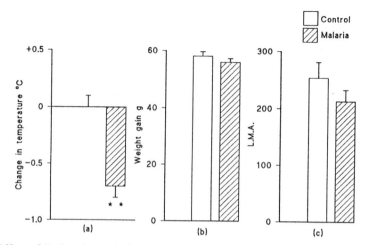

Figure 2 Effect of *P. berghei* malaria in rats on (a) colonic temperature on Day 7 compared with value on Day 0 (Change in temperature, °C), (b) gain in body weight from Day 0 to Day 7 (Weight gain, g) and (c) locomotor activity on Day 7, measured over 5 min (L.M.A. {LKB Animex Activity Meter scalar units}). Bars show the mean and vertical lines the s.e.m. for n = 12 (a, b) or n = 8 rats (c). ** P < 0.01.

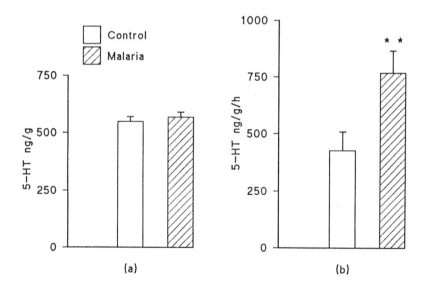

Figure 3 Effect of *P. berghei* malaria in rats 7 days after inoculation on (a) the concentration of 5-HT in wet brain tissue (5-HT ng/g) and (b) the turnover rate of 5-HT in the rat brain (5-HT ng/g/h). Bars show the mean and vertical lines the s.e.m. for (a) n = 8 and (b) n = 6. ** P < 0.01.

P. berghei infection on Day 7 had no effect on the brain concentration of 5-HT in rats, but the turnover rate of the monoamine in malarial rats was significantly higher than in control animals (Figure 3). Hypothermia, associated with shivering, piloerection and peripheral vasoconstriction, developed in rats treated with pargyline and L-tryptophan. One hour after drug treatment, colonic temperature had fallen by $-1.8 \pm 0.7°C$ in control rats and by $-3.0 \pm 0.6°C$ in malarial rats (n = 6), the difference between the two groups was significant at $P < 0.1$.

Discussion

P. berghei infection induced hypothermia in rats on Day 7 when parasitaemia was about 19%, although rats can respond to heat-killed *P. berghei*, as well as bacterial endotoxin, with a rise in body temperature (4). The absence of fever and the fall in temperature associated with *P. berghei* infection could be due to a general debilitating effect (e.g. anaemia) of an active infection on heat production and/or heat conservation in small rodents, which, with large body surface area:body mass ratios, have a tendency to lose body heat. However, *P. berghei* infection in mice is accompanied by a rise in thermogenic oxygen consumption (10). In this study, infected rats on Day 7 did not appear debilitated. They did not display marked behavioural changes, exploratory locomotor activity was similar to that of control rats, and shivering, piloerection and peripheral vasoconstriction were not evident. Food intake was reduced during the period of high parasitaemia (Days 5, 6, 7), but not earlier and there was no effect on either weight gain or body weight measured at the end of the study (Day 7). A reduction in food intake could, however, contribute to the development of hypothermia.

5-HT is a putative central neurotransmitter implicated in many physiological processes, including the reduction of food intake (5) and the promotion of heat loss (6,7) in the rat. The activity of serotonergic neurones in the brain, measured by the turnover of 5-HT, is increased, for example, by exposure of rats to high (32°C) ambient temperature (11). The data presented here show that the hypothermia in rats infected with *P. berghei* is also associated with an increased turnover of 5-HT in the brain, indicating that the absence of fever and the development of hypothermia in rats infected with *P. berghei* may be due in part to the increased activity in central 5-HT pathways regulating food intake and/or body heat loss.

References

1.   Knell AJ. Malaria (Oxford University Press, Oxford 1991).
2.   White NJ, Warrell DA. The management of severe malaria. In: Malaria: principles and practice of malariology (Churchill Livingstone, Edinburgh 1988) pp 865-88.
3.   Cordeiro RSB, Cunha FQ, Filho JA, Flores CA, Vasconcelos HN, Martins MA. *Plasmodium berghei*: physiopathological changes during infections in mice. Ann Trop Med Parasitol 1983;77:455-65.
4.   Oyewo EA, Dascombe MJ. Malaria (*Plasmodium berghei*) in rats: effects on body temperature and response to salicylate therapy. In: Homeostasis and thermal stress: experimental and therapeutic advances (Karger, Basel 1986) pp 92-5.
5.   Leibowitz SF, Weiss GF, Walsh UA, Viswanath D. Medial hypothalamic serotonin: role in circadian patterns of feeding and macronutrient selection. Brain Res 1989;503:132-40.
6.   Bligh J. Central transmitter substances and thermoregulation. In: Temperature regulation in mammals and other vertebrates (North-Holland, Amsterdam 1973) pp 133-52.
7.   Dascombe MJ. The pharmacology of fever. Prog Neurobiol 1985;25:327-73.
8.   Snyder SH, Axelrod J, Zweig M. A sensitive and specific assay for tissue serotonin Biochem Pharmacol 1965;14:831-5.
9.   Neff NH, Tozer TN. *In vivo* measurement of brain serotonin turnover. Adv Pharmacol 1968;6A:97-109.
10.   Cooper AL, Dascombe MJ, Rothwell NJ, Vale MJ. Effects of malaria on O_2 consumption and brown adipose tissue activity in mice. J Appl Physiol 1989;67:1020-3.
11.   Simmonds MA. Effect of environmental temperature on the turnover of 5-hydroxytryptamine in various areas of rat brain. J Physiol 1969;211:93-108.

PROSTAGLANDINS AND THE ACUTE PHASE IMMUNE RESPONSE

Dino Rotondo,

Division of Biochemical Sciences, Rowett Research Institute,
Bucksburn, Aberdeen AB2 9SB, Scotland, U.K.

Introduction: The acute phase of the immune response occurs in response to infectious agents. This initial phase occurs within minutes of the challenge which results in a cascade of immune interactions and the most prominent and easily measured manifestation of this immunostimulation is fever. Experimentally the acute phase response, especially fever, can be induced in animals by administering the purified components of the infectious agents which stimulate this response. The purified component of bacterial cell walls, endotoxin (lipopolysaccharide, LPS), can mimic the immunological activation observed during infection and can be used to determine immune responsiveness. Similarly it is possible to mimic the immunoactivation observed during viral infection by using synthetic agents which are similar to viral components and in this respect the synthetic double-stranded ribonucleic acid polyinosinic: polycytidylic acid (poly I:C) has been extremely useful. Fever and the acute phase in general occur in a series of steps in response to the sequential production of mediators within the body. The initial challenge leads to the induction and release of interleukin-1 (IL-1) [1] which is synthesised by various cells in the body especially peripheral blood monocytes or tissue macrophages. IL-1 concomitantly induces the production of prostaglandins from many cells, if not all, in the body and this increase in the level of prostaglandins [1], specifically prostaglandin E_2, (PGE_2) is responsible for ultimately producing a febrile response (Fig. 1).

Acute phase reactants: Many acute phase reactants and responses have been described these are summarised in table 1. However, prostaglandins do not appear to have been included among these.

responses	reactants
nitrogen catabolism	fibrinogen
neutrophilia	serum amyloid A
hypozincaemia	C-reactive protein
hypoferraemia	haptoglobin
anorexia	ceruloplasmin
sleep	α_2-macroglobulin
fever	complement complex proteins

Table 1. Major documented acute phase responses and reactants. adapted from [2]

It was shown by Rotondo *et al.* [3] that blood levels of PGE_2 also increase following the administration of immunostimulatory agents such as LPS, poly I:C and IL-1. The increase in blood levels of PGE_2 is accompanied by a febrile response indicating that these changes are associated with the acute phase and that PGE_2 may play an important role.

Source of the increased levels of PGE_2 in blood: The increase in circulating levels of PGE_2 following administration of LPS is in the order of 8- to 10-fold which represents a massive upregulation of biosynthesis and/ or downregulation of catabolism.   In all likelihood both mechanisms operate as a mere reduction in catabolism with a consequent accumulation of basal levels is unlikely to account for the massive increase in the balance of circulating PGE_2. During the acute phase response following intravenous administration of LPS the half-life of PGE_2 through the lungs is increased [4] indicating that LPS itself can increase blood levels of PGE_2 by both mechanisms. In addition, LPS, poly I:C and IL-1 can stimulate PGE_2 release from monocytes

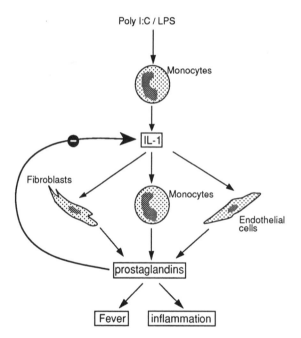

Figure 1. Schematic representation of the cascade of mediator release following administration of exogenous immunostimulatory agents (LPS/ poly I:C) showing the activation of monocytes and the release of IL-1 with the subsequent production of PGE_2 from multiple cell types. PGE_2 at localised sites participates in inflammatory reactions and in the brain is thought to result in a febrile response. PGE_2 is also thought to provide an autoregulatory negative feedback loop which controls the activation of monocytes/ macrophages to limit the release of cytokines particularly IL-1 [8].

[5] and blood vessels [6] indicating that cells within the circulation (monocytes) and the cells of the vasculature possess the potential to greatly contribute to the increased blood levels of PGE_2.

The possible role of PGE_2 during acute phase responses: It is not certain what role the increased levels of PGE_2 play during the acute phase, however, it is well established that PGE_2 can modulate a wide variety of immunological responses [7]. A particularly important action of PGE_2 which could potentially modulate the acute phase response is the inhibition of IL-1 release from monocytes. The stimulation of IL-1 release from monocytes by LPS *in vitro* was shown by Knudsen et al. [8] to be decreased in the presence of 10 nM PGE_2 and was increased by the cyclooxygenase inhibitor indomethacin. This indicates that both exogenous PGE_2 and PGE_2 released from monocytes in response to LPS stimulation is inhibitory to further IL-1 release. This represents an important autoregulatory negative feedback loop to limit the activation cascade following the stimulation of IL-1 release by LPS as IL-1 has been shown to induce its own release from monocytes and the vasculature. The stimulation of IL-1 release by LPS in the absence of PGE_2 should, therefore, result in an amplification and prolongation of cytokine release from monocytes with consequently protracted target cell responses which could be detrimental and result in damage to host tissues. PGE_2 may therefore act locally either on the cell from which it is released or adjacent cells as a modulator of immune cell activation (see Fig. 1) although it may also act on other organs. The changes in blood levels of PGE_2 in response to LPS or IL-1 are mirrored by changes in the febrile response which would indicate that it is possibly involved in the genesis of the febrile response.

Relationship between body temperature changes and blood levels of PGE_2: It was suggested in a recent review by Kluger [9] that changes in blood flow could possibly alter the PGE_2 recovered in each sample during the febrile response to immunostimulatory agents especially from the marginal ear vein of the rabbit which undergoes enormous changes in blood flow during thermoregulatory vascular responses. This would imply that changes in thermoregulatory status would determine the plasma level of PGE_2, however, it was shown in the study of Rotondo *et al.* [3] that animals which were placed in an environmental chamber at 33 °C and developed a hyperthermia did not show any changes in plasma levels of PGE_2 indicating that the increase in body temperature *per se* is not responsible for the increase in plasma PGE_2. Similarly conscious rabbits which were placed in an environmental chamber at 8 °C and maintained normal body temperature, showed that poly I:C, which in conscious rabbits produces a biphasic increase in body temperature with the first peak occurring after 90 minutes, produced a normal biphasic fever and an increase in blood levels of PGE_2 after 90 minutes exactly as observed at ambient temperatures of 22 °C (Fig. 2). During blood sampling the ear blood vessels are always pre-dilated using the heat from a lamp, therefore, even although there is a generalised thermoregulatory vasoconstriction during fever development there remains a high blood flow to the ear from which blood is sampled. Other experiments also show that plasma levels of PGE_2 are stimulus-dependent but independent of body temperature and that body temperature appears to follow changes in PGE_2 levels. Immunostimulator-induced (LPS, poly I:C, IL-1α/ IL-1β and TNF-α) changes in body temperature always appear to parallel changes in plasma levels of PGE_2. This also correlates with the defervescent actions (reduction of an existing fever) of non-steroidal anti-inflammatory agents such as ketoprofen and the antipyretic actions (attenuation of fever following pretreatment) of dexamethasone and ketoprofen [3].

Recent experiments with rabbits in which thermoregulatory capacity was removed by urethane anaesthesia showed that a steady decrease in body temperature occurred, except the increases which were manually induced using a heated table when rectal temperature began to drop below 37 °C i.e. approximately 2.5 °C below normal body temperature for rabbits. In these animals, although no changes in body temperature were observed following administration of poly I:C a normal increase in plasma levels of PGE_2 could be detected 90 minutes after i.v. administration of poly I:C with no changes in basal PGE_2 levels in saline controls (Fig. 2). Blood was collected from cannulated jugular veins not from the marginal ear vein. This clearly shows that the biochemical phenomena associated with the changes in PGE_2 biosynthesis occur independently of thermoregulation. However, changes in the production of PGE_2 in the circulation may modulate thermoregulatory responses by entering the CNS from peripheral sites, possibly the brain microvasculature [6], resulting in a febrile response. Measurement of the penetration of $[^{125}I]$-labelled PGE_2 into the csf, by push-pull perfusion of the 3rd cerebral ventricle, following i.v. administration showed that radioactivity entered the brain but only if it was given at the peak of the LPS or poly I:C-induced fever. Little or no radiolabel was detected in the csf of animals (unpublished observations). This demonstrates that PGE_2 can enter the brain from the peripheral circulation.

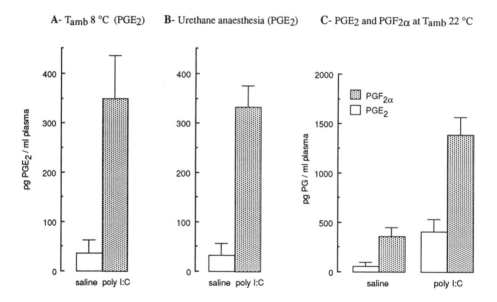

Figure 2. Plasma levels of prostaglandins measured from rabbits given different treatments under different conditions. (A) Plasma levels of PGE_2 were measured 90 minutes after either saline (open column) or 5 μg/kg poly I:C (shaded column) i.v. in conscious rabbits which were placed in an environmental chamber at 8 °C. These animals maintained normal body temperature and showed a normal biphasic increase in body temperature with the first peak occurring 90 minutes after poly I:C. (B) Plasma levels of PGE_2 were measured 90 minutes after either saline (open column) or 5 μg/kg poly I:C (shaded column) i.v. from cannulated jugular veins in urethane-anaesthetised rabbits which were unable to thermoregulate. (C) Plasma levels of PGE_2 (open columns) and $PGF_{2\alpha}$ (shaded columns) were measured 90 minutes after either saline or poly I:C (5 μg/kg) i.v. in conscious rabbits at an ambient temperature of 22 °C. Prostaglandins were measured by radioimmunoassay as described previously [3] and values are the means of n = 5 ± s.d.

The involvement of PGE_2 from peripheral sources in the febrile response would also correlate with the actions of indomethacin, which is thought to be unable to cross the blood-brain-barrier. In several studies indomethacin given peripherally is an effective antipyretic and also decreases csf levels of PGE_2 whereas when given centrally it is not so effective in attenuating fever in response to intravenous pyrogens [10; 11] indicating that the inhibition of prostaglandins in the periphery is important for antipyresis .

Are other eicosanoids associated with the acute phase response ?   The question also arises of whether other arachidonic acid-derived metabolites in addition to PGE_2 are involved in the acute phase. In the study of Davidson *et al.* [12] it was shown that no changes in blood levels of the thromboxane A_2 (TXA_2) metabolite, TXB_2, or leukotriene B_4 occurred during poly I:C fever whereas PGE_2 increased in the normal manner suggesting that leukotrienes and TXA_2 play no role in the acute phase response. Blood levels of $PGF_{2\alpha}$ are generally much higher than PGE_2 levels and also change during poly I:C fever, however, the increase is only in the order of 2-3-fold (Fig. 2). It is not certain whether this indicates that $PGF_{2\alpha}$ is also involved in some aspect of the acute phase response although the metabolism of PGE_2 in rabbit blood appears to yield a PGF metabolite [13] and the changes in blood levels of $PGF_{2\alpha}$ may, therefore, reflect the metabolism of PGE_2. This would suggest that PGE_2 is unique among eicosanoids in its involvement during the acute phase response to immunostimulatory agents. This is in agreement with the observations of Bishai *et al.* [6] who showed that PGE_2 but not TXB_2 changed during fever.

Conclusion: It is not certain what role PGE_2 plays in the acute phase response to exogenous immunostimulatory agents such as LPS and poly I:C or a wide variety of cytokines. However, the level of PGE_2 can change up to 10-fold following LPS stimulation which would, at least, qualify PGE_2, as a major acute phase reactant. PGE_2 could be involved in the localised control of cytokine release from activated monocytes and endothelial cells or act in the brain, subsequent to its biosynthesis in the brain vasculature, to modulate the thermoregulatory responses during fever.

References:

[1]   DINARELLO, C. A. (1991). Interleukin-1 and Interleukin-1 antagonism. *Blood.*, **77**, 1627-1652.

[2]   DINARELLO, C. A. (1984). Interleukin-1. *Rev. Infect. Dis.*, **6**, 51-95.

[3]   ROTONDO, D., ABUL, H. T., MILTON, A. S. AND DAVIDSON, J. (1988). Pyrogenic immunomodulators increase blood levels of PGE_2 simultaneously with the increase in body temperature. *Eur. J. Pharmac.*, **154**, 145-152.

[4]   IZUMI, T. AND BAHKLE, Y. S. (1988). Modification by steroids of pulmonary oedema and prostaglandin E_2 kinetics induced by endotoxin in rats. *Br. J. Pharmacol.*, **93**, 955-963.

[5]   HARTUNG, K., SCHLICK, E., STEVENSON, H. C. AND CHIRIGAS, M. A. (1983). Prostaglandin E synthesis and release by murine macrophages and human monocytes after *in vitro* treatment with biological response modifiers. *J. Immunopharmacol.*, **5**, 129-146.

[6]   BISHAI, I., DINARELLO, C. A. AND COCEANI, F. (1987). Prostaglandin formation in feline cerebral microvessels -effect of endotoxin and inteleukin-1. *Canadian J. Phsiol. Pharmacol.*, **65**, 2225-2230.

[7]   NINNEMANN, J. L. (1988). Prostaglandins, Leukotrienes and the Immune Response. Ed. J. L. Ninnemann. New York, Cambridge University Press.

[8]   KNUDSEN, P. J., DINARELLO, C. A. AND STROM, T. B. (1986). Prostaglandins posttranscriptionally inhibit monocyte expression of interleukin-1 activity by increasing intracellular cyclic adenosine monophosphate. *J. Immunol.*, **137**, 3189-3194.

[9]   KLUGER, M. J. (1991). Fever: Role of pyrogens and cryogens. *Physiol. Rev.*, **71**, 93-127.

[10] STITT, J. T. AND BERNHEIM, H. A. (1985). Differences in endogenous pyrogens induced by i.v. and i.c.v. routes in rabbits. *J. Appl. Physiol.*, **59**, 342-347.

[11] MURAKAMI, N. (1992). Function of the OVLT as an entrance into the brain for endogenous pyrogens. in Neuro-Immunolofy of fever. Ed. T. Bartfai and D. Ottoson, Pergamon, Oxford, pp. 107-114.

[12] DAVIDSON, J., MILTON, A. S., ROTONDO, D., SALMON, J. and WATT, G. (1990). The Effect of the Lipoxygenase Inhibitor BW A797C on Eicosanoid Formation following Pyrogen Administration. *Br. J. Pharmacol.*, **100**, 447P.

[13] GRANSTROM, E., FITZPATRICK, F. A. AND KINDAHL, H. (1982). Radioimmunologic determination of 15-Keto-13,14-dihydro-11β,16ξ-cyclo-PGE$_2$. in Methods in Enzymology, Eds. W. E. M. Lands and W. L. Smith, Academic Press, New York, pp. 306-320.

Temperature Regulation
Advances in Pharmacological Sciences
© 1994 Birkhäuser Verlag Basel

PYROGENIC IMMUNOMODULATOR STIMULATION OF A NOVEL THYMIC CORTICOTROPHIN-RELEASING FACTOR *IN VITRO*.

N.G.N. Milton, E. Swanton & E.W. Hillhouse.

Department of Clinical Biochemistry, University of Newcastle-upon-Tyne, UK.

SUMMARY:

A novel rat thymic corticotrophin releasing factor (CRF) has been identified which is activated by interferon and polyribonucleotide interferon inducers. Thymic CRF is released as a dimer of two 15 kDa proteins, is chromatographically distinct from the hypothalamic CRF-41 peptide, and is a biologically active CRF. These results suggest that thymic CRF is a candidate for the pyrogen induced peripheral CRF which modifies febrile responses *in vivo*.

INTRODUCTION:

Recent studies have suggested that the both the immune system and the stress response are activated by pyrogens during fever. Our *in vivo* studies[1,2] have suggested that a peripherally acting corticotrophin-releasing factor-41 (CRF-41)-like compound is activated by the pyrogenic interferon (INF) inducer polyinosinic:polycytidylic acid [poly (I):poly (C)] and plays a modulatory role in both the febrile and prostaglandin responses activated by this pyrogen. The peripheral CRF-41-like compound does not appear to have a hypothalamic origin since poly (I):poly (C) was unable to stimulate the release of immunoreactive CRF-41 from the rat hypothalamus *in vitro*[3]. However, this peripheral CRF-41-like compound does share immunological and pharmacological characteristics with hypothalamic CRF-41.

The thymus contains endocrine cells which produce a number of polypeptide hormones[4]; some of which are specific to the thymus (thymosins, thymopoitin and thymulin) whilst others are identical or similar to polypeptides associated with other endocrine tissues (neurophysins and luteinizing hormone releasing hormone). The thymus also produces many cytokines including the endogenous pyrogens interleukin-1 (IL-1), tumour necrosis factor (TNF) and INFs. Activation of the HPA axis causes changes within the thymus, including a transient atrophy. The thymus and hormones associated with the hypothalamo-pituitary-adrenocortical (HPA) axis show complex interactions and provide a pivotal link between stress and immunity.

CRF-41 has been shown to act both centrally and peripherally to regulate immune function. The source and nature of the peripheral CRF which regulates the immune system is currently unknown. In 1973 Brodish proposed that there was a blood borne tissue CRF which could activate the HPA axis independently of hypothalamic CRF[5] and more recently a biologically active thymic CRF has been proposed which is immunologically and chromatographically different from hypothalamic CRF-41[6]. The aim of this study was to determine whether the pyrogenic INF inducer poly (I):poly (C) plus related and unrelated immunomodulators could activate the synthesis and release of such a thymic CRF *in vitro*.

MATERIALS & METHODS:

Thymus glands were removed from freshly killed rats and mechanically dissociated in HANKS balanced salt solution. After centrifugation (500 g: 5 min) cells were resuspended in RPMI 1640 culture medium and washed three times. Single cell suspensions of thymic cells were plated (10 million cells/ml) into 24-well plates and incubated at 37°C in RPMI 1640 medium for 18 hrs in the presence of test immunomodulators. The culture medium was harvested for subsequent analysis. Immunomodulators tested were poly (I):poly (C); poly (A):poly (U); poly (C):poly (G); rat interferon (INF), lipopolysaccharide (LPS), Lipid-A, muramyl dipeptide (MDP), human recombinant interleukin-1α (IL-1α) and IL-1β. To study the synthesis of thymic-CRF *in vitro* thymic cell cultures were maintained in RPMI 1640 culture medium containing ^{3}[H]-Leucine (Leu). Thymic-CRF released into the medium was immunoprecipitated with rabbit anti-CRF-41 antiserum and levels of ^{3}[H]-Leu in the immunoprecipitates measured with a beta counter. Immunoreactive CRF-41-like material released into culture medium was assayed using ultra-sensitive 'two-site' enzyme amplified immunometric assays (EAIA's) for rat CRF-41[7].

The irCRF-41-like material was precipitated with ammonium sulphate (45%), desalted, using columns with a 5 kDa molecular weight cut-off, and eluted in pyrogen free water prior to immunoprecipitation, with either rabbit anti-CRF-41 antiserum or murine monoclonal anti-CRF-41 antibodies, to generate a partially purified form of thymic-CRF. Purified thymic-CRF was subjected to reversed phase (C_{18}) chromatography, size exclusion HPLC or SDS-PAGE analysis.

Purified thymic-CRF was also tested for adrenocorticotrophin (ACTH) and prostaglandin E_2 (PGE$_2$) releasing activities using a rat anterior pituitary fragment bioassay, with measurement of ACTH and PGE$_2$ release by EAIA and radioimmunoassay respectively.

RESULTS:

Poly (I):poly (C) and INF both stimulated incorporation of ^{3}[H]-Leu into immunoprecipitated thymic CRF-like material. The basal release of CRF-41-like immunoreactivity from thymic cells was < 20 pg/ml and LPS, MDP, Lipid-A IL-1α and IL-1β had no effect on CRF-41 release. CRF-41-like immunoreactivity was increased, in a dose dependent manner, by poly (I):poly (C); poly (A):poly (U); poly (C):poly (G) and rat INF (Fig 1).

Fig 1 - EFFECT OF INTERFERON AND INTERFERON INDUCERS ON THE RELEASE OF CRF-LIKE IMMUNOREACTIVITY FROM THYMIC CELLS *IN VITRO*.

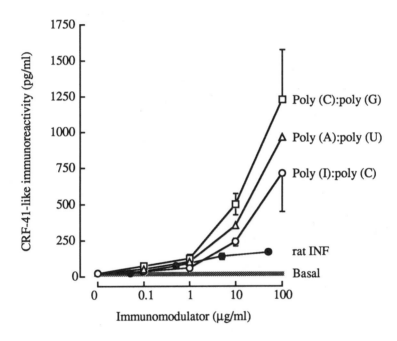

Analysis were carried out on thymic-CRF preparations purified from poly (I):poly (C) stimulated cells. Dilution curves of synthetic rat hypothalamic CRF-41, Sep-Pak C_{18} extracted hypothalamic CRF-41, immunopurified thymic CRF and Sep-Pak C_{18} extracted thymic CRF showed significant differences between hypothalamic CRF-41 and thymic CRF. Size exclusion HPLC of immunopurified thymic-CRF from synthesis experiments revealed the presence of a single protein peak with a molecular weight of 25-35 kDa. When the HPLC was carried out in the presence of β-mercaptoethanol a single peak with a molecular weight of 10-20 kDa was observed. SDS-PAGE analysis, under reducing conditions, of the HPLC purified 25-35 kDa form of thymic CRF revealed a minor 31 kDa protein band and a major 15 kDa protein band. Both the 15 kDa and 31 kDa bands were shown to incorporate [3][H]-Leu.

When incubated with rat anterior pituitary tissue *in vitro* the thymic CRF stimulated a three fold increase in ACTH release and a two fold increase in PGE_2 release (Fig 2).

Fig 2 - EFFECT OF THYMIC CRF AND HYPOTHALAMIC CRF-41 ON ACTH AND PGE_2 RELEASE FROM THE RAT ANTERIOR PITUITARY GLAND *IN VITRO*.

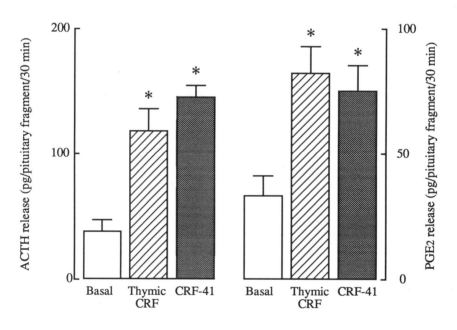

DISCUSSION:

The results from *in vitro* incubations of thymic cells in the presence of pyrogens suggest that INF and INF inducers activate both the synthesis and release of thymic CRF. Thymic CRF is not released in response to LPS or IL-1, suggesting that the regulation of its synthesis and release may be INF dependent.

A characteristic of hypothalamic CRF-41 is that it can be purified from tissue extracts and biological fluids using Sep-Pak C_{18} reversed phase columns[8]. Hypothalamic CRF-41 is retained on such columns and can be eluted with acetonitrile. However, when the immunopurified thymic CRF was subjected to Sep-Pak C_{18} extraction no immunoreactive CRF was recovered in the acetonitrile eluates. The differences between hypothalamic CRF-41 and thymic CRF observed in the two-site immunometric assays suggest that the proteins are not identical but do contain some similar structures.

Chromatographic characterization of affinity purified thymic CRF revealed the presence of a 31 kDa protein which could be reduced to yield a 15 kDa protein. This suggests that the 31 kDa form may in fact be a dimer of two 15 kDa proteins, linked via disulphide bridges, presumably between cysteine (Cys) residues in each 15 kDa protein. The cross-reactivity of the 31 kDa protein with antibodies which bind the C-terminal residues of hypothalamic CRF-41 suggests that the protein contains a CRF-41-like sequence. The detection of the 31 kDa form using two-site assays for hypothalamic CRF-41 backs up this suggestion. The precursor for hypothalamic CRF-41 is derived from a 187 amino-acid protein[9]. The amino-acid sequence for rat pre-pro-CRF-41 contains two Cys residues at positions 19 and 22. However, both these Cys residues are within the 24-27 amino acid signal sequence which is thought to be cleaved from the 187 precursor prior to secretion. The CRF-41 sequence comprises residues 144-185 of pre-pro-CRF-41. The molecular weight of pre-pro-CRF-41 is thought to be 22-25 kDa, with the pro-CRF-41 having a predicted molecular weight of 19-22 kDa. This pre-pro-CRF-41 form, containing both the Cys residues and CRF-41 sequence, is significantly larger than the 15 kDa form of thymic CRF. CRF-41 has been shown to activate both ACTH and PGE_2 release from the anterior pituitary gland *in vitro*[10], whilst precursor forms of CRF-41 have been suggested to lack such activity. Thymic CRF was shown to stimulate both ACTH and PGE_2 release further strengthening the hypothesis that this material is indeed a corticotrophin-releasing factor.

In conclusion, we have identified a novel rat thymic CRF which shows immunological, chromatographic and structural differences to hypothalamic CRF-41. Despite these differences thymic CRF shares similar adrenocorticotrophin and prostaglandin releasing activities with hypothalamic CRF-41. This biologically active thymic CRF is released and synthesized by thymic cells in culture when they are stimulated with either interferon or interferon inducers and provides us with a potential candidate for the pyrogen induced peripheral CRF which modifies febrile responses.

ACKNOWLEDGMENTS:

This work was supported by a project grant from the Wellcome Trust.

REFERENCES:

[1] Milton NGN, Hillhouse EW, Milton AS. A possible role for peripheral corticotrophin-releasing factor-41 in the febrile response of conscious rabbits. J Physiol (Lond) 1993; 465: 419-30.

[2] Milton NGN, Hillhouse EW, Milton AS. Modulation of the prostaglandin responses of conscious rabbits to the pyrogen polyinosinic:polycytidylic acid by corticotrophin-releasing factor-41. J Endocrinol 1993; 138; 7-11.

[3] Milton NGN, Self CH, Hillhouse EW. Effects of pyrogenic immunomodulators on the release of corticotrophin-releasing factor-41 and prostaglandin E2 from the intact rat hypothalamus *in vitro*. Br J Pharmacol 1993; 109: 88-93.

[4] Ritter MA, Crispe IN. The Thymus, Oxford University Press, Oxford, 1992.

[5] Brodish A. Hypothalamic and extrahypothalamic corticotrophin-releasing factors in peripheral blood. In Brodish A & Redgate ES (eds) Brain-pituitary-adrenal interrelationships. Karger, Basal, 1973, pp 128-51.

[6] Healy DL, Hodgen GD, Schute HM, Chrousos GP, Loriaux DL, Hall NR, Goldstein AL. The thymus-adrenal connection: thymosin has corticotropin-releasing activity in primates. Science 1983; 222: 1353-5.

[7] Milton NGN, Hillhouse EW, Fuller JQ, Self CH. Corticotrophin releasing factor (CRF-41) in the human and rat - utility of a highly sensitive enzyme amplified immunometric assay. J Neuroendocrinol 1990; 2: 889-95.

[8] Rivier J, Spiess J, Vale W. Characterization of rat hypothalamic corticotropin-releasing factor. Proc Natl Acad Sci USA 1983; 80: 4851-5.

[9] Jingami H, Mizuno N, Takahashi H, Shibahara S, Furutani Y, Imura H, Numa S. Cloning and sequence analysis of cDNA for rat corticotropin-releasing factor precursor. FEBS Lett 1985; 191: 63-6.

[10] Vlaskovska MA, Hertting G, Knepel W. Adrenocorticotropin and β-endorphin release from rat adenohypophysis *in vitro*: inhibition by prostaglandin E2 formed locally in response to vasopressin and corticotrophin-releasing factor. Endocrinology 1984; 115: 895-903.

Temperature Regulation
Advances in Pharmacological Sciences
© 1994 Birkhäuser Verlag Basel

APPARENT DISSOCIATION BETWEEN LIPOPOLYSACCHARIDE-INDUCED

INTRAPREOPTIC RELEASE OF PROSTAGLANDIN E_2 AND FEVER IN

GUINEA PIGS

M. Szekély, E. Sehic, V. Menon, and C.M. Blatteis
University of Tennessee, Memphis, TN 38163, USA

Summary
The i.v. administration of a lipopolysaccharide (LPS) regularly produced increases in both core temperatures (T_c) and levels of prostaglandin E_2 (PGE_2) in the intrapreoptic (iPO) extracellular fluid of guinea pigs with unilaterally implanted microdialysis probes, but these changes were not temporally related. The increase in PGE_2 was greater in guinea pigs with bilaterally implanted microdialysis probes, although in half of the animals LPS caused a fall in T_c. The bilateral iPO microdialysis of cyclooxygenase inhibitors abrogated both the T_c and PGE_2 changes induced by LPS. These results suggest that PGE_2 may be produced in the preoptic area of conscious guinea pigs in response to LPS, but that its release and fever may be dissociated responses.

Introduction

It is generally believed that PGE_2 produced in the preoptic area (POA) of the anterior hypothalamus has a modulatory role in the febrile response to systemic pyrogens (1); the POA is the putative site of the primary, but not exclusive (2), fever-driving controller (3). However, lately, the question has arisen whether PGE_2 may, in fact, be formed outside of the brain substance (4), then diffuse into it through, in particular, the circumventricular organ *organum vasculosum laminae terminalis* (OVLT), and then on to the POA and other hypothalamic nuclei (5). It has also been suggested that PGE_2 may be formed within the OVLT itself (6) and act locally (7) or on neurons of the adjacent neuropil (8). If PGE_2 were produced in the POA in response to circulating pyrogens, inhibitors of PGE_2 synthesis delivered directly into this site should prevent its local release and also abrogate the febrile response. But, if fevers were inhibited by this local blockade yet PGE_2 appeared in the POA, it would infer that this PGE_2 diffused into the POA from outside this region and had no direct connection to the febrile process. Such a result, thus, would suggest that the development of fever is dependent on PGE_2 specifically produced and released locally in the POA. Conversely, if the administration of PGE_2 synthesis inhibitors into the POA blocked the appearance of PGE_2 in the POA but did not prevent fever production, it would suggest that fever and the generation of PGE_2 in the POA are dissociated phenomena. The present study was undertaken to determine which result would obtain.

Materials and Methods

Male Hartley guinea pigs weighing 300 - 350 g were used in these experiments. Artificial cerebrospinal fluid (aCSF) was microdialyzed either uni- or bilaterally at 2 μl/min for 6 h into the POA of conscious guinea pigs prepared 60 h previously (9). *S. enteritidis* lipopolysaccharide (LPS, 2 μg/kg, i.v.) was administered 80 min after the onset of microdialysis. In other experiments, sodium indomethacin trihydrate (indo, 1 μg/μl aCSF, pH 7.4) or sodium meclofenamate (meclo, 1.2 μg/μl aCSF, pH 7.8) was similarly microdialyzed bilaterally, beginning 80 min after the onset of aCSF microdialysis and 30 min before the injection of LPS. Appropriate control groups were included in all the experiments. Core temperature (T_c) was monitored at 2-min intervals and microdialysate effluents collected over 30-min periods for the duration of these experiments; the samples were analyzed for their PGE_2 content by radioimmunoassay (irPGE_2). The ambient temperature was $24 \pm 1°$ C. Only experiments in which the intrapreoptic localization of the microdialysis probes was confirmed histologically were included in the results.

Results

Neither the unilateral nor the bilateral microdialysis of aCSF into the POA affected the T_c of these guinea pigs; irPGE_2 levels in the microdialysate effluents from the POA declined steadily over the 6-h duration of the experiments.

LPS induced characteristic, biphasic, *ca*. 1.4° C febrile rises in the animals with unilaterally implanted microdialysis probes; mean iPO irPGE_2 levels changed in these guinea pigs in apparently close correlation with the febrile course. However, examination of individual experiments revealed that the patterns of PGE_2 and T_c changes did not correspond in all cases. Thus, in 4 of 11 animals, the rise in irPGE_2 occurred only during the first rising phase of fever. In 4 other animals, it was associated with both rising phases, but became maximal during the second rising phase. Finally, in 3 animals, the rise was sustained at high levels during both phases of fever. These results are depicted in Figs. 1A - D.

In 3 of 6 guinea pigs with bilaterally implanted microdialysis probes, LPS failed to produce fevers. To the contrary, T_c fell significantly in these animals, beginning 90 min after LPS injection. In the other 3 animals, LPS evoked biphasic, *ca*. 1.4° C fevers, but these were protracted and had not abated when the experiments were ended. However, increases in iPO irPGE_2 levels occurred in all 6 guinea pigs over the entire post-LPS period. Indeed, the increases in irPGE_2 were larger in the animals in which T_c fell than in those in which it rose. These data are illustrated in Figs. 2A and B.

Neither the bilateral microdialysis of indo nor of meclo *per se* affected T_c or irPGE_2 levels (n=11). However, both inhibited the increase in iPO irPGE_2 induced by LPS in these animals. Indo (n=4) blocked and meclo (n=6) attenuated the protracted febrile rises LPS induced in half of

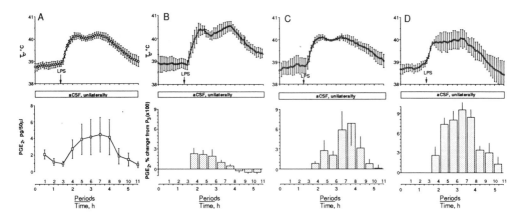

Fig. 1. Effects of LPS (2 µg/kg, i.v.; arrow) on T_c and iPO PGE_2 levels of guinea pigs with unilaterally implanted microdialysis probes. A: Mean (± SE) courses of the responses of all the animals. B-D: Mean (± SE) changes representative of the three different types of responses (see text); PGE_2 levels are expressed as percent changes from pre-LPS values.

the controls, and both drugs prevented the T_c falls LPS caused in the other half. These results are presented in Figs. 2C and D.

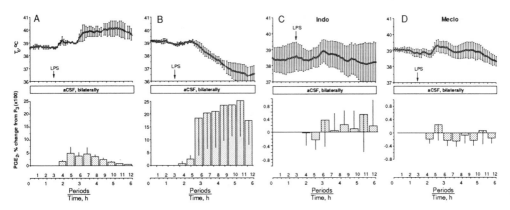

Fig.2. A-B: Differential effects of LPS (2 µg/kg, i.v.; arrow) on T_c and iPO PGE_2 levels (expressed as percent changes from pre-LPS values) of guinea pigs with bilaterally implanted microdialysis probes (see text). C-D: Effects of the bilateral iPO microdialysis of indo (1 µg/µl) and meclo (1.2 µg/µl) on the T_c and iPO PGE_2 changes induced by LPS, i.v.

A teardrop-shaped, variably-sized region of diffuse gliosis was seen surrounding the dialysis probe tip sites of most guinea pigs after these dialysis sessions.

Discussion

The present results show that, in all cases, the i.v. administration of a presumptively pyrogenic dose of LPS consistently induced an increase in the level of irPGE$_2$ assayed in microdialysate effluents from the POA of conscious guinea pigs. Furthermore, the local, bilateral, iPO microdialysis of two different cyclooxygenase inhibitors always abrogated the iPO extracellular fluid PGE$_2$ rises induced by i.v. LPS. The effects of LPS or of the inhibitors on PGE$_2$ levels, however, did not correlate, in most cases, with corresponding changes in the courses of the LPS-induced febrile responses. Thus, although increased amounts of irPGE$_2$ generally appeared in the POA of guinea pigs with unilaterally implanted dialysis probes after LPS administration, the temporal patterns of these increases were not coincident with the courses of the associated febrile responses. This dissociation between LPS-induced PGE$_2$ and T_c changes was particularly apparent in guinea pigs with bilaterally implanted probes; *viz.*, in half of these, iPO PGE$_2$ levels rose markedly while T_c decreased. On the other hand, both the PGE$_2$ and T_c changes induced by LPS in these animals were prevented by the iPO microdialysis of indo or meclo. According to our original hypothesis, the latter results would suggest that the induction of LPS fever depends on the production and release of PGE$_2$ in the POA, while the former results would imply that fever and the occurrence of PGE$_2$ in the POA are dissociated phenomena.

Intracerebral microdialysis has become an established technique for investigating changes in the extracellular levels of putative neuromediators in the brains of conscious animals (10). It offers various advantages over other methods of sampling (11). However, as in other methods, it also involves the implantation of a device, with attendant trauma. Unavoidably in this process, neurons are destroyed, fibers are damaged, and the integrity of the blood-brain barrier is broken. Inflammatory cells promptly migrate to the injured site, astrocytes begin to proliferate locally, and a zone of diffuse gliosis gradually develops, the degree of tissue reactions depending on the extent of the implantation damage. Since local blood flow and metabolism reportedly return to essentially normal conditions within 1 d, and reactive gliosis does not become prominent until 4 days after probe insertion (12), we chose to conduct our dialysis experiments beginning 2 1/2 d after probe implantation (9). However, it is apparent from the histological evidence of the present study that variable gliotic, and probably also vascular, changes affected tissue integrity around the probe tips sooner than expected. To what extent these tissue reactions affected, in turn, our neurochemical measurements is uncertain, but we would surmise that they accounted importantly for the variable results obtained. Thus, we suspect that the PGE$_2$ levels we detected in the POA effluents were contaminated by variable contributions from invaded inflammatory cells and local astrocytes, activated by blood-borne LPS (and cytokines it presumably also induced) that diffused into the POA was included in the samples. The different patterns of PGE$_2$ we observed following LPS administration may be due, therefore, to the consequences of implantation trauma, *e.g.*, the distance and path that LPS may have to traverse to reach the probe site after exiting ruptured vessels, the variable density and distribution of the gliotic region around the probe, etc., different in each case. The broken morphological and neurochemical integrity of the POA, by the same token, likely also may account for the hypothermic response to LPS of some of the guinea pigs with bilaterally implanted microdialysis probes.

The findings that indo and meclo prevented the PGE_2 and T_c responses to LPS of these doubly implanted animals would support the argument, on the other hand, that the production of fever depends on the presence of PGE_2 in the POA . Since it is probable that the PGE_2 detected in the extracellular fluid from the POA was derived, in variable amounts, from sources not integral to this region, and that all PG synthesis, irrespective of cell source, would presumably be blocked equally by the cyclooxygenase inhibitors delivered into this site, it cannot be determined from the present data which cell types in the POA would normally be activated to produce PGE_2 in response to circulating LPS. To the extent that dialysis was performed in this study mostly in gliotic tissue, it might be conjectured that the PGE_2 that we collected originated largely in this tissue. Whether it was artifactitiously stimulated in this case by infiltrated LPS or is the normal source of circulating LPS-induced iPO PGE_2 remains to be clarified.

It should be emphasized that the traumatic injury and consequent tissue reactions caused by the introduction of a microdialysis probe into the brain substance also attend the insertion of other devices in different sampling/perfusion methods (11). These technique- dependent artifacts probably account, at least in part, for the conflicting data in the literature regarding PGE_2 as a fever mediator. Better definition of its role in the modulation of fever, therefore, awaits resolution of these technical problems.

Acknowledgements

Supported by NIH grant NS-22716.

References

1.   Milton AS. Thermoregulatory actions of eicosanoids in the central nervous system with particular regard to the pathogenesis of fever. Ann NY Acad Sci 1989; 559: 392-410.
2.   Blatteis, CM, Banet M. Autonomic thermoregulation after separation of the preoptic area from the hypothalamus in rats. Pflügers Arch 1986; 406: 480-4.
3.   Blatteis CM. The pyrogenic action of cytokines. In: Rothwell N, Dantzer R, editors. Interleukin-1 in the brain. New York: Pergamon, 1992:93-114.
4.   Rotondo D, Abul HT, Milton AS, Davidson J. Pyrogenic immunomodulators increase the level of prostaglandin E_2 in the blood simultaneously with the onset of fever. Eur J Pharmacol 1988; 154:145-52.
5.   Katsuura G, Arimura A, Koves K, Gottschall PE. Involvement of *organum vasculosum of the lamina terminalis* and preoptic area in interleukin-1β-induced ACTH release. Am J Physiol 1990;258:E163-71.

6.   Hashimoto M, Ishikawa T, Tokata S, Goto F, Bando T, Sakakibara Y, Iriki M. Action site of circulating interleukin-1 on the rabbit brain. Brain Res 1991; 540: 217-23.
7.   Matsumura K, Watanabe Y, Onoe H, Watanabe Y, Hayaishi O. High density of prostaglandin E_2 binding sites in the anterior wall of the 3rd ventricle: a possible site of its hyperthermic action. Brain Res 1990; 553:147-51.
8.   Stitt JT. Prostaglandins, the OVLT and fever. In: Bartfai T, Ottoson D, editors. Neuroimmunology of fever. Oxford: Pergamon, 1992:155-65.
9.   Quan N, Xin L, Ungar AL, Hunter WS, Blatteis CM. Validation of the hypothermic action of preoptic norepinephrine in guinea pigs. Brain Res Bull 1992;28:537-42.
10.   Robinson TE, Justice JB Jr. Microdialysis in the neurosciences. Amsterdam: Elsevier, 1991.
11.   Blatteis CM. Methods for evaluating neural mechanisms in the pyrogenic and other acute-phase responses to cytokines. In: de Souza E, editor. Biology of cytokines. Orlando: Academic, in press.
12.   Benveniste H, Drejer J, Schousboe A, Diemer NH. Regional cerebral glucose phosphorylation and blood flow after insertion of a microdialysis fiber through the dorsal hippocampus in the rat. J Neurochem 1987; 49:729-34.

Temperature Regulation
Advances in Pharmacological Sciences
© 1994 Birkhäuser Verlag Basel

THERMOREGULATORY RESPONSES OF RABBITS

TO COMBINED HEAT EXPOSURE, PYROGEN AND DEHYDRATION

C. M. Blatteis, A.L. Ungar, and R. B. Howell
University of Tennessee, Memphis, TN 38163 USA

SUMMARY
To determine how the competing demands of combined stressors for shared regulatory systems are met, rabbits were exposed to 37° C, or injected with an exogenous pyrogen, or water-deprived for 48 h, in separate experiments. In other experiments, two or all three of these treatments were combined. Core (T_c) and ear skin (T_{sk}) temperatures were monitored continuously. Based on the incidence of mortality, dehydration was the chief contributor to decompensation when it and other stressors were applied together. Core heating *per se* did not appear to be a significant factor in the outcomes.

INTRODUCTION

In nature, animals are continually exposed to a wide variety of stimuli which introduce multiple perturbations that challenge the constancy of their internal environment. To prevent undue deviations from homeostatic levels, appropriate and coordinated corrective measures are constantly instituted. These adjustments involve the interweaving of several regulatory mechanisms whose combined effect, thus, is to maintain equilibrium. While well appreciated in concept, in fact relatively few studies have as yet systematically examined the interrelationships among different regulatory systems when two or more physiological drives occur concurrently. Knowledge of the nature of homeostatic interactions under such conditions is of interest because, in many instances, the responses to different drives utilize common effector mechanisms.

The thermoregulatory system is of particular interest in this regard because, alone among homeostatic systems, it possesses, apart from thermosensors and the sweat glands in certain species, no exclusively proprietary effectors. Rather, the various autonomic and behavioral effectors utilized to maintain core temperature (T_c) are components of other regulatory systems, *viz.*, the cardiovascular, respiratory, neuromuscular, endocrine, etc. systems. It must, therefore, compete for these resources when confronted with stimuli that simultaneously disturb T_c and other regulated variables, the homeostatic defense of which also requires these systems. Examples of situations for which some experimental data are available include combined heat stress and exercise (T_c and body fluids [1, 2]), combined cold and hypoxia (T_c and tissue energy supply [3]), and combined heat and CO_2 (T_c and respiration [4, 5]). The results suggest that the eventual compensatory response depends partly on the relative strengths of the stimuli and partly on the immediate importance for survival of the functions being defended. On this basis, we would postulate that the regulation of T_c occupies a rank in a hierarchy of homeostatic controls; a rank, however, that may change depending on coexistent conditions.

To better understand the place of T_c regulation in complex physiological situations, we have assessed, first, the effects of two concurrent disturbances that impose a load on the thermoregulatory system but evoke opposite thermoeffector responses, *viz.*, an exogenous pyrogen (fever, *i.e.*, increased heat production and decreased heat loss) and heat exposure (no increase in heat production but increased heat loss); then, the responses to the application of a third, simultaneous disturbance, the compensation of which utilizes regulatory systems also used in the defense of T_c, *viz.*, dehydration. It is evident that the potential for morbidity and even mortality exists in situations in which combined disturbances exceed the limits that can be tolerated; *e.g.*, dehydration during heat stress accelerates heat exhaustion (6). To assess, at the outset, the extent of the competition for shared effectors, we elected in this first study to activate heat defense responses as strongly as possible.

MATERIALS AND METHODS

Fully conscious, *Pasteurella*- and coccidia-free male, New Zealand White rabbits at weights of 3.0 - 3.5 kg were used in these experiments. They were housed in individual cages at $22 \pm 2°$ C with light and darkness alternating every 12 h, and provided with Ralston-Purina Rabbit chow and tap water *ad libitum*. For an experiment, the rabbits were held in a locally fabricated, webbed cloth supporter that minimized movement but did not cause restraint stress. They were habituated to the experimental procedures by being placed in their supporters and handled for 3 h for 5 consecutive days before an experiment.

We measured the T_c and ear skin temperatures (T_{sk}) of 4 groups of 6 rabbits exposed for 5 h to $37 \pm 1°$ C ambient temperature (heat), administered *S. enteritidis* lipopolysaccharide (LPS, 2 µg/kg, i.v.) in $26.5 \pm 1.1°$ C (thermoneutrality, T_n), or water-deprived (no water, NW) for 2 days in T_n. The corresponding controls were exposed to T_n for 5 h, received pyrogen-free saline (PFS, i.v.) in T_n, or had normal access to water (W) in T_n, respectively. In other experiments, two of the above conditions were combined, *viz.*, heat exposure and LPS, LPS and dehydration, and heat exposure and dehydration; or all three stressors, *i.e.*, heat exposure, LPS and dehydration were applied concurrently. In the relevant experiments, LPS (or PFS) was administered just before initiating the exposure to heat so that the two drives would coincide closely; water deprivation was instituted 48 h before these treatments. Each animal in a group was used twice such that two of the treatments remained constant while the third changed; thus, a rabbit served as its own control. At least 1 week elapsed between tests. To obviate possible, superimposed effects of circadian rhythms, all the experiments were begun at 8:30 a.m.

RESULTS

The data are summarized in Table 1.

The T_c and T_{sk} of the rabbits exposed to T_n varied insignificantly over the duration of the experiments. Acute exposure to heat elevated these T's. LPS evoked monophasic fevers in T_n. There were no significant differences between the T_c and T_{sk} of the dehydrated and euhydrated rabbits in T_n. When heat exposure and LPS were combined, the time to reach the pre-fixed peak value of T_c when heat exposure was ended was significantly shorter than during exposure to 37° C alone. Moreover, the elevated T_c was sustained without abatement for another 2.0 ± 0.2 h after return to T_n; T_{sk}, on the other hand, decreased immediately upon ending the heat exposure (by 8.6 $\pm$ 0.5° C in 1.1 ± 0.1 h), then stabilized for the remainder of the experimental period. One of the animals succumbed during the night following this experiment. The administration of LPS to

Table 1. Maximal core (T_c) and simultaneous ear skin (T_{sk}) temperatures (means $\pm$ SD) of 4 groups of 6 rabbits in response to the treatments shown.

Treatments	T_c °C	T_{sk}, °C	t, hours[a]	Mortality
T_n + PFS + W	39.2 ± 0.2	32.6 ± 3.5		
Tn + PFS + NW	39.2 ± 0.3	32.6 ± 4.4		
Heat + PFS + W	41.7 ± 0.4[b]	39.5 ± 1.3	2.8 ± 0.5	
Heat + PFS + NW	42.1 ± 0.8[b]	37.2 ± 2.8	3.2 ± 0.8	3/6
T_n + LPS + W	41.6 ± 0.3	31.7 ± 5.1	3.1 ± 0.1	
T_n + LPS + NW	40.3 ± 0.5	26.3 ± 2.3	2.2 ± 0.6	3/6
Heat + LPS + W	41.7 ± 0.2[b]	39.9 ± 1.1	1.4 ± 0.0	1/6
Heat + LPS + NW	42.5 ± 0.1[b]	38.3 ± 2.8	1.4 ± 0.5	5/5

[a]Time to reach maximum T_c. [b]For humane reasons, it was set *a priori* that heat exposure would end when $T_c \cong 41.5 - 42.0°$ C. Abbreviations: T_n, thermoneutrality; PFS, pyrogen-free saline; W, euhydrated; NW, dehydrated; LPS, lipoplysaccharide.

dehydrated rabbits in T_n produced protracted monophasic fevers that were lower than those of the euhydrated rabbits and, therefore, peaked earlier; their T_{sk}, however, were not different. Neither T abated over the following 4.5 h. Three animals died during the following night; the others were moribund the next morning. Exposure to heat of the dehydrated rabbits did not affect them differently than the euhydrated animals; however, 3 of the animals subsequently died in T_n. Finally, the combination of all three stressors resulted in a rapid T_c rise, which continued more slowly after the heat exposure was ended. T_{sk} was elevated in the heat, but fell immediately on reentry to T_n (by $1.2 \pm 0.0°$ C in 0.8 ± 0.3 h). None of these animals survived more than 3 h after exposure.

DISCUSSION

Heat exposure, fever and dehydration are stressors that, individually, produce disturbances in more than one regulated system, calling for diverse compensatory responses. A potential for conflict is created when these stimuli are combined and impose concurrent stressful loads on the body because the various defense mechanisms recruited to meet the demands of each stressor are also partly common to the responses to the other stressors. The present results illustrate how, under the stressful conditions of these experiments, the conflicting demands of competing stressors for shared regulatory mechanisms impacted synergistically on these systems, such that their concomitant needs could not be adequately met, in some cases with fatal consequences. The mechanistic bases for these deleterious effects remain to be precisely clarified, but it would appear that the animals' dehydration was the most aggravating factor in these outcomes while core heating did not seem to contribute significantly to the decompensation.

ACKNOWLEDGMENTS

Supported by NIH Grant HL 47650.

REFERENCES

1.   Nadel ER. Body fluid and electrolyte balance during exercise: competing demands with temperature regulation. In: Hales JRS, editor. Thermal Physiology. New York: Raven, 1984:365-76.

2.   Horowitz M, Samueloff S. Dehydration stress and heat acclimation. In: Yousef MV, editor. Milestones in Environmental Physiology. The Hague:SPB Academic, 1989: 91-100.

3.   Blatteis C M. Shivering and nonshivering thermogenesis during hypoxia. In: Smith RE, editor. Proc Internat Symp Environmental Physiology (Bioenergetics). Bethesda: FASEB, 1972:151-68.

4.   See WR. Interactions between chemical and thermal drives to respiration during heat stress. In: Hales JRS, editor. Thermal Physiology. New York: Raven, 1984:353-8.

5.   Leigh J. Dynamic mathematical models of the interaction between the thermoregulatory system and the chemical respiratory control system in mammals. In: Hales JRS, editor. Thermal Physiology. New York: Raven, 1984:359-64.

6.   Pitts GC, Johnson RE, Consolazio FC. Work in the heat as affected by intake of water, salt and glucose. Am J Physiol 1944; 142: 253-9.

Temperature Regulation
Advances in Pharmacological Sciences
© 1994 Birkhäuser Verlag Basel

FEVER AND THE ORGANUM VASCULOSUM LAMINAE TERMINALIS: ANOTHER LOOK

W. S. Hunter[a], E. Sehic[b], and C. M. Blatteis[b]
[a]Southern Illinois University, Carbondale, IL, and [b]University of Tennessee, Memphis, TN 38163 USA

SUMMARY

Lesions were made to compare in guinea pigs the previously demonstrated antipyretic effectiveness of ablation of the anteroventral third ventricle (AV3V) brain region, which contains the *organum vasculosum laminae terminalis* (OVLT), with that of the rostroventral *lamina terminalis*. Small lesions discretely localized to this site were reported to enhance the febrile responses of rats and rabbits. We found that both types of lesions blocked LPS-induced fever so long as the OVLT was completely ablated. Fever was not increased by the more discrete lesions in these guinea pigs. The present findings are consistent with the hypothesis that the pyrogenic signal is transduced from blood to brain via the OVLT.

INTRODUCTION

How circulating pyrogens signal the brain to initiate fever has been a long standing and vexing question. In 1983, we reported that electrolytic ablation of the AV3V region, which includes the OVLT, prevented the febrile response of conscious guinea pigs to the i.p. injection of *S. enteritidis* (LPS) (1). In the same study, we showed that this lesion did not impair the pyrogenic sensitivity of the surrounding medial preoptic area (MPO), as reflected by its continued capacity to induce fever in response to the local microinjection of a semi-purified, homologous EP; the febrigenic controller is thought to be located in the MPO (2). We concluded, therefore, that the AV3V region contained structures critical for the normal development of fever in guinea pigs. We demonstrated later that AV3V lesions did not affect the thermoregulatory ability of guinea pigs, as assessed by the animals' capacity to mount appropriate and effective thermoeffector responses to acute exposures to ambient temperatures ranging from 12 to 32° C (3). We found subsequently that the fever induced in sheep by i.v. LPS was prevented only if the vascular plexus of the OVLT was destroyed and suggested, therefore, that it was the critical interface between circulating pyrogens and the brain (4). In 1985, Stitt (5) came to the same interpretation, but from different results. He reported that the febrile response of rabbits and rats to the i. v. injection of a semi-purified preparation of human EP was transiently increased, rather than suppressed, following electrolytic lesions of the ventrorostral portion of the *lamina terminalis* (LT). This effect culminated in 6 - 11 days post-lesion, then gradually declined again over the subsequent 10 days. Although it is possible that the divergent results were due to species differences and the fact that different pyrogens and routes of administration were used, another possibility may be the size and localization of the lesions produced in the two studies; *viz*, Stitt's LT lesions were smaller and more discrete than our AV3V lesions. To investigate this possibility, we have compared the effects of electrolytic ablation of the OVLT region of guinea pigs by the two approaches.

MATERIALS AND METHODS

Male Hartley guinea pigs weighing 350 - 400 g were used in these experiments. To ablate the AV3V region, a 0.5-mm diameter, blunt-tipped tungsten wire electrode insulated with epoxy resin except for 0.5 mm at the tip was inserted stereotaxically at coordinates A = 11.6 mm, L = 0.0 mm, and V = 9.5 mm (1, 6); the lesioning current was anodal, 3 mA DC for 20 s. To ablate the ventrorostral LT, a 0.3-mm diameter, stainless steel wire electrode tapered to a 100 μm tip and polyurethane-insulated except for 0.5 mm at the tip was inserted at coordinates A = 11.8 mm, L = 0.0 mm, and V = 9.8 mm; the lesioning current was 2 mA DC for 10 s. In other animals, the relevant electrodes were inserted at the same, corresponding A and L coordinates but only to V = 6.0 mm, and no current was passed (sham-operated). Ten or eleven days after this surgery, the guinea pigs, fully conscious, were injected with 2 μg of *S. enteritidis* LPS/kg, i.v. Core temperature (T_C) was recorded at 2-min intervals from 90 min before to 3 h after this injection. During the measurements, the animals were lightly restrained in wire-mesh confiners to which they had been habituated for 2-3 h/d for 3 d before an experiment; the ambient temperature was 23 ± 1° C. The locations of the lesions were subsequently verified histologically, classified according to the extent of destruction of the OVLT, and matched with the animals' corresponding changes in T_c in response to the LPS treatment.

RESULTS

The results indicated that neither the method of approach nor the size of the lesion demonstrably influenced the outcomes. Rather, the determining factor was the localization of the ablated tissue. Thus, lesions placed precisely on the midline by either approach consistently blocked fever (T_C rise < 0.5° C), provided the ablated region included the anterior aspect of the LT, extending 0.5 mm posteriorly on the rostrodorsal surface of the optic chiasm, in the middle ventralmost portion of the 3V; this region contains the OVLT. If any part of the OVLT remained after the lesion by either method of ablation, fevers indistinguishable from those in sham-operated animals were elicited. Fig. 1 illustrates coronal sections of the brain of an AV3V- and an OVLT-ablated guinea pig that did not develop fever after i.v. LPS. Fig. 2 illustrates the courses of the thermal responses to LPS of guinea pigs with completely and incompletely ablated OVLTs. Ten animals sustained complete lesions and did not develop fever; 7 of these underwent the AV3V and 3 the OVLT approach. Eleven guinea pigs had incomplete ablations of the OVLT and developed fever; these included 5 lesions directed to the AV3V and 6 to the OVLT.

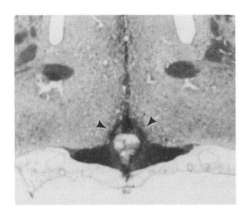 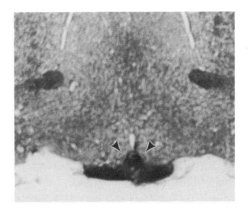

Fig. 1 Coronal sections of guinea pig brains showing complete AV3V (left panel) and OVLT (right panel) lesions. Arrows indicate ablated areas.

The lesions localized to the AV3V region were always more extensive than those to the rostroventral LT (Fig. 1). The likelihood, therefore, that the critical region would be included in the ablated area was greater in animals with AV3V lesions. Hence, the "success" rate was higher in the AV3V (7 of 12 attempts)- than in the LT (3 of 9)-ablated animals. But in no instance were enhanced febrile responses to LPS observed in animals with either lesion.

DISCUSSION

The present results confirm our findings of 10 years ago (1) that the integrity of the ventral portion of the AV3V, i.e., the brain region that contains the OVLT, is critical for the development of LPS-induced fever in guinea pigs. They extend those results by substantiating in this species our earlier findings in sheep (3) that destruction of the vascular plexus of the LT, i.e., of the *organum vasculosum* itself, is necessary in order for the suppression of the febrile response to circulating LPS to become manifest. The present results also show that the method of ablation, *viz.*, the electrode size and type, the strength and duration of the lesioning current, and, consequently, the lesion size, does not influence this outcome; i.e., so long as the ablated region includes the OVLT, fever is consistently prevented.

There was no indication in this study that small lesions confined to the LT affect the febrile response differently than larger lesions directed to the AV3V. The reported transiently enhanced

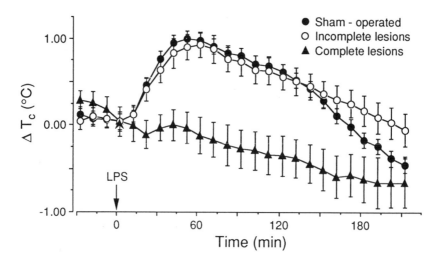

Fig. 2. Changes in colonic temperatures (ΔT_c, means $\pm$ SE) of guinea pigs following the administration of LPS (2 µg/kg, i.v.) to sham-operated ($T_{c\text{initial}} = 38.7 \pm 0.2°$ C; n =12), incomplete OVLT lesions ($T_{c_i} = 38.9 \pm 0.2°$ C; n=10), and complete OVLT lesions ($T_{c_i} = 38.8 \pm 0.1°$ C; n=11) guinea pigs.

fevers of rats and rabbits with localized LT lesions (5), therefore, could largely be due to species differences.

The specific events that occur in the OVLT and transduce the circulating pyrogenic messages into signals that trigger fever production remain to be elucidated (7 - 9).

ACKNOWLEDGMENTS

Supported by NIH grant NS22716 and the Illinois Fraternal Order of the Eagles.

REFERENCES

1.   Blatteis CM, Bealer SL, Hunter WS, Llanos-Q J, Ahokas RA, Mashburn TA Jr. Suppression of fever after lesions of the anteroventral third ventricle in guinea pigs. Brain Res Bull 1983;11:519-26.
2.   Blatteis CM. The pyrogenic action of cytokines. In: Rothwell N, Dantzer R, editors. Interleukin-1 in the brain. New York: Pergamon, 1992:93-114.
3.   Blatteis CM, Hunter WS, Wright JM, Ahokas RA, Llanos-Q J, Mashburn TA Jr. Thermoregulatory responses of guinea pigs with anteroventral third ventricle lesions. Can J Physiol Pharmacol 1987; 65:1261-6.

4.   Blatteis CM, Hales JRS, McKinley MJ, Fawcett AA. Role of the anteroventral third ventricle region in fever in sheep. Can J Physiol Pharmacol 1987;65:1255-60.
5.   Stitt JT. Evidence for the involvement of the *organum vasculosum laminae terminalis* in the febrile response of rabbits and rats. J Physiol (Lond) 1985;368:501-11.
6.   Luparello TJ. Stereotaxic atlas of the forebrain of the guinea pig. Basel:Karger, 1967.
7.   Blatteis CM, Dinarello CA, Shibata M, Llanos-Q J, Quan N, Busija DW. Does circulating interleukin-1 enter the brain? In: Mercer JB, editor. Thermal Physiology 1989. Amsterdam:Elsevier, 1989:385-90.
8.   Blatteis CM. Role of the OVLT in the febrile response to circulating pyrogens. Progr Brain Res 1992; 91:409-12.
9.   Stitt JT. Postaglandins, the OVLT and fever. In: Bartfai T, Ottoson D, editors. Neuroimmunology of fever. Oxford: Pergamon, 1992:155-65.

Temperature Regulation
Advances in Pharmacological Sciences
© 1994 Birkhäuser Verlag Basel

LIPOPOLYSACCHARIDE (LPS)-INDUCED FOS EXPRESSION IN THE BRAINS OF

FEBRILE RATS

A. Oladehin, J. A. Barriga-Briceno, and C. M. Blatteis
University of Tennessee, Memphis, TN 38163, USA

SUMMARY

We have mapped the distribution of Fos-like immunoreactivity (FLI) in the brains of restraint-trained, LPS-treated rats. Fos was visualized in the septum, preoptic area, endocrine hypothalamus, bed nuclei of the stria terminalis, thalamus, amygdala, and nuclei of the solitary tract. Only diffusely scattered FLI neurons were observed in these regions of trained pyrogen-free saline (PFS)-treated controls. But FLI neurons were abundant in these regions of untrained PFS-treated rats, and also in other regions thought to be involved in stress. Thus, discrete hypothalamic and brainstem nuclei are activated by LPS and overlap to some extent with those stimulated by stress.

INTRODUCTION

The entrance of pathogenic microorganisms or their products into the body evokes an array of systemic responses, *e.g.*, fever, acute-phase proteinemia, neutrophilic leucocytosis, increased slow-wave sleep, reduced food intake, and changes in the circulating levels of various hormones, in defense of the invaded host against the potentially deleterious effects of these agents. These responses are collectively termed the acute-phase reaction (APR). It is now generally recognized that many of these responses are modulated by the brain. Thus, the microinjection of putative endogenous mediators of the APR into various, discrete brain loci evokes pertinent reactions, *e.g.*, fever and acute-phase proteinemia in the preoptic area of the anterior hypothalamus (1), reduced food intake in the lateral hypothalamus (2). These brain regions are presumed to be interconnected into a neural network that becomes activated in a coherent, organized concatenation of events (3). However, the extent of this network has not yet been fully documented during the APR.

Immediate early genes, such as *c-fos*, are acutely induced in the nuclei of many cells by various stimuli; they function in the mediation of genomic responses to cellular stimulation (4). As applied to the CNS, the expression of the Fos protein of the *c-fos* protooncogene is thought, therefore, to correlate with the functional activation of discrete neurons (5). We have used this feature to examine the distribution of activated neurons in the brains of rats following the administration of LPS, an exogenous inducer of the APR.

MATERIALS AND METHODS

Male Sprague-Dawley rats weighing 230-500 g were used in this study. The animals were housed three to a cage and maintained on a 12-12 h light-dark cycle at an ambient temperature of 22 $\pm 2°C$. Tap water and rodent chow were provided *ad libitum*. The rats were randomly assigned into trained LPS (*E. coli*, Difco; 10 µg/kg at 1 ml/kg, i.v.)-, trained pyrogen-free saline (PFS; 1 ml/kg, i.v.)-, and untrained PFS (1 ml/kg, i.v.)-treated groups; *i.e.*, all but the latter group were habituated for 4 h/day for 4 consecutive days prior to experimental use to restraint in individual plexiglass restrainers, to handling and to insertion of a colonic thermistor. The untrained rats were controls for possible, nonspecific, stress-related Fos expression. Core temperatures (T_c) were measured at 15-min intervals from 90 min before until 90 min after drug administration. At this time, the T_c of the LPS-treated animals had usually risen to *ca.* 1.2°C above its initial value; this febrile response was used as the index of the efficacy of LPS to induce the APR. The rats were anesthetized and perfused transcardially with heparinized PFS, followed by 4% paraformaldehyde and 0.2% picric acid in 0.15 M phosphate-buffered PFS (PBS). The brains were removed, blocked, and postfixed overnight. The next day, tissue blocks containing the hypothalamus and the medulla oblongata were sectioned into 50 µm serial sections, using a vibratome. After rinsing with PBS, the sections were treated with 1% H_2O_2, rinsed, and incubated in: 1) 10% normal goat serum (NGS, Sigma) in PBS containing 0.4% Triton X-100 for 1 h, 2) *c-fos* (ab-2, rabbit polyclonal antibody, Oncogene Science) diluted at 1:1000 in PBS containing 2% NGS and 0.4% Triton X-100 for 48 h at 4 °C, 3) biotinylated goat anti-rabbit IgG (heavy and light chains, Vector Lab) diluted at 1:100 in PBS for 3 h, and 4) standard avidin-biotin complex solution (Vector Lab), using the manufacturer's recommended dilution. The sections were subsequently stained with 3,3'-diamino-benzidine (DAB, Sigma; 0.5 mg/ml), 0.02% nickel ammonium sulphate, and 0.002% H_2O_2 in PBS, mounted on gelatin/chrom alum-coated glass slides, air dried, dehydrated through graded ethanol solutions, cleared in xylene, and coverslipped with permount. In every case, tissues from a control and an experimental animal were processed together for the same period of time and with the same reagents. The sections were examined using a Nikon Labphot microscope at a magnification of 200X. The presence of Fos was evident as dark black reaction products in the cell nuclei. Fos-like immunoreactive (FLI) cells were plotted with a camera lucida and selected sections photographed with a Nikon N2020 automatic camera system. The localization and distribution of stained nuclei were compared in identical brain regions among the three groups.

RESULTS

The data are summarized in Table 1. Representative sections are illustrated in Fig. 1.

Fos was visualized in the following regions of the LPS-treated rats: septum (medial and lateral nuclei), preoptic area (medial and lateral), hypothalamus (paraventricular, arcuate, supraoptic and posterior nuclei), bed nuclei of the stria terminalis, thalamus (paraventricular, paratenial, anteromedial, rhomboid and reuniens nuclei), central amygdaloid nuclei, and nuclei of the solitary

Table 1. Fos expression in various brain regions of LPS- and PFS-treated rats.

Regions	Trained PFS	Trained LPS	Untrained PFS
Septum	+	+++	++++
Preoptic area	+	+++	++++
Hypothalamus	+	+++	++++
Thalamus	+	+++	++++
Amygdala	-	++	++++
Supraoptic nuclei	-	-	+++
Bed nuclei of the stria terminalis	+	+++	++++
Nuclei of the solitary tract	+	++	++++
Cortex	-	-	+++
Limbic system	-	-	+++

tract. The circumventricular organs *organum vasculosum laminae terminalis*, subfornical organ, and *area postrema* were not labeled *per se*, but their adjacent regions were. Neurons in the anterior hypothalamic area, hypothalamic suprachiasmatic, ventromedial and dorsomedial nuclei, medial forebrain bundle, and median eminence did not express FLI at this 90-min post LPS time. Only diffusely scattered FLI neurons were observed within the brains of the trained PFS-treated rats; they were located mostly in the lateral septum, hypothalamic paraventricular nuclei, midline nuclei of the thalamus, and nuclei of the solitary tract. But Fos-positive cells were abundant in these regions of the untrained PFS-treated (stressed) rats, and also in their supraoptic nuclei, amygdala, periventricular regions, frontal, parietal and entorhinal cortices, corpus callosum, globus pallidus, claustrum, and nuclei of the olfactory tracts.

DISCUSSION

In general, the distribution of Fos-positive cells in the LPS-treated rats was consistent with that of the brain regions thought to modulate the APR (3). For example, abundant FLI neurons were

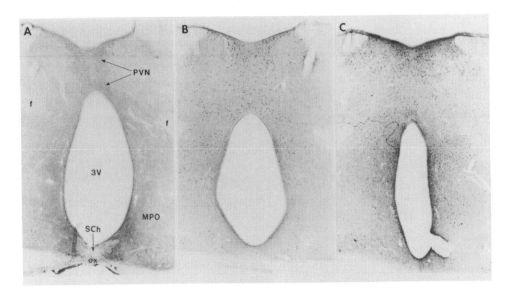

Fig. 1. Photomicrographs showing the hypothalamus of trained PFS (A)-, trained LPS (B)-, and untrained PFS (C)-treated rats. Abbreviations: PVN, paraventricular nucleus; MPO, medial preoptic area; f, fornix; ox, optic chiasm; 3V, third ventricle; SCh, suprachiasmatic nucleus.

observed in the preoptic and septal areas, the paraventricular, arcuate and supraoptic nuclei, and the bed nuclei of the stria terminalis, all sites previously implicated in the neuromodulation of fever production and lysis (6, 7).

Interestingly, Fos was also visualized in these animals in neurons adjacent to the subfornical organ and *area postrema*, two circumventricular organs that have not been previously associated with the central activation of the APR (8); but Fos was not observed in the immediate vicinity of the median eminence, which has been previously involved in the endocrine APR (8). In addition, Fos was unexpectedly expressed in these LPS-treated rats in some brain regions that were found also to be labeled in the present untrained PFS-treated rats, e.g., the amygdala, the midline nuclei of the thalamus, and the supraoptic nuclei. By contrast, all these regions were only lightly and diffusely labeled in the trained PFS-treated rats. Since the latter brain sites have been reported to become activated during nonspecific stress (9, 10), these results would suggest that the APR to LPS may include a neural component similar to that produced by stress ("sickness behavior"?).

ACKNOWLEDGMENTS

Supported by NIH grants NS-22716 and HL-47650.

REFERENCES

1. Blatteis CM, Hunter WS, Llanos-Q J, Ahokas RA, Mashburn TA Jr. Activation of acute-phase responses by intrapreoptic injections of endogenous pyrogen in guinea pigs. Brain Res Bull 1984;12:689-95.
2. Plata-Salaman CR, Oomura Y, Kai Y. Tumor necrosis factor and interleukin-1β: suppression of food intake by direct action in the central nervous system. Brain Res 1988;448:106-14.
3. Blatteis CM. Neuromodulative actions of cytokines. Yale J Biol Med 1990;63:133-46.
4. Sagar SM, Sharp FR, Curran T. Expression of *c-fos* protein in brain: metabolic mapping at the cellular level. Science 1988;240:1328-31.
5. Dragunow M, Faull R. The use of *c-fos* as a metabolic marker in neuronal pathway tracing. J Neurosci Methods 1989;29:261-5.
6. Blatteis CM. The pyrogenic action of cytokines. In: Rothwell N, Dantzer R, editors. Interleukin-l in the brain. New York: Pergamon, 1992:93-114.
7. Kasting NW. Criteria for establishing a physiological role for brain peptides. A case in point: the role of vasopressin in thermoregulation during fever and antipyresis. Brain Res Rev 1989;14:143-55.
8. Katsuura G, Arimura A, Koves K, Gottschall PE. Involvement of organum vasculosum of the lamina terminalis and preoptic area in interleukin-1β-induced ACTH release. Am J Physiol 1990;258:E163-71.
9. Palkovits M. Neuroanatomical overview of brain neurotransmitters in stress. In: Loon AR, Kvetnansky R, McCarty R, Axelrod J, editors. Neurochemical and humoral mechanisms, vol 1. New York: Gordon and Breach, 1989:31-42.
10. Smith MA, Banerjee S, Gold PW, Glowa J. Induction of *c-fos* mRNA in rat brain by conditioned and unconditioned stressors. Brain Res 1992;578:135-41.

Temperature Regulation
Advances in Pharmacological Sciences
© 1994 Birkhäuser Verlag Basel

ANALYSIS OF BODY TEMPERATURE AND BLOOD PROTEIN IN HYPOTHERMIC SYRIAN HAMSTERS AND RATS

Nobu Ohwatari, Masaki Yamauchi, Munenori Shimazu, Takaaki Matsumoto, Klaus Pleschka* and Mitsuo Kosaka
Department of Environmental Physiology, Institute of Tropical Medicine, Nagasaki University, Sakamoto 1-12-4, Nagasaki 852, Japan; *Max-Planck-Institut für Physiologishe und Klinische Forschung, W.G. Kerckhoff-Institut, D-61231 Bad Nauheim, Federal Republic of Germany

Summary

To clarify the ability of cold tolerance of Syrian hamsters (Mesocricetus auratus) during the nonhibernating state as compared with Wistar rats, hypothermia was induced by acute cold exposure at 0°C with ice packs. On skin insulation, the back skin was better than the abdominal skin in both species, and Wistar rats was superior to Syrian hamsters. Minimal temperature in the abdominal cavity of Syrian hamsters in nonhibernation (7.22 ± 0.31°C) was significantly lower than 17.11 ± 0.39°C of Wistar rats. Blood proteins on SDS-PAGE 18hr after acute cold exposure changed as follows: 17.5, 19, 35 and 43 kDa proteins increased and 32 and 40 kDa proteins decreased. A little change occurred in proteins of the brain and kidney of Syrian hamsters.

Introduction

The hypothermia in hibernation is actively controlled by the physiological functions of the mammalian hibernator. On the other hand, hypothermia induced by acute cold exposure in the non-hibernating state is a passive phenomenon for the hibernator. Therefore, many differences between the two types of hypothermias may exist in terms of physiological functions and responses. The hibernators recover from body temperature below 6°C, whereas nonhibernating endotherms, made hypothermic, die at body temperature in the range of 12-20°C (Adolph, 1951, Spector, 1956). However, hypothermic golden hamster (rectal temperature, 7°C) were made hypothermia by the technique of Musacchia (1972) using a gas mixture containing 90% helium; 10% oxygen at 0°C. In the present study, Syrian hamster recovered from body temperature below 8°C without the technique by Musacchia. This study was performed in order to clarify the ability for cold tolerance of Syrian hamsters (Mesocricetus auratus) in the passive hypothermic state induced by acute cold exposure during the nonhibernating state.

Materials and Methods

Fifteen Syrian hamsters and six Wistar rats were used in this study, and body weights were 182.7±25.1g and 347.5± 41.1g, respectively. Animals were reared at an ambient temperature of 22.0±2.0°C with a relatve humidity of 60±5% and in a photo-period of 12:12 (Light: 6:00-18:00). Food and water were supplied ad libitum.

Experiments were performed in the environmental chamber at 23°C and 60% rh. After animal was anesthetized with sodium pentobarbital (25 mg/kg i.p.), thermistor thermosensors (0.8 mm in diameter) were inserted into the abdominal cavity and the interscapular brown adipose tissue. Thermosensors were subcutaneously set on the abdomen and the back. A small balloon connected to a pressure transducer was subcutaneously implanted on the chest to pick up the respiratory rate. The limbs of animal were fixed to a wooden frame (170×270 mm) loosely as not to block the movement for cold shivering.

After arousal from anesthesia, the whole body surface except the face was covered with ice packs at 0°C for acute cold exposure. Cold exposure was continued until bradypnea or bradycardia and high amplitude in ECG occurred during hypothermia. After cessation of cold exposure, the animal was allowed to recover from hypothermia without any heating procedures in the same environmental condition.

Temperatures in the abdominal cavity (Tab), under the abdominal skin (Tabs), under the back skin (Tbs) and in the interscapular brown adipose tissue (Tbat) were simultaneously measured with a thermistor thermometer every minute. Respiratory rate, ECG and EMG in the musculus scalenus medius were simultaneously recorded with a data recorder.

Animals were sacrificed at 0, 1, 3, 6 and 18 hours after the end of acute cold exposure and tissue proteins were analyzed by 9.0 and 12.5% SDS-PAGE.

Body temperatures were fitted into regression equations for mathematical analysis of thermal transition with time as follows: [T: temperature (°C), t: time (min)]

$$T = C \cdot \exp(-k \cdot t) + a \cdot t + b \tag{1}$$

Formula (1) was used to analyze the decreasing phase of body temperatures due to cold exposure. The term $C \cdot \exp(-k \cdot t)$ reflects the dynamic change in the initial period during acute cold exposure, and $a \cdot t + b$ mainly reflects the static change after the initial period.

$$T = a \cdot t^3 + b \cdot t^2 + c \cdot t + d \tag{2}$$

Formula (2) represents a cubic curves and was useful in analyzing the recovery phase after the cold exposure. The pattern and increase rate of core temperatures were analyzed from the inflection point of these cubic curves. Values are represented as mean±SE, and all statistical comparisons were performed with Fisher PLSD.

Results

Body temperatures of Syrian hamsters and Wistar rats decreased significantly during cold exposure (Table 1). The cooling times were 65.9±5.8 minutes for Syrian hamsters and 54.4±6.3 minutes for Wistar rats. Tab of Syrian hamsters quickly increased just after acute cold exposure, whereas Tab of Wistar rats further decreased. Difference between minimal temperatures (Syrian hamster: 7.22±0.31°C, Wistar rats: 17.11±0.39°C) was significant (p<0.01).

Table. 1. Body temperatures of Syrian hamsters and Wistar rats before and at the end of cold exposure
H: Syrian hamster (N=14), R: Wistar rat (N=6), Tab: Temperature in the abdominal cavity, Tbat: Temperature in the interscapular brown adipose tissue Tre: Rectal temperature, Tabs: Temperature under the abdominal skin, Tbs: Temperature under the back skin

		Tab	Tbat	Tre	Tabs	Tbs
Before	H	37.24±0.26	37.51±0.16	36.87±0.38	36.23±0.37	36.07±0.22
	R	36.35±0.50	36.31±0.26	36.09±0.88	36.13±0.59	35.68±0.59
End	H	7.40±0.35a	11.15±0.76	6.56±0.37a	3.53±0.44	8.44±0.44b
	R	18.77±0.80ac	21.37±0.49c	18.99±0.53ac	8.39±0.66c	18.99±0.32bc

There are significant differences between Before and End in same column (p<0.001). Mean ± SE, a: p<0.01 as compared with Tbat in the same line, b: p<0.01 as compared with Tabs in the same line, c: p<0.01 as compared with Syrian hamster in the same stage.

Tbs maintained a higher level than Tabs during acute cold exposure and was higher than Tab at the end of cold exposure in both species. Regression equations in decreasing phase during acute cold exposure are shown in Table 2. This equation modified with Newton's cooling formula, was best fit to analyze the decrease of temperature in this study. Temperatures of Syrian hamsters decreased more quickly compared with those of Wistar rats in the initial term of cold exposure. For the last half term of cold exposure, decrease rates of body temperatures in Syrian hamsters were lesser than those in Wistar rats.

On Tab and Tre in recovery phase after the cold exposure, biphasic rise in Syrian hamster and monophasic rise in Wistar rat were observed. To get the point of the minimal increase rate of Tab in Syrian hamsters, Tab data excluding the last portion of the recovery phase were fitted to the regression equation of a cubic curve. This point is equal to an inflection point calculated from the

Table. 2. Coefficients of regression equations on body temperatures in decreasing phase during acute cold exposure.

$T = C \cdot \exp(-k \cdot t) + a \cdot t + b$ [T: temperature (°C), t: time (min)]

H: Syrian hamster, R: Wistar rat, Tab: Temperatures in the abdominal cavity, Tbat: Temperatures in the interscapular brown adipose tissue, Tre: Rectal temperature, Tabs: Temperatures under the abdominal skin, Tbs: Temperatures under the back skin

		C	$k(10^{-2})$	$a(10^{-2})$	b
Tab	H	16.30	9.915	-21.53	20.18
	R	5.130	14.88	-33.19	31.88
Tbat	H	9.842	9.759	-29.05	28.04
	R	12.43	4.633	-8.013	24.00
Tre	H	16.71	7.611	-26.43	19.65
	R	4.080	4.110	-32.96	33.35
Tabs	H	29.17	40.78	-6.116	7.27
	R	23.80	21.87	-12.35	13.58
Tbs	H	17.07	12.40	-19.63	18.89
	R	9.436	25.12	-19.86	27.31

(Syrian hamster: t<76 min, Wistar rat: t<48 min)

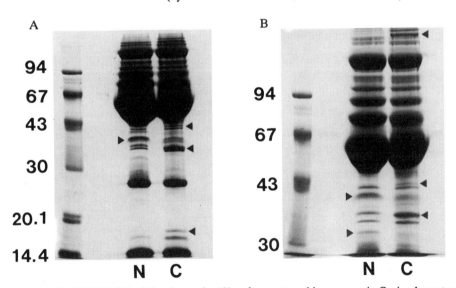

Fig.1. Changes in SDS-PAGE of blood protein 18hr after acute cold exposure in Syrian hamster.
A: 12.5% SDS-PAGE, B: 9.0% SDS-PAGE, N: Normal control hamster, C: Cold exposed hamster, Number on the left side: Scale (kDa).

cubic curve. The recovery time was differed for each animal in both species, therefore fitting to the regression equation was individualized. Regression coefficients were over 0.999 in each animal. For example on Tab, coefficients a, b, c and d in regression equations ($T = a \cdot t^3 + b \cdot t^2 + c \cdot t + d$) were 0.00003, -0.00473, 0.33737 and 6.10387 for a Syrian hamster and -2.088×10^{-6}, 0.00085, 0.00579 and 18.144 for a Wistar rat, respectively. The value of factor "a" was positive in Syrian hamsters, but negative in Wistar rats. The inflection point calculated from the regression equation in a Syrian hamster was 15.12°C at 52.6 minutes after acute cold exposure. Tab at inflection points was 14.58±0.34 °C. Spikes in EMG due to cold shivering disappeared in hypothermia, and reappeared again after the inflection point in Tab of each Syrian hamster.

Acute cold exposure produced changes in proteins of blood, brain and kidney in Syrian hamsters. Changes in SDS-PAGE pattern of blood proteins 18hrs after the acute cold exposure were as follows: Proteins of 17.5, 19, 35 and 43 kDa increased and proteins of 32 and 40 kDa decreased (Fig.1). A little change occurred in proteins of the brain and kidney of Syrian hamsters.

Discussion

Body temperatures of Syrian hamsters decreased more quickly during the initial term of cold exposure as compared with those of Wistar rats (Table 2). Body weights might be attributed to this difference.

Tabs decreased more quickly than Tbs during acute cold exposure (Table 1, 2). This result suggests that the back skin insulation is better than the abdominal skin insulation for both species. It is caused by facts which the back skin was thicker than the abdominal skin and hair length of the back skin was much longer than that of the abdominal skin in both species. The difference between back skin and abdominal skin insulations suggests that the posture to curl itself up in a cold environment is useful for suppression of heat loss from the body surface and the spinal cord, which is an important organ, is located under the back skin at the midline. The skin insulation in Wistar rats was superior to that of Syrian hamsters from Table. 2.

Tbat was always higher than Tab and Tre during and after acute cold exposure in both species. In the latter half of cold exposure, decrease rates of core temperatures in Syrian hamster were lesser than those in Wistar rat. After acute cold exposure, Tab of Syrian hamsters quickly increased at once, whereas Tab in Wistar rats further decreased more than 10 minutes. These results suggest that the non-shivering thermogenesis disturbing the decrease of Tab elevated Tab quickly due to cessation of cold exposure and heat was further supplied from brown adipose tissues, and the capacity the non-shivering thermogenesis in Syrian hamster is superior to Wistar rats in hypothermia.

In the recovery phase after acute cold exposure, the biphasic rise in Tab and Tre of the Syrian hamster were fitted into a positive cubic curve, and the monophasic rise in Tab and Tre of the Wistar rat and in Tbat of both species were fitted into a negative cubic curve. Inflection point was at the point of minimal increase rate for a positive cubic curve and was at the point of maximal increase rate for a negative one in the range of 0°C to 40°C. Tab was 14.58±0.34°C at the inflection point in Syrian hamsters. This temperature approximates the threshold temperature wherein spikes in EMG due to cold shivering appear again. These results suggest that cold shivering thermogenesis contributes to the increase of body temperatures mainly in the second phase after the inflection point, and the threshold for Tab is near 15°C. Changes in heart rate and Tre were similar to the report from Temple et. al. (1977).

Acute cold exposure produced changes of proteins in the blood, and a little change of proteins in the brain and the kidney of Syrian hamsters. On the other hand, Kondo and Kondo (1992) reported changes of blood protein in Asian chipmunks even in the nonhibernation in winter season. Further experiments should be done to confirm the change of proteins using various methods.

It is well known mammalian hibernators can recover from core temperature at 6°C in hibernation. In this study, the Syrian hamster as a hibernator recovered from core temperature at 7.22±0.31°C in nonhibernation. This temperature is very low as compared to that of nonhibernator and approximates the core temperature of hibernating mammals. It is considered that thermoregulatory and other regulatory functions for cold tolerance is active even in the hypothermia not only in hibernation but in nonhibernation, and concomitant functions are additionally enhanced for long term hypothermia in hibernation.

References

Adolph, E. F. (1951): Responses to hypothermia in several species of infant mammals. Am. J. Physiol. 166: 75-91.
Musacchia, X. J. (1972): Heat and cold acclimation in helium cold hypothermia in the hamster. Am. J. Physiol. 222: 495-498.
Spector, W. S. (Editor), (1956): Handbook of Biological Data. Philadelphia, PA: Saunders.
Temple, G. E., Musacchia, X. J. & Jones, S.B. (1977): Mechanisms responsible for decreased glomerular filtration in hibernation and hypothermia. J. Appl. Physiol. 42(3): 420-425.
Kondo N. & Kondo J. (1992): Identification of novel blood proteins specific for mammalian hibernation. J. Biol. Chem. 267, 473-478.

Temperature Regulation
Advances in Pharmacological Sciences
© 1994 Birkhäuser Verlag Basel

NEUROPHYSIOLOGY OF THERMOREGULATION:
ROLE OF HYPOTHALAMIC NEURONAL NETWORKS

Jack A. Boulant
The Ohio State University
Columbus, Ohio 43210, U.S.A.

SUMMARY

This paper reviews hypothalamic neuronal thermosensitivity. It concentrates on recent studies using extracellular and intracellular recording techniques. It examines three basic types of neurons in the rostral hypothalamus: 1) warm sensitive neurons that integrate local thermal information with peripheral inputs and endogenous factors; 2) cold sensitive neurons, many of which are synaptically driven by warm sensitive neurons; and 3) temperature insensitive neurons that form the mortar used to build local neuronal networks.

INTRODUCTION

The neural control of body temperarture depends on the ability of central neurons to sense and convey information. This applies, in particular, to rostral hypothalamic neurons; neurons that sense changes in central temperature, integrate this information with other information (e.g., peripheral temperature, pyrogens and other endogenous factors), and synatically relay this information to effector neurons to produce heat loss, heat retention or heat production responses. The limited goal of this paper is to review some mechanisms that allow hypothalamic neurons to sense and relay thermal information. Admittedly, this approach is overly simplistic; it implies that the neural control of a complex system, like thermoregulation, can be appreciated by understanding the characteristics of three basic types of hypothalamic neurons. This review is also quite personal; it focuses on the author's own collaborative experiments with several investigators. The author is extremely grateful to these colleagues and appreciates the opportunity to work with these creative individuals.

While a thermosensitive area in the rostral brain stem had been suggested in earlier studies, it was, perhaps, the classic 1938 study by Magoun, Harrison, Brobeck and Ranson [1] that best provided the early delineation of the preoptic area and anterior hypothalamus (PO/AH) as a temperature sensitive region that was important in controlling a thermoregulatory response. Subsequent studies, particularly by Hammel, Hardy and their coworkers, showed that many

appropriate responses could be evoked by local changes in PO/AH temperature, using animals
implanted with water-perfused hypothalamic thermodes. Fig. 1A shows a sagittal view of a rabbit
brain with two thermodes implanted in the septum and PO/AH. Studies in several species have
shown that local warming of this area can produce proportional increases in panting, sweating,
skin blood flow and various heat loss behaviors; while local cooling of this area can produce
proportional increases in shivering, non-shivering thermogenesis and various heat retention
behaviors [reviewed in 2-4].

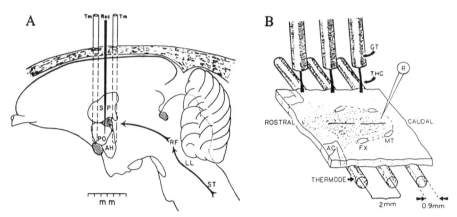

Fig. 1. *In vivo* and *in vitro* preparations used in electrophysiological studies of hypothalamic
thermosensitive neurons. **A**. Sagittal view of a rabbit's brain showing the locations of implanted
thermodes (Tm). Rec, recording microelectrode; PO, preoptic region; AH, anterior hypothalamus;
SP, septum; RF, midbrain reticular formation; LL, lateral lemniscus; ST, lateral spinothalamic
tract. (Modified from [8].) **B**. Horizontal tissue slice through the rat diencephalon showing 3
underlying thermodes and a recording microelectrode (R). THC, thermocouples in guide tubes
(GT); AC, anterior commissure; FX, fornix; MT, mammillothalamic tract. (From [9].)

EXTRACELLULAR RECORDING STUDIES

Neuronal types. Starting in the early 1960s with the first single unit recordings by Nakayama
and coworkers [5], a host of electrophysiological studies have used extracellular microelectrodes to
identify different types of hypothalamic and septal neurons, based on their firing rate responses to
thermode-induced changes in local temperature (see Fig. 1A). As shown in Fig. 1B, similar
types of neurons can also be recorded *in vitro*, using hypothalamic tissue slices. Fig. 2 shows
examples of the three basic neuronal types. Warm sensitive neurons (Fig. 2B) increase their firing
rates (impulses/sec; imp·s^{-1}) when the local temperature is increased, and cold-sensitive neurons
(Fig. 2C) decrease their firing rates when the local temperature is increased. A neuron's
thermosensitivity or thermal coefficient is the slope of the firing rate plotted as a function of
temperature. In many studies, the criterion for warm-sensitivity is a positive thermal coefficient of

at least +0.8 imp·s^{-1}·°C^{-1}, and the criterion for cold-sensitivity is a negative thermal coefficient of at least -0.6 imp·s^{-1}·°C^{-1}. If a neuron's thermal coefficient (or slope) does not met these criteria, it is classified a temperature insensitive (Fig. 2A). Temperature insensitive neurons may also be divided into 2 sub-groups: moderate-slope temperature insensitive neurons (+0.8 to +0.1 imp·s^{-1}·°C^{-1}) and low-slope temperature insensitive neurons (+0.1 to -0.1 imp·s^{-1}·°C^{-1}).

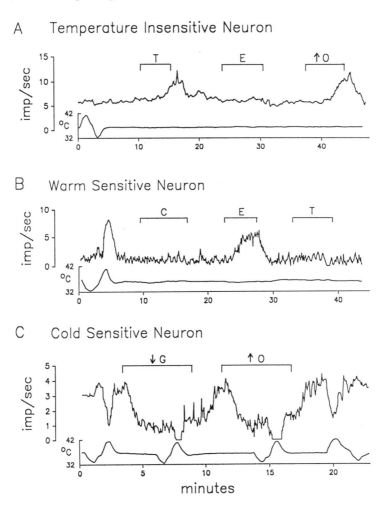

Fig. 2. Effects of endogenous factors on 3 different preoptic neurons recorded in rat tissue slices. Each record shows firing rate (imp/sec) and tissue temperature. **A**. temperature insensitive neuron. **B**. warm sensitive neuron. **C**. cold sensitive neuron. Bars indicate when the perfusion with normal medium was switched to an experimental medium containing: T, testosterone; E, estradiol; C, ethanol control; O, hyperosmotic (309 mosmol.); G, low gluscose (1 mM). (From [10].)

In the PO/AH, the proportions of thermosensitive neurons remain fairly consistent in both *in vivo* and *in vitro* electrophysiological studies [reviewed in 4]. Usually, about 30% of the PO/AH neurons are considered warm sensitive; less than 5-10% are cold sensitive; and the remainder (i.e., >60%) are temperature insensitive. Using the horizontal tissue slice preparation (Fig. 1B), it has also been shown that these proportions are similar in other diencephalic regions [6]. This indicates that neuronal thermosensitivity, *per se*, does not necessarily imply that a neuron has a role in temperature regulation. The whole-animal thermode studies would suggest, however, that at least some of the PO/AH neurons have taken advantage of their thermosensitivity and have made the appropriate synaptic connections to actually control thermoregulatory responses.

In addition to defining local thermosensitivity, the above-mentioned criteria also best delineate the PO/AH neurons that respond to afferent projections from remote or peripheral thermosensitive neurons. *In vivo* studies (Fig. 1A) have characterized afferent input to hypothalamic neurons, by studying neuronal responses to changes in skin or spinal temperatures [7] or electrical stimulation of projections to the hypothalamus [8]. In general, it is the warm sensitive and cold sensitive neurons, not the temperature insensitive neurons, that respond to these afferent inputs.

Similarly, recent studies have used hypothalamic tissue slices to better understand the properties of individual neurons and their synaptic networks. Fig. 1B shows how a microelectrode can record at several different locations in a horizontal slice through the entire rat hypothalamus. Tissue temperature can be changed by warming or cooling the perfusion medium that bathes the slice. Alternatively, regional changes in tissue temperature can be accomplished using the three thermodes underneath the slice. Each thermode permits separate and independent thermal stimulation in a confined region of the slice. In this way, a neuron's own thermosensitivity can be identified, using a single thermode directly under the microelectrode. Following this, the local temperature can be maintained constant while the temperature is changed in a remote part of the slice, using one of the other thermodes. This identifies the types and locations of neurons that receive axonal projections from thermosensitive neurons in remote parts of the hypothalamus. Like the *in vivo* studies depicted in Fig. 1A, the *in vitro* studies in Fig. 1B suggest that it is the locally thermosensitive neurons that receive the projections from thermosensitive neurons in other regions of the hypothalamus [9]. Neurons that are insensitive to local temperature are also unaffected by remote temperatures. These studies suggest that the thermosensitive neurons, particularly the warm sensitive neurons, serve to integrate local thermal information with afferent thermal information coming from remote sources. These studies also indicate that thermosensitive neurons form a synaptic network that extends throughout the diencephalon.

Neuronal integration of thermal and non-thermal information. One advantage of *in vitro* studies is that hypothalamic neurons can easily be tested for their responsiveness to pyrogens and other endogenous factors that may influence neuronal activity. In addition to temperature regulation, the

PO/AH has been implicated in the control of several homeostatic systems, including the regulation body water, glucose and reproduction. Before the use of hypothalamic slices, the concept of "neuronal functional specificity" suggested that a neuron's functional role in these systems may be determined by its sensitivity to a particular endogenous factor. For example, the most likely role of a PO/AH thermosensitive neuron might be temperature regulation, while the most likely role of an osmosensitive neuron might be body water regulation, and the most likely role of a testosterone-sensitive neuron might be reproduction. A corollary to "functional specificity" suggested that most of the PO/AH osmosensitive, glucosensitive, or steroid-sensitive neurons would be confined to the population of temperature insensitive neurons. Fig. 2 summarizes a series of *in vitro* experiments that contradicted this "functional specificity" hypothesis [10]. Of the preoptic neurons tested with both low glucose and hyperosmotic media, responsiveness was observed in 44% of the thermosensitive neurons, but in only 27% of the temperature insensitive neurons. Similarly, of the neurons tested with estradiol and testosterone, steroid-sensitivity was observed in 52% of the thermosensitive neurons, but in only 35% of the temperature insensitive neurons. These studies suggest that even at the level of the single neuron, there is a basis for interactions between different regulatory systems; changes in nonthermal endogenous factors can affect thermoregulation, and similarly, changes in temperature can affect each of the other regulatory systems. Perhaps most importantly, these neurons with multiple-sensitivities suggest that within the hypothalamus, there are interactive networks concerned with homeostasis in general, which is the composite of all regulatory systems.

As indicated above, temperature sensitive neurons are located throughout the hypothalamus [6]. A recent study [11] has found thermosensitive neurons in the hypothalamic suprachiasmatic nucleus (SCN), an area producing the circadian pacemaker that affects several regulatory systems, including the diurnal changes in body temperature and sleep. In addition, sleep and other circadian events can be affected by temperature, suggesting that the thermosensitive SCN neurons may contribute to interactions between temperature and circadian rhythms. Perhaps the most interesting characteristic of certain SCN neurons is that they display circadian changes both in their firing rates and in their thermosensitivities [11]. This occurs in tissue slices from the rat (a nocturnal animal) where SCN neurons show circadian firing rate changes, with highest firing rates occurring during the middle of the day. In contrast, these SCN neurons show circadian changes in neuronal thermosensitivity, with the greatest warm-sensitivity occurring at night. During the day, few temperature sensitive neurons are recorded in the SCN; in fact, the proportion of thermosensitive neurons is less than any other hypothalamic region. On the other hand, the proportion of SCN thermosensitive neurons increases dramatically during the night. In certain parts of the SCN, more than 40% of the neurons are warm sensitive at night. These changes in SCN neuronal thermosensitivity may account for interactions between body temperature and circadian rhythms.

INTRACELLULAR RECORDING STUDIES

Warm sensitive neurons. Using extracellular recordings in tissue slices, it is possible to record the activity of warm sensitive neurons before, during, and after reversible synaptic blockade with high [Mg] - low [Ca] perfusion. PO/AH warm sensitive neurons tend to retain their thermosensitivity during synaptic blockade, suggesting that warm-sensitivity is an intrinsic property of these neurons. To understand the cellular basis of neuronal thermosensitivity, recent studies have employed sharp-tip and whole-cell patch microelectrodes to record the intracellular activity of PO/AH neurons in tissue slices [12-13].

PO/AH neuronal thermosensitivity cannot be explained simply by the effect of temperature on the resting membrane potential or on a cell's input resistance. Firing rate changes in warm sensitive neurons are not due to depolarizations during warming or hyperpolarizations during cooling. Moreover, like most neurons, the input resistance of almost all PO/AH neurons (i.e., warm sensitive, cold sensitive, and temperature insensitive neurons) decreases with warming and increases with cooling.

As indicated in Fig. 3, PO/AH warm sensitive neurons display rhythmic firing rates with constant interspike intervals. Much of the basis of warm sensitivity in these neurons resides in the depolarizing prepotential (or pacemaker potential) that precedes each action potential. Fig. 3 shows that the rate of rise of this depolarizing prepotential increases with an increase in temperature. This, in turn, leds to a decreased interspike interval and increased firing rate with warming.

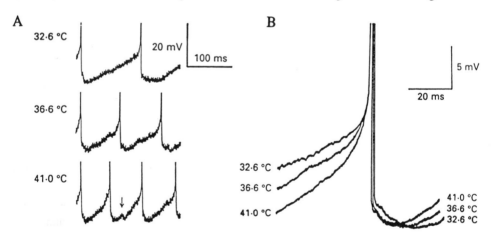

Fig. 3. Effects of temperature on the intracellular activity of a preoptic warm sensitive neuron in a rat hypothalamic tissue slice. A. records showing spontaneous action potentials (truncated) preceded by slow, depolarizing prepotentials (or pacemaker potentials). Arrow indicates putative IPSP. B. averaged pre- and post-spike activity showing that warming increases the rate of rise of the depolarizing prepotential. (From [12].)

By using intracellular filling with Lucifer yellow, the morphology of recorded neurons can be identified [13]. Warm sensitive neurons have branching dentritic patterns that could allow them to receive afferent input from at least two different projection pathways. This supports the concept that these neurons act to integrate peripheral information with local information.

As indicated by the 41°C record in Fig. 3A, PO/AH warm sensitive neurons also display postsynaptic potentials. Even though neuronal warm sensitivity is not dependent on synaptic input, intracellular recordings in tissue slices show that these neurons display considerable excitatory and inhibitory synaptic input from local sources. In general, however, the frequency of these postsynaptic potentials is not dependent on temperature, suggesting that much of this inhibitory and excitatory input comes from nearby temperature insensitive neurons. Temperature can, nevertheless, alter the "effectiveness" of this local synaptic activity. Cooling, for example, increases the amplitude of presynaptic action potentials which, in turn, increases the amount of neurotransmitter released. In addition, cooling increases the input resistance of the postsynaptic membrane which can increase IPSP and EPSP amplitudes. Therefore, despite intrinsic mechanisms responsible for neuronal warm-sensitivity, the effect of temperature on "synaptic effectiveness" can contribute to the thermosensitivity of local hypothalamic networks [12]

Cold sensitive neurons. Many of the early neuronal models suggested that cold-sensitivity in hypothalamic neurons was not inherent to those neurons, but was due, instead, to synaptic inhibition from nearby warm sensitive neurons. This concept is supported by some in vitro extracellular studies showing that PO/AH cold sensitivity is lost during synaptic blockade [14-15]. This remains equivocal, however, since: 1) in some hypothalamic areas, neuronal cold-sensitivity is retained during synaptic blockade [15]; and 2) it is known that reduction in extracellular [Ca^{++}] can not only block synaptic activity, but may directly alter intrinsic neuronal activity as well.

Support for the early neuronal models comes from some intracellular studies [12,16]. Intracellular recordings suggest that the firing rate activity and thermosensitivity of PO/AH cold sensitive neurons is highly dependent on IPSP and EPSP activity. Since temperature can change the frequency of some of these postsynaptic potentials, it appears that these neurons are synaptically driven by nearby PO/AH thermosensitive neurons.

In other hypothalamic areas, particularly in the posterior hypothalamus, neuronal cold sensitivity is retained during synaptic blockade and during ouabain-blockade of the Na-K pump [17-18]. Neurons in these areas also possess delayed cold-sensitivity; i.e., they are unresponsive to brief cooling, but if the cooling is maintained, after 1-8 minutes there if an increased firing rate. This suggests that temperature can alter intracellular modulators or second messengers which, in turn, cause delayed and prolonged changes in the firing rate.

Temperature insensitive neurons. In all hypothalamic regions, including the PO/AH, the majority of neurons are temperature insensitive. By their synaptic links with both temperature sensitive and insensitive neurons, the temperature insensitive neurons form the foundation of most neuronal networks. Some of these neurons have moderate thermal coefficients, but others are classified as low-slope temperature insensitive neurons (having thermal coefficients less than 0.1 imp·s$^{-1.}$°C^{-1}). In some respects, these low-slope temperature insensitive neurons are extremely interesting, since their firing rates remain absolutely unchanged even if temperature is changed over a 10°C range.

One study suggests that the temperature insensitivity of these neurons is dependent on the Na-K pump [17]. Ouabain blockade of the Na-K pump causes dramatic increases in both the firing rate and thermosensitivity of these neurons. Since the hyperpolarizing Na-K pump is temperature dependent, one might expect warming to increase this activity. Accordingly, these neurons may employ this pump to counteract other membrane processes that might produce either increased excitability or depolarization during warming.

Intracellular recordings show that, like warm sensitive neurons, temperature insensitive neurons display depolarizing prepotentials or pacemaker potentials [12-13]. In the low-slope temperature insensitive neurons, however, the pacemaker potential's rate of rise is not affected by temperature. This suggests that there are ionic conductance differences, when compared to warm sensitive neurons. In some cases, artificial depolarization can increase the thermosensitivity of these neurons, and this is accompanied by thermal-dependent changes in the pacemaker potential's rate of rise. It is possible, therefore, that these neurons strongly rely on the Na-K pump to stabilize resting membrane potential and insure a temperature independence in their pacemaker potential.

ACKNOWLEDGEMENTS

The author's work cited here has been supported by National Institutes of Health grant NS-14644 and by the F.A. Hitchcock Professorship for Environmental Physiology. The author wishes to express appreciation to Casa Ybel and to his students and collaborators.

REFERENCES

1. Magoun, H.W., F. Harrison, J.R. Brobeck, and S.W. Ranson. Activation of heat loss mechanisms by local heating of the brain. J. Neurophysiol. 1: 101-114, 1938.

2. Boulant, J.A. Hypothalamic control of thermoregulation: neurophysiological basis. In: Handbook of the Hypothalamus: Behavioral Studies of the Hypothalamus, edited by P.J. Morgane and J. Panksepp. New York: Dekker, 1980, vol. 3, pt A, p. 1-82.

3. Boulant, J.A., M.C. Curras, and J.B. Dean. Neurophysiological aspects of thermoregulation. In: Advances in Comparative and Environmental Physiology, 4, Animal Adaptation to Cold, edited by L.C.H. Wang. Berlin: Springer-Verlag, 1989, pp. 117-160.

4. Boulant, J.A. and J.B. Dean. Temperature receptors in the central nervous system. Ann. Rev. Physiol. 48: 639-654, 1986.

5. Nakayama, T., Eisenman, J. S., and Hardy, J. D. Single unit activity of anterior hypothalamus during local heating. Science 134, 560-561, 1961.

6. Dean, J.B. and J.A. Boulant. In vitro localization of thermosensitive neurons in the rat diencephalon. Am. J. Physiol. 257: R57-R64, 1989.

7. Boulant, J.A. and J.D. Hardy. The effect of spinal and skin temperatures on the firing rate and thermosensitivity of preoptic neurons. J. Physiol. 240: 639-660, 1974.

8. Boulant, J.A. and H.N. Demieville. Responses of thermosensitive preoptic and septal neurons to hippocampal and brain stem stimulation. J. Neurophysiol. 40: 1356-1368, 1977.

9. Dean, J.B., M.L. Kaple and J.A. Boulant. Regional interactions between thermosensitive neurons in diencephalic slices. Am. J. Physiol. 263: R670-R678, 1992.

10. Boulant, J.A. and N.L. Silva. Multisensory hypothalamic neurons may explain interactions among regulatory systems. NIPS 4: 245-248, 1989.

11. Derambure, P.S. and J.A. Boulant. Temperature sensitivity of suprachiasmatic neurons: circadian rhythms in rat hypothalamic tissue slices. Sleep Res. 22: 617, 1993.

12. Curras, M.C., S.R. Kelso and J.A. Boulant. Intracellular analysis of inherent and synaptic activity in hypothalamic thermosensitive neurones in the rat. J. Physiol. 440: 257-271, 1991.

13. Griffin, J.D. and J.A. Boulant. Thermosensitive characteristics of hypothalamic neurons determined by whole-cell recording. Soc. Neurosci. Absts. 17: 835, 1991.

14. Kelso, S.R. and J.A. Boulant. Effect of synaptic blockade on thermosensitive neurons in hypothalamic tissue slices. Am. J. Physiol. 243: R480-R490, 1982.

15. Dean, J.B. and J.A. Boulant. Effects of synaptic blockade on thermosensitive neurons in rat diencephalon in vitro. Am. J. Physiol. 257: R65-R73, 1989.

16. Nelson, D.O. and C.L. Prosser. Intracellular recordings from thermosensitive preoptic neurons. Science 213: 787-789, 1981.

17. Curras, M.C. and J.A. Boulant. Effects of ouabain on neuronal thermosensitivity in hypothalamic tissue slices. Am. J. Physiol. 257: R21-R28, 1989.

18. Dean, J.B. and J.A. Boulant. Delayed firing rate responses to temperature in diencephalic slices. Am. J. Physiol. 263: R679-R684, 1992.

EFFECT OF 5-HT RECEPTOR AGONISTS OR ANTAGONISTS ON HYPOTHALAMIC 5-HT RELEASE OR COLONIC TEMPERATURE IN RATS

Lin, M.T. and Liu, H.J.*
Department of Physiology, National Cheng Kung University, Tainan city, Taiwan, ROC; and *
Department of Physiology, Taipei Medical College, Taipei, Taiwan, ROC

Summary
Intraperitoneal administration of either 1-(3-chlorophenyl)-piperazine (5-HT1 receptor agonist), 8-hydroxy-DPAT (5-HT1A receptor agonist), CGS-12066B (5-HT1B receptor agonist), DOI (5-HT2 receptor agonist) or 2-methyl-serotonin (5-HT3 receptor agonist) produced dose-dependent decrease in both the colonic temperature and the hypothalamic 5-HT release measured by *in vivo* voltammetry in unanaesthetized rats. On the other hand, intraperitoneal administration of propranolol, S(-) (5-HT1 receptor antagonist)or ketanserine (5-HT2 receptor antagonist) caused dose-dependent increase in both the colonic temperature and the hypothalamic 5-HT release in rats.

Introduction

Evidence is provided from a variety of biochemical, physiological and behavioural studies that multiple serotoninergic (5-HT) receptors exist in the central nervous system. At the present time, at least five distinct 5-HT binding site subtypes have been differentiated by radioligand techniques in brain homogenates [1]. Other lines of evidence has accumulated to suggest that 5-HT1 receptors mediate the hypothermia responses in rats and mice [2]. However, little information is available about the interaction between the 5-HT receptor agonists or antagonists, the hypothalamic 5-HT release and body temperature.

The present study was an attempt to explore the possible involvement of the hypothalamic 5-HT release in the thermal responses to 5-HT receptor agonists or antagonists in rats. Voltammetry at carbon fibre electrodes implanted in the hypothalamus would provide a means of measuring endogenous 5-HT release *in vivo* in unanaesthetized rats restrained in a rat stock in the present study [3]. Such a technique would permit repeated measurements in the same animal without the need to radiolabel transmitter stores or kill the animal.

Materials & methods

Male Albino rats weighing 250 - 300 g were housed singlely under a 12/12 h light/dark cycle (light on at 06:00 h) with free access of food and water. For surgery, rats were anaesthetized with sodium pentobarbital (6 mg/100 g,i.p.) and placed in a Kopf stereotaxic frame in the flat skull position. The frontal and parietal bones were exposed by a midline incision on the scalp. After appropriately located craniotomy had been trephined, two self-tapping screws were inserted into the parietal or frontal bones. After connection to the polarograph (Biopulse, Tacussel, France), the nafion-coated carbon fibre electrodes were stereotaxically implanted in the anterior hypothalamus according to the following coordinates: A 8.7 mm, L±0.8 mm, and V 8.7 mm [4].Before implantation, these electrodes (with 12 µm thick and 500 µm long) were subjected to an electrical treatment and a nafion coating and then tested *in vitro* as previously described (Crespi et al. 1988; Lin et al. 1992). This allowed us to record every 2 s, an oxidation current at 300 mV (the 5-hydroxyindole oxidation current) by using differential pulse amperometry. Electrochemical signals were quantified by measuring the height of the oxidation peaks as described previously [5]. Both the screws and the electrodes were anchored with acrylic cement to the calvarian surface, which had been scraped clean of periosteum. The reflected skin was replaced around the acrylic mound containing the screws and the electrodes and were sutured with chromic gut (000). The animals were allowed to recover for 2 days before they were used. At the end of experiments, a current (DC; 5 V for 10 s) was passed through the carbon fibre electrode, producing an electrolytic lesion. The lesion site was verified on 20-µm-thick serial coronal sections stained with cresyl violet technique.

The effects of 5-HT receptor agonists or antagonists on both colonic temperature (Tco) and 5-HT release in the hypothalamus in unanaesthetized rats acclimated to a rat restraining stock. Colonic temperature was measured by using copper-constantan thermocouples. All drug solutions were prepared in pyrogen-free glasswares and containers. All solutions were sterile, non-pyrogenic and, as an added precaution, were passed through 0.22 µm Millipore bacterial filters immediately prior to use. Drugs administered intraperitoneally included 1-(3-chlorophenyl) piperazine (5-HT1 agonist, RBI), DOI (5-HT2 agonist, RBI), 2-methyl-serotonin (5-HT3 agonist, RBI) ketanserine(5-HT2 antagonist; RBI, Natik, MA),and S (-) propranolol(5-HT1 antagonist; RBI). Doses refer to the salt of the drugs. Injection volume was 1 ml/kg for all i.p. injections.

Results

Rats acclimated to a rat restraining stock were equilibrated in an ambient temperature (Ta) of $22^{\circ}C$ for a period of at least 90 min. Intraperitoneal injection of either piperazine (0.5 - 5.0 µg/kg), DOI (0.5 - 5.0 µg/kg), or 2-methyl-serotonin (0.5 - 5.0 µg/kg) caused dose-dependent fall in Tco in rats with implanted hypothalamic carbon fibre electrodes. The Tco fall almost immediately after injection and reached its bottom level of about 1.26 for a dose of 5 µg/kg of DOI or $1.12^{\circ}C$ for a dose of 5 µg/kg of piperazine at 45 min. The Tco returned its pre-injection level about 210 min and 180 min, respectively, for DOI and piperazine. Generally, the hypothalamic 5-HT release value followed the time course of Tco after DOI or piperazine injection. The data are summarized in both Table 1 and Fig. 1.

Table 1. Effects of intraperitoneal administration of serotoninergic receptor agonist and antagonists on colonic temperature (Tco) of rats

Treatment	Maximal changes in Tco(^{o}C)
1. 0.9%Saline	-0.18±0.05
2. 1-(3-chlorophenyl)-piperazine:	
0.5 µg/kg	-0.65±0.07*
1.0 µg/kg	-0.88±0.09*
5.0 µg/kg	-1.15±0.15*
3. DOI:	
0.5 µg/kg	-0.71±0.08*
1.0 µg/kg	-0.92±0.07*
5.0 µg/kg	-1.26±0.10*
4. 2-Mithyl-serotonin:	
0.5 µg/kg	-0.57±0.07*
1.0 µg/kg	-0.68±0.07*
5.0 µg/kg	-0.84±0.10*
5. Propranolol, S (-):	
0.5 µg/kg	+0.53±0.07*
1.0 µg/kg	+0.64±0.09*
5.0 µg/kg	+1.15±0.17*
6. Ketanserine:	
0.5 µg/kg	+0.72±0.08*
1.0 µg/kg	+0.90±0.12*
5.0 µg/kg	+1.23±0.13*

The values are expressed as the means±SEM of 8 rats for each group.
The delta values denote the difference between the control values before the injection and 1 h after injection.
* $P < 0.05$, Significantly different from the corresponding control values (Student's t-test).

It can be seen from the figure that the changes in the hypothalamic 5-HT release occurred about 15 min preceded the changes of Tco.

On the other hand, i.p. injection of s (-) propranolol (0.5 - 5.0 µg/kg) or ketanserine (0.5 - 5.0 µg.kg) caused dose-dependent rise in Tco in rats with hypothalamic electrodes (Table 1). Again, it can be seen from Fig. 2 that, after injection, the changes of the hypothalamic 5-HT release occurred about 30 min prior to the changes of Tco.

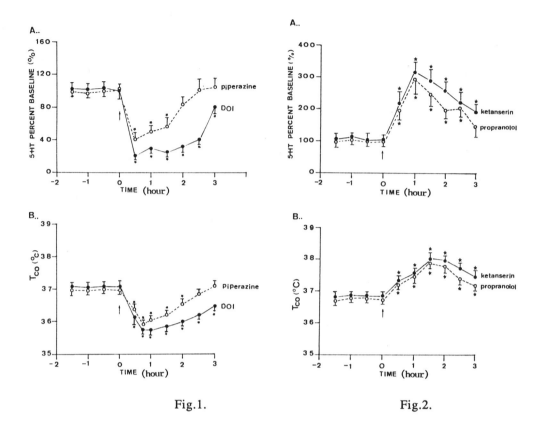

Fig.1.   Fig.2.

Fig. 1. Effects of i.p. administration g piperazine (5 µg/kg) or DOI (5 µg/kg) on both hypothalamic 5-HT release and colonic temperature (Tco) in unanaesthetized rats. The values are means±SEM of 8 rats for each point. * P < 0.05, significantly different from coresponding control values (pre-jection), Student's t-test. Basal 5-HT levels were considered stable after observation of at least 90 min with no upward a downward trend. The basal values durig the last 30 min were averaged and all data were expresses as a percent of the mean baseline.

Fig. 2. Effects of i.p. administration of ketanserin (5 µg/kg) or propranolol (5 µg/kg) on both hypothalamic 5-HT release and Tco in rats. The values are means±SEM of 8 rats. * P < 0.05, significantly different control values (pre-injection), Student's t-test).

Discussion

The present results showed that systemic administration of either piperazine (5-HT1 agonist), DOI (5-HT2 agonist) or 2-methyl-sterotonin (5-HT3 agonist) [6], produced dose-dependent hypothermia in rats. For a given dose, either piperazine produced the same degree of hypothermia as those of DOI. As compared to those of 2-methyl-serotonin, for a given dose, either piperazine or DOI produced a higher degree of colonic temperature reduction. By contrast, either propranolol (5-HT1 antagonist) or ketanserine (5-HT2 antagonist) [6]. produced dose-dependent hyperthermia in rats. From a given dose, propranolol produced the same degree of hypothermia as those of ketanserine. In addition, it was found that systemic administration of 8-OH-DPAT (5-HT1 agonist) or ipsapirone (5-HT1 agonist) in rats also produced dose-dependent hypothermia [7]. Furthermore, both the present and previous results demonstrated that the hypothermic respones to 5-HT1 or 5-HT2 agonists in the rat were antagonized by 5-HT1 or 5-HT2 antagonists [2]. Thus, there did not appear to be any association between 5-HT receptor subtypes and thermoregulation. However, these results suggest that , at least in rodents, activation of 5-HT receptors decreases body temperature, whereas inhibition of 5-HT receptors increases body temperature [8].

As mentioned earlier, several 5-HT receptor agonists such as buspirone, ipsapirone or 8-OH-DPAT were shown to inhibit raphe cell firing . The inhibitory effect of ipsapirone or 8-OH-DPAT on raphe cell firing was blocked by pretreatment with propranolol [9]. Systemic administration of 8-OH-DPAT was shown to decrease the extracellular 5-HT in the hypothalamus measured by *in vivo* dialysis coupled to HPLC-ECD [10]. The present results provide new evidence that the 5-HT release in the hypothalamus measured by *in vivo* voltammetry was reduced by systemic administration of either 5-HT1 or 5-HT2 receptor agonists, but enhanced by systemic administration of 5-HT1 or 5-HT2 receptor antagonists in rats. Other lines of evidence also showed that repeated treatment with either electroconvulsive shock or antidepressant drugs (such as MAOI or 5-HT reuptake inhibitors) blunted the hypothermic effect to 8-OH-DPAT [1,11,12]. The authors claim that antidepressants, which enhance serotonergic transmission [11], all attenuated the hypothermic response to 8-OH-DPAT. Hensler et al. [1] have demonstrated that treatment with phenelzine and clorgyline attenuated the hypothermic response to 8-OH-DPAT and was accompanied by diminished binding of [^{3}H]8-OH-DPAT in the hypothalamus. However, other drugs, such as tranylcypromine, citalopram or sertraline, diminished the hypothermic response but did not alter 5-HT1 receptor binding in these hypothalamic nuclei [1], suggesting that antidepressant drugs may not attenuate the hypothermic response by acting directly at the hypothalamus. These observations prompted us to think that systemic administration of 5-HT receptor agonists or antagonists may elicit hypothermic or hyperthermic response in rats, respectively, by acting presynaptically at 5-HT receptors in the raphe neurons.

Acknowledgements The work reported here was supported by grants from the national Science Council of Republic of China.

References
1. Hensler JG, Kovachich GB, Frazer A. A quantitative autoradiographic study of serotonin-1A receptor regulation; Effect of 5,7-dihydroxytryptamine and antidepressant treatments. Neuropsychopharmacology 1991;4:131-44.

2. Moser P. The effect of putative 5-HT-1A receptor antagonists on 8-OH-DPAT-induced hypothermia in rats and mice. Eur J Pharmacol 1991;193:165-72.

3. Marsden CA, Joseph MH, Kruk ZL, Maidment NT, O'Neill RD. IN vivo voltammetry - present electrodes and methods. Neuroscience 1988;25: 389-400.

4. Paxinos G, Watson C. The rat brain in stereotaxic coordinates, Academic press, New York, 1982.

5. Crespi F, Martin KF, Marsden CA. Measurement of extracellular basal levels of serotonin in vivo using nafion-coated carbon fiber electrodes combined with differential pulse amperometry. Neuroscience 1988;27:885-96.

6. Peroutka SJ. 5-Hydroxytryptamine receptor subtypes. Ann Rev Neurosci 1988;11:45-60.

7. Koenig JI, Meltzer HY,Gudelsky GA. 5-Hydroxytryptamine-1A receptor mediated effects of buspirone, gepirone and ipsapirone. Pharmac Biochem Beahv 1988;29:711-5.

8. Won SJ,Lin MT. 5-Hydroxytryptamine receptors in the hypothalamus mediate thermoregulatory responses in rabbits. Naunyn-Schmiedeberg's Arch pharmacol. 1988;338:256-61.

9. Hinter BC, Roth HL, Peroutka SJ. Antimigraine drug interactions with 5-hydroxytryptamine-1A receptors. Ann. Neurol. 1986;19:511-3.

10. Auerbach SB, Minzenberg MJ, Wilkinson LO. Extracellular serotonin and 5-hydroxyindoleacetic acid in hypothalamus of the unanesthetized rat measured by in vivo dialysis coupled to high-performance liquid chromatography with electrochemical detection: dialysate serotonin reflects neuronal release. Brain Res 1989;499:281-90.

11. Blier P, de Montigney C, Chaput Y. A role for the serotonin system in the mechanism of action of antidepressant treatment: preclinical evidence. J clin Psychiat 1990;4:14-20.

12. Goodwin GM, DeSouza RJ, Green AR. Attenuation by electroconvulsive shock and antidepressant drugs of the HT1A receptor-mediated hypothermia and serotonin syndrome produced by 8-OH-DPAT in the rat. Psychopharmacology 1987;91:500-5.

Temperature Regulation
Advances in Pharmacological Sciences
© 1994 Birkhäuser Verlag Basel

TEMPERATURE SENSITIVTY OF RAT SPINAL CORD NEURONS RECORDED IN VITRO

Schmid, H.A., U. Pehl and E. Simon

Max-Planck-Institut für physiologische und klinische Forschung, W.G. Kerckhoff-Institut, Parkstr. 1, 61231 Bad Nauheim, Germany; Supported by DFG Si 230/8-1.

Summary

The *in vitro* slice technique was employed in order to characterize the temperature sensitivity and pharmacological responsiveness of spontaneously active neurons from the dorsal horn of the rat spinal cord. 54% of the recorded neurons (n=79) were warm-sensitive, 3% cold-sensitive, the remaining temperature insensitive. The numbers of warm-sensitive vs. temperature-insensitive neurons and their responsiveness to substance P (SP) and Glutamat (Glu) were evenly distributed throughout the dorsal horn, except for warm-sensitive neurons in lamina II, which were not influenced by SP. In general, the temperature-sensitive neurons recorded from the dorsal horn can be regarded as the cellular basis for the *in vivo* unequivocally characterized temperature sensory function of the spinal cord.

Introduction

The local temperature sensitivity of the spinal cord (SC) of various homeothermic species has clearly been established in *in vivo* experiments. Local temperature displacements of the spinal cord elicit adequate heat loss and heat gain mechanisms, like vasoconstriction and shivering during cooling and panting or vasodilation during warming (1,2,3). The existence of temperature sensitive neurons within the spinal cord (4) and their functional connection to peripheral temperature receptors has been demonstrated in *in vivo* experiments (5), by recording from temperature-sensitive fibers in the ascending anterolateral tract of the cervical spinal cord after local temperature stimulation of the lumbar spinal cord and/or relevant skin areas. Numerous studies have shown that the peripheral pain and temperature receptors convey their information via slowly conducting C- and Aδ- fibers to neurons located in the superficial part of the dorsal horn (6,7,8). Furthermore it has been shown that substance P (SP) and glutamate (Glu) are colocalized in the majority of slowly conducting fibers (9) and that SP is a mediator for potentiation of the postsynaptic response of dorsal horn neurons caused by Glu (10). The rationale for choosing these two neurotransmitters was that they might also be used by temperature receptors, thus mimicking specific temperature input from the periphery.

The aim of the present study was to investigate electrophysiologically the inherent temperature-sensitivity of spinal cord neurons, their local distribution within the dorsal horn and their responsiveness to SP and Glu in a slice preparation. This *in vitro* approach allowed the physiological and pharmacological characterization of temperature-sensitive neurons in the absence of their peripheral and most of their central input.

Material and Methods

Adult male Wistar rats (150 - 320 g) were decapitated and a section of the upper lumbar spinal cord region (segments L_2 to L_4) was quickly removed and superfused with ice-cold artificial cerebrospinal fluid (aCSF) of the following composition (in mM): NaCl, 124; KCl, 5; NaH_2PO_2, 1.2; $MgSO_4$, 1.3; $CaCl_2$, 1.2; $NaHCO_3$, 26; glucose, 10; pH: 7.4, equilibrated with 95% O_2 and 5% CO_2, 290 mOsmol/kg). Several coronal slices (500 μm thick) were cut with a custom-made tissue slicer and preincubated at 35°C for at least 1 hour, before the first slice was transferred to the temperature controlled recording chamber. The chamber was continuously perfused with prewarmed (37.0°C) aCSF at a rate of 1.6 ml/min. Extracellular recordings were made with glass-coated platinum/iridium (Pt-Ir) electrodes from various layers of the spinal cord. Spontaneously active neurons from the superficial dorsal horn (laminae I and II) and around the central canal (lamina X) were investigated for their local temperature sensitivity. Only in those cases where no neurons could be found in these two regions, spontaneously active neurons were investigated in other regions (laminae III-VI) of the lumbar spinal cord as well. The temperature was changed within 7 min either sinusoidally between 33 and 41°C, or slow (0.02°C/s) temperature ramps were applied to record the static temperature dependencies. Fast step-like temperature changes (0.5°C/s) were also performed in order to detect possible dynamic temperature responses. A statistical program, which either fits one or two regression lines (11) to a set of data calculated the set-point of a temperature response without requiring any further preconditions. Cells were classified as being warm- or cold-sensitive, when either the one (linear regression) or the steeper of the two regression lines (piecewise regression) had a slope (i.e. Temperature Coefficient, TC) exceeding +0.6 or -0.6 imp/s/°C, respectively. The spontaneous activity of each neuron was evaluated by averaging its activity for 60 seconds prior to the first temperature stimulus. Mean values given in the text are presented ±SEM. Statistical significance was determined using t-test.

Results

43 of the neurons recorded from coronal slices (n=79) were warm-, 2 cold-sensitive, the remaining were temperature-insensitive. The average spontaneous firing rate of warm-sensitive neurons (10.5±1.5 imp/s) was significantly (p<0.001) higher than of temperature-insensitive neurons (2.3±0.5 imp/s). Both cold-sensitive neurons were recorded in lamina X

and had a FR below 1 imp/s. The average TC of all warm-sensitive neurons was 1.9±0.2 imp/s/°C and was not significantly different in individual laminae.

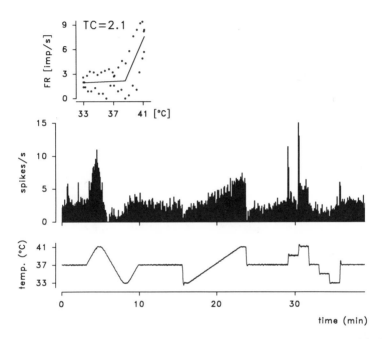

Fig. 1. Continuous rate-meter recording of a warm-sensitive neuron recorded in lamina II of the dorsal horn from a spinal cord slice. Inset: frequency/temperature relationship for the time of the sinusoidal temperature stimulus. Temperature coefficient (TC) is 2.1 imp/s/°C for the temperature range between 38.5 and 41.0°C.

Fig. 1 displays an example of a warm-sensitive neuron recorded from the lumbar dorsal horn. This neuron is characterized by a relatively low spontaneous firing rate under control conditions, as it was typical for the majority of warm-sensitive neurons in the region of lamina II. This rate-meter recording also displays the applied stimulus protocol. The temperature coefficient of this cell was determined from the slope of the regression line from the sinusoidal temperature change as shown in the inset of Fig. 1. The ramp-like and the step-like temperature stimuli both resulted in a reduced temperature responsiveness as compared to the sinus. This phenomenon was generally seen when comparing the temperature sensitivity of a given cell in response to different stimuli. A clear phasic response could be observed in the neuron of Fig. 1 and in 63% of all warm-sensitive neurons during rapid temperature steps of 2°C in the hyperthermic direction. No striking difference has been observed in the number of warm-, cold-sensitive and temperature insensitive neurons in laminae II and X. Due to the method of searching for spontaneously active neurons more neurons have been recorded from these two regions as compared to laminae III-VI.

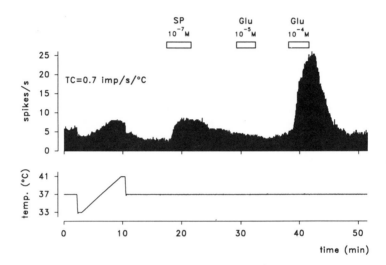

Fig. 2 Continuous rate meter recording of a warm-sensitive neuron (TC = Temperature coefficient) recorded in lamina X from a transversal spinal cord slice. Superfusion with substance P (SP) and Glutamate (Glu) cause an excitation at the indicated concentrations.

After characterizing their temperature sensitivity a total of 23 neurons were tested for their responsiveness to SP and/or Glu (Fig. 2). None of the neurons investigated decreased their activity in response to any of the two substances in the indicated concentrations (table 1). Table 1 summarizes the results and subdivides the transmitter-responses according to the temperature sensitivity and the laminae in which the individual neurons have been found. The only notable exception so far from the relatively even distributions of temperature-sensitive neurons, transmitter sensitivity and location (the numbers of neurons from other laminae may still be too small for more far reaching conclusions) is the absence of warm-sensitive neurons in lamina II responding to SP.

Table 1. Effect of substance P (10^{-7}-10^{-6}M) and glutamate (10^{-5}-3×10^{-4}M) on the activity of warm-sensitive and temperature-insensitive neurons in different laminae of rat spinal cord slices.

		substance P				glutamate			
		+	-	0	tested	+	-	0	tested
warm-	lamina II	0	0	6	6	1	0	4	5
sensitive	laminae III-VI	2	0	0	2	2	0	0	2
neurons	lamina X	2	0	0	2	2	0	0	2
temp.-	lamina II	4	0	3	7	3	0	3	6
insens.	laminae III-VI	2	0	0	2	1	0	0	1
neurons	lamina X	1	0	1	2	2	0	0	2
total		11	0	10	21	11	0	7	18

Discussion

This study provides evidence, that the number and temperature sensitivity of warm- and cold-sensitive neurons from SC slices is comparable to the number and temperature sensitivity of preoptic area and anterior hypothalamus (PO/AH) neurons recorded in slices (12,13,14). Acknowledging a functional role for the PO/AH in thermoregulation, the similarity of the results supports the significance of the spinal cord for temperature regulation. With the exception that the number of cold-sensitive fibers found in ascending neurons of the anterolateral tract in cats (4) exceeded the numbers of cold-sensitive neurons in the recent study on SC slices, the properties of the recorded responses after local stimulation of the spinal cord were similar.

Aim of the pharmacological approach of this study was to find additional criteria by which the specificity of local temperature-sensitive SC neurons could be further be characterized. With regard to the specificity of the peripheral input to local temperature-sensitive SC neurons, Simon (5) has shown that the functional overlap is virtually 100%, i.e. temperature-sensitive SC neurons receive specific input from peripheral temperature receptors. So far no correlation between transmitter sensitivity, temperature sensitivity and location in the dorsal horn emerged, except for the absence of an effect of SP on warm-sensitive neurons in lamina II. The ineffectiveness of SP to activate these neurons is noteworthy with regard to the well documented input of thinly- or un- myelinated Aδ- and C- fibers conveying pain and temperature signals to neurons in laminae I and II (8,15). Many C- and Aδ- fibers are known to use SP as a transmitter and direct evidence has been provided for SP conveying pain signals to dorsal horn neurons (16). Following this argumentation one would expect that SP affects temperature-sensitive SC neurons in these laminae, a conclusion not supported by our results. On the other hand recent studies suggest (17) that the majority of SP containing fibers in the dorsal roots innervate visceral structures, thus not belonging to the functional pathway monitoring skin temperature. Following this evidence, the absence of an effect of SP on warm-sensitive SC neurons in lamina II is in line with the separation of visceral and somatic pathways. Not mentioning the other known central sources of SP input (15) to the dorsal horn, the described peripheral SP input to these cells does so far neither support, nor oppose a functional specificity of the temperature-sensitive SC neurons. This result is similar to studies on PO/AH neurons (18 for ref.), where also no clear correlation could be observed between temperature- and transmitter-sensitivity in various species.

In summary, these data underline the functional significance of the local temperature sensitivity of SC neurons. Although specific criteria for temperature-sensitive SC neurons are, like in the PO/AH, still missing, the spinal cord offers a better clue to characterize temperature-sensitive neurons, due to its segmental structure and due to the known morphological separation of neurons belonging to the afferent or efferent signal transduction pathway.

References
(1) Simon, E., Rev Physiol Biochem Pharmacol 1974; 71:1-76.
(2) Simon, E., Pierau, Fr.-K. and Taylor, D.C.M., Physiol Rev 1986; 66:235-300.
(3) Jessen, C. and Simon-Oppermann, C., Experientia 1976; 32:484-5.
(4) Simon, E. and Iriki, M., Pflügers Arch 1971; 328:103-20.
(5) Simon, E., Pflügers Arch 1972; 337:323-32.
(6) Hellon, R.F. and Misra, N.K., J Physiol 1973; 232:375-88.
(7) Hensel, H., Thermoreception and temperature regulation. 1981; London: Academic Press.
(8) Perl, E.R., In: Thermoreception and temperature regulation. 1990; Berlin:Springer:89-106.
(9) De Biasi, S. and Rustioni, A., Proc Natl Acad Sci USA 1988; 85:7820-24
(10) Randic, M., Hecimovic, H. and Ryu, P.D., Neurosci Lett 1990; 117:74-80.
(11) Vieth, E., J Appl Physiol 1989; 67:390-6.
(12) Boulant, J.A. and Dean, J.B., Annu Rev Physiol 1986; 48:639-54.
(13) Nakashima, T., Pierau, Fr.-K., Simon, E. and Hori, T., Pflügers Arch 1987; 409:236-43.
(14) Schmid, H.A. and Pierau, F.-K., Am J Physiol (Regulatory Integrative Comp Physiol 33) 1993; 264:R440-8.
(15) Jessel, T.M., In: Substance P in the nervous system. 1982; London: Pitman:225-48.
(16) De Konink, Y., Ribeiro-Da-Silva, A., Henry,J.L. and Cuello, A.C., Proc Natl Acad Sci USA 1992; 89:5073-7.
(17) Lawson, S.N., Prabhakar, E., Perry, M.J. and McCarthy, P.W., Neuropeptides 1992; 22 (N1):39-40.
(18) Sato, H. and Simon, E., Pflügers Arch 1988; 411:34-41.

THE EFFECT OF AMBIENT TEMPERATURE ON THE MODULATION OF THERMOREGULATORY MECHANISMS BY SELECTIVE OPIOID PEPTIDES

Cynthia M. Handler, Thomas C. Piliero, Ellen B. Geller and Martin W. Adler,
Department of Pharmacology, Temple Univ. School of Medicine, Philadelphia, PA, USA

SUMMARY

We have characterized the effects of highly selective opioid agonists and antagonists on body temperature, heat exchange and oxygen consumption at ambient temperatures above and below the thermoneutral zone. The thermoregulatory effects of these selective agonists differ from those seen at thermoneutral ambient. Such differences are to be expected when a system involved in thermoregulation is perturbed from its normal state.

Control and maintenance of body temperature over a wide range of environmental temperatures is characteristic of homeotherms. Activation and integration of autonomic and behavioral thermoregulatory mechanisms provide for the relatively constant core temperature. Thermoregulatory responses do not occur in isolation but are intricately interrelated to other homeostatic systems. Within the central nervous system, neurons with opioid receptors are widely distributed and have multiple physiological roles, including a role in thermoregulation.[1,2] Alterations in body temperature may be induced by administration of opioid ligands, with the effect dependent upon species, receptor type on which the ligand acts, dosage, route of administration, and ambient temperature.[2]

In studies carried out in rats at 20° C, Lynch et al.,[3] Zwil et al.[4] and Handler et al.[5] demonstrated that opioid agonists produce distinct patterns in oxygen consumption and heat exchange that are correlated to changes in body temperature. In contrast, Thornhill and Desautels[6] did not find post-injection increases in oxygen consumption following acute central or peripheral morphine administration. They did report dose-dependent increases in rectal body temperature and tail skin temperature (indicative of vasodilation) but found no temperature increase within the brown adipose tissue (BAT), nor was there an increase in [^{3}H]GDP binding to BAT mitochondria isolated during peak hyperthermia. Using highly selective opioid agonists, Handler et al.[5] showed that changes in oxygen consumption and heat exchange occur during the first 15-30 min post-injection. It is likely that the timing of the isolation of mitochondria and measurement of Tb account for the differences between these studies.

Brain surface temperature (Tb), oxygen consumption (VO$_2$) and heat exchange (Q) were measured at ambient temperatures of 5° and 30° C in a gradient-layer calorimeter following central administration of PL-017 (1.86 nmol), dynorphin A$_{1-17}$ (4.86 nmol) and DPDPE (4.64 nmol), opioid peptides selective for μ, κ and δ receptors, respectively. The results of this study provide information on ambient-temperature-induced modifications of the thermoregulatory process and the interactive neuromodulator network, giving further insight into the role of the opioid system in thermoregulation.

METHODS

Male albino Sprague-Dawley rats (Zivic-Miller) were housed 4-5 per cage in an animal room at 22° ± 2°C and 50 ±10% relative humidity on a 12 hr L/12 hr D cycle. Seven days prior to testing, a polyethylene cannula was implanted into the right lateral ventricle and a YSI thermistor was implanted on the surface of the left lateral cortex under ketamine anesthesia (150 mg/kg i.p.). A gradient-layer calorimeter was flushed with air (20% O$_2$) and brought to testing temperature (5° or 30°±0.5°C) prior to placing the animal in the calorimeter. Data were continuously recorded, with the last 60 min constituting a pre-injection control for comparisons with continuous recordings 180 min post-injection. Doses of drugs were selected from full dose-response curves for each agonist (ambient temperature 20° C), using rectal body temperature as the end point.[7] Results are reported as ΔTb (°C), ΔVO$_2$ (ml O$_2$/g body wt/hr), and ΔQ (cal/g body wt/hr) from baseline controls for each time interval. Data were analyzed for each agonist or agonist/antagonist combination using a one-way analysis of variance with a repeated measure variable of time, followed by a post-hoc Fisher's test. Results are reported as means of each measured variable ± SEM. Levels of significance were p ≤ 0.05.

RESULTS/DISCUSSION

Baseline Tb values at 5°, 20° and 30° C did not differ significantly from each other, while at 5° and 30° C, baseline Q and VO$_2$ values differed significantly from those at 20° C. This consistency of body temperature is a function of the integrated role of metabolic rate and heat loss. At 30° C, heat loss increased 12% above the rate for 20° C, and VO$_2$ is more than twice the rate seen at 20° C. At an ambient temperature of 5° C,

VO_2 was nearly three times the rate at 20° C while heat loss increased 23%. These data provide insight into the balance maintained between heat production and heat loss at ambient temperatures outside the thermoneutral zone and the mechanisms that contribute to the maintenance of homeothermy.

The thermoregulatory effect of a drug may be altered by changing the ambient temperature. Mu-receptor stimulation by opioid ligands, given centrally at low doses to rats, induces hyperthermia through increased metabolic rate at ambient temperatures within the thermoneutral zone.[5,7] PL-017(icv), given at an ambient temperature of 30° C, increased Tb 1.39°±0.48° C, 30-75 min post-injection. This hyperthermia resulted from an immediate increase in VO_2 with a concurrent reduction in heat loss (Table 1). The conservation of metabolic heat from increased cellular respiration produces hyperthermia. Tb returned to baseline levels as a result of an increase in heat loss which followed a return of metabolic rate to baseline values. Environmental heat stimulation has a variety of effects on the thermoregulatory processes and the interactive neuronal network. Biswas and Poddar[8] demonstrated that increases in Tb were inducible by acute exposure to high environmental temperatures, suggesting that central inhibitory GABA neurons ,as well as opioid receptors, have an active role in the thermoregulatory process at high ambient temperatures.

Much less is known about the role of κ receptors at elevated ambient temperatures. It is clear that within the thermoneutral zone administration of κ-selective agonists to rats results in hypothermia.[2,5,7,9,10] Ambient temperatures of 30° C abolished dynorphin-induced hypothermia seen at 20° C[5] (Table 1). In agreement with our findings, Mandenoff et al.[11] have reported that the decreased level of VO_2 and the resulting hypothermia, induced by central administration of κ-selective agonists at 20° C is eliminated at higher ambient temperatures (29° C). Nor-BNI, a κ-selective antagonist, had no effect on Tb, Q or VO_2, alone or in combination with dynorphin at 30° C.

The role of the δ receptor in temperature regulation is unclear. Spencer et al.[10] induced hypothermia using 100 μg DPDPE in restrained animals, and hypothermia followed by hyperthermia in free-moving animals (no ambient temperature specified). Lower doses of DPDPE (30 μg; icv) induced only hypothermia. Handler et al.[5] used a low dose (3 μg, icv) of DPDPE and found no changes in VO_2 or Q and thus, no alterations in Tb at 20° C. Central injection of D-ala-deltorphin II, a $δ_2$-selective agonist, had no statistically significant effect on Tb in rats except at 4° C [12] or with a high dose (50 μg/rat) at 22° C. DPDPE, alone or in combination with the δ-selective antagonist, naltrindole, given icv to rats at 30° C, produced no statistically significant changes in Q, VO_2 or Tb (Table 1). These results reinforce the hypothesis that the δ receptor has only a minor role in thermoregulation.

TABLE 1

		5°C	20°C	30° C
	ΔTb	-	+	+
PL-017	ΔVO_2	-	+	+
	ΔQ	-	-	-
	ΔTb	-	-	±
DYNORPHIN	ΔVO_2	-	-	±
	ΔQ	-	-	±
	ΔTb	±	±	±
DPDPE	ΔVO_2	±	±	±
	ΔQ	-	±	±

Summary of data on post-injection effects of selective opioid agonists on body temperature (Tb), oxygen consumption (VO_2) and heat exchange (Q). A significant increase is indicated by +, a significant decrease by -; no change is indicated by ±.

Marked differences in the role of the opioid receptors in thermoregulation appear when the ambient temperature is lowered. Drugs which suppress the thermogenic response to cold may produce a change in Tb only at ambient temperatures below the thermoneutral zone.[13] At 5° C, icv PL-017 induced hypothermia (Table 1). This hypothermia (-1.01 ± 0.46 ° C) resulted from a decrease in VO_2 occurring 15-30 min post-injection. Conservation of metabolic heat followed, and in conjunction with a significant rise in VO_2, effected the return of Tb to baseline values at 105 min post-injection. These results are in contrast to those of Bloom and Tseng.[14] They reported that central administration of low doses (<1 μg) of the less μ-selective opioid agonist, β-endorphin, to mice produced hyperthermia over a wide range of ambient temperatures while higher doses of β-endorphin produced either hyperthermia or hypothermia, depending on the ambient temperature. Paolino and Bernard[15] reported hypothermia in rats given high doses of morphine (systemic administration) at an ambient temperature of 5° C. Rosow et al.[16] demonstrated that the dose-dependent responses to various opiates in mice were influenced by the environmental temperature. The results of the present study and others clearly demonstrate that lower ambient temperature may reverse the action of μ agonists on VO_2, thus altering Tb. At both ambient temperatures used in this study, the PL-017-induced changes in VO_2 and Q were blocked by the μ-selective antagonist, CTAP.

Central administration of dynorphin A_{1-17} at 5° C induced statistically significant hypothermia (-1.05±0.48° C) for the entire post-injection period. The hypothermia resulted from a decrease in VO_2, which returned to pre-injection levels, with the continuous reduction in Q, prevented the animals from becoming poikilothermic (Table 1). These results clearly indicate an opioid-induced change in set point. However, one-third of the animals tested did not shiver, either pre- or post-injection and were unable to thermoregulate following administration of dynorphin and died within 60 min (31% of total animals tested). Further studies, measuring rectal Tb, were conducted in order to account for the dynorphin-induced deaths. We observed that icv dynorphin induced either barrel-rolling or clonic seizures (behavior that is not seen with this dose of dynorphin at 20° C) during the first 30 min post-injection. Tb fell an average of 4°-5° C. We believe that the seizures, the lack of handling and the more confined space within the calorimeter contributed to the dynorphin-induced deaths. Nor-BNI partially attenuated changes in Q and VO_2 induced by dynorphin.

At an ambient of 5° C, DPDPE did not produce statistically significant changes in Tb or VO_2 (Table 1). DPDPE did decrease in Q which was reversed by naltrindole.

We have shown that environmental conditions have a modulatory effect on the role of the opioid system in thermoregulation. Our data demonstrate that stimulation of central opioid receptors by selective μ and κ ligands over a range of ambient temperatures results in changes in oxidative metabolism which cause alterations in Tb. The increase or decrease in metabolic rate is responsible for the quantity of heat available to regulate Tb via heat exchange mechanisms. The use of receptor-selective antagonists over a wide range of environmental temperatures has been shown to block the changes in VO_2 and Q and thus in Tb, indicating that these changes are mediated via specific opioid receptors. Studies of similar interactions between selective agonists and antagonists, as well as those which employ combinations with other neuromodulators, promise to provide important new information on the interrelationships between systems which are utilized by homeotherms to maintain a stable internal temperature needed for optimal functioning of the multiple systems in the body when confronted by thermal stress.

ACKNOWLEDGEMENTS

Support provided by grants T32DA07237 and DA00376 from The National Institute on Drug Abuse.

REFERENCES

1. Clark WG. Influence of opioids on central thermoregulatory mechanisms. Pharmacol Biochem Behav 1979; 10: 609-613.

2. Adler MW, Geller EB, Rosow CE, Cochin J. The opioid system and temperature regulation. Annu Rev Pharmacol Toxicol 1988; 28: 429-449.

3. Lynch TJ, Martinez RP, Furman MB, Geller EB, Adler MW. A calorimetric analysis of body temperature changes produced in rats by morphine, methadone, and U50,488H. In Problems of Drug Dependence 1986, NIDA Research Monogr. 76, ed. LS Harris, p. 82, U.S.Dept.Health & Human Services, Washington,D.C.; 1987.

4. Zwil AS, Lynch TJ, Martinez RP, Geller EB, Adler MW. Calorimetric analysis of ICV morphine in the rat. In Problems of Drug Dependence 1987, NIDA Res. Monogr. 81, ed. LS Harris, p. 285, Gov't Printing Office, Washington,D.C., 1988.

5. Handler CM, Geller EB, Adler MW. Effect of μ-, κ-, and δ-selective opioid agonists on thermoregulation in the rat. Pharmacol Biochem Behav 1992; 43:1209-121.

6. Thornhill, JA, Desautels, M. Is acute morphine hyperthermia of unrestrained rats due to selective activation of brown adipose tissue thermogenesis? J Pharmacol Exp Ther 1984; 231: 422-429.

7. Adler MW, Geller EB. Physiological functions of opioids: Temperature regulation. In Handbook of Experimental Pharmacology, Vol. 104/II, Opioids II, ed. by A. Herz, H. Akil and E. J. Simon, pp. 205-238, Springer-Verlag, Berlin, 1992.

8. Biswas S, Poddar MK. Involvement of GABA in environmental temperature-induced change in body temperature. Meth Find Exp Clin Pharmacol 1988; 10: 747-749.

9. Spencer RL, Hruby VJ, Burks TF. Body temperature response profiles for selective mu, delta, and kappa opioid agonists in restrained and unrestrained rats. J Pharmacol Exp Ther 1988; 246: 92-101.

10. Cavicchini E, Candeletti S, Ferri S. Effects of dynorphins on body temperature of rats. Pharmacol Res Commun 1988; 20: 603-604.

11. Mandenoff A, Seyrig JA, Betoulle D, Brigant L, Melchior JC, Apfelbaum M. A kappa opiate agonist, U50,488H, enhances energy expenditure in rats. Pharmacol Biochem Behav 1991; 39:215-217.

12. Broccardo M, Improta G. Hypothermic effect of D-Ala-deltorphin II, a selective δ opioid receptor agonist. Neurosci Lett 1992; 139: 209-212.

13. Shemano I, Nickerson M. Effect of ambient temperature on thermal responses to drugs. Can J Biochem Physiol 1958; 36: 1243-1249.

14. Bloom AS Tseng LF. Effects of ß-endorphin on body temperature in mice at different ambient temperatures. Peptides 1981; 2: 293-297.

15. Paolino RM, Bernard BK. Environmental temperature effects on the thermoregulatory response to systemic and hypothalamic administration of morphine. Life Sci 1968; 7: 857-863.

16. Rosow CE, Miller JM, Pelikan EW, Cochin J. Opiates and thermoregulation in mice. I.Agonists. J Pharmacol Exp Ther 1980; 213: 273-283.

Temperature Regulation
Advances in Pharmacological Sciences
© 1994 Birkhäuser Verlag Basel

REGULATION OF BODY TEMPERATURE : INVOLVEMENT OF OPIOID AND HYPOTHALAMIC GABA

Suchandra Ghosh (née Biswas) and Mrinal K. Poddar
Department of Biochmistry, University of Calcutta,
35, B. C. Road, Calcutta 700 019, India.

SUMMARY : *Exposure (2 h) to HET (40°C) increased the BT of rats. The BT of normal (28°C) and HET exposed rats was increased with bicuculline or morphine .Treatment with naloxone or muscimol reduced the BT of both normal and HET exposed rats. The bicuculline-induced rise in BT of HET exposed rat was potentiated following cotreatment with morphine, but not due to cotreatment with naloxone. The exposure of normal rat to HET and/or morphine treatment reduced the hypothalamic GABAergic activity. Naloxone at 28°C increased the hypothalamic GABA by increasing its synthesis and release; but exposure of naloxone treated rat to HET reduced hypothalamic GABAergic activity. These results, thus, suggest that central opioidergic system regulates the BT under both normal and HET exposed conditions through the interaction with hypothalamic GABA.*

INTRODUCTION

The primary centre of thermoregulatory control in mammals is the hypothalamic nuclei of the brain, with the array of thermosensitive neurons that receive afferent neuronal imput from thermoreceptors in both the periphery and other parts of the central nervous system (CNS) (1,2). In homeotherms, a constant body temperature (BT) is maintained by the balance between the heat loss and heat gain mechanisms (3). Since the balance between the heat production and the heat loss may be deranged by changing the ambient temperature, the output of thermoreceptors are subject to constant control (4) provided by neurons distributed in CNS. The physiological balance between the heat production and heat loss is under the regulation of central neurotransmitters. Several studies have shown the involvement of neurotransmitters including choline, dopamine, serotonin and GABA systems in thermoregulation (5-7). In recent past, the role of opioid in thermoregulation (8) and the influence of environmental temperature on the thermic responses to opioids have also been known (9).Several evidences have suggested that opioid interacts with different neurotransmitters in different functions of the brain (10). Since higher environmental temperature (HET) increases the BT and changes the hypothalamic GABA (11), it is not unlikely that there may be some interaction between the opioid and neurotransmittes systems in thermoregulation. In the present pharmacological and biochmical investigation we have studied the possible interaction of opioid with hypothalamic GABA in the regulation of normal BT and HET-induced change in BT.

MATERIALS AND METHODS

GABA, Na-glutamate, pyridoxal phosphate, ninhydrin, α-ketoglutarate, methylbenzothiazolinohydrazine and ethanolamine-O-sulfate (EOS), muscimol, bicuculline, atropine sulfate, physostigmine and naloxone were purchased from Sigma chemical Company (St. Louis, MO, USA). Morphine was obtained as a gift from Medical College, Calcutta, India. All other reagents used in the present study were of analytical grade.

Adult male albino rats (120-140 g, b.wt.) of Charles Foster strain, kept in a 12 h dark/12h light cycle room having temperature $28^{\circ} \pm 1^{\circ}$C and constant relative humidity (80±5%), were maintained with standard laboratory diet and water *ad libitum.*.

Experimental Conditions(s) of Drug Treatment and/or Exposure to HET for Pharmacological and Biochemical Studies :

(a) *Pharmacological study* : Animals were divided into different groups of six animals each. The animals of groups 1-7 were treated with vehicle (saline), morphine, naloxone, muscimol and bicuculline individually or their different combinations and were exposed to HET $(40^{\circ}+1^{\circ}$C) in a thermostatically controlled ventilated chamber of a fixed relative humidity $(80 \pm 5\%)$ for exactly 2 h. The animals of groups 8-14 kept in a chamber at room temperature $(28^{\sigma} \pm 1^{\circ}$C) with constant relative humidity $(80 \pm 5\%)$, were considered as control and were treated with vehicle or above mentioned individual drug or their different combinations under similar conditions as described in experimental animals. Doses and durations of treatments of different drugs with and without HET exposure prior to measurement of BT are described in Table2.

(b) *Biochemical study* : In the present study, the animals of groups 1-3, kept at room temperature $(28^{\circ} \pm 1^{\circ}$C) with constant relative humidity $(80 \pm 5\%)$, were treated with vehicle (saline), morphine (1 mg/kg, i.p.), and naloxone (1 mg/kg, i.p.) respectively. The animals (Group 1) treated with vehicle were considered as control. Morphine and naloxone were injected 60 min and 30 min respectively prior to sacrifice. The animals of groups 4-6, exposed to HET $(40^{\circ} \pm 1^{\circ}$C) for exactly 2 h in a thermostatically controlled chamber with a constant relative humidity $(80 \pm 5\%)$, were treated with saline, morphine and naloxone respectively. Morphine was injected 60 min and naloxone 30 min before the end of 2 h following HET exposure. At the end of 2 h following HET exposure, the animals were killed by sharp decapitation between 10.00 - 11.00 h. Heads were immediately thrown into liquid nitrogen for the estimation of GABA levels. For the estimation of enzyme activities the brains were collected under ice cold $(0^{\circ} - 4^{\circ}$C) condition.

Dissection of Hypothalamic Region of Brain : The hypothalamic region of brain of both control and experimental animals were dissected out according to the method as described by Poddar and Dewey (12).

Biochemical Assay : GABA content in hypothalamus was estimated following the method of Lowe *et al.* (13). Activities of glutamate decarboxylase (GAD, E.C. 4.1.1.15) and GABA-transaminase (GABA-T, E.C. 2.6.1.19) were measured according to the method of MacDonnell and Greengard (14) and Sytinsky *et al.* (15) respectively. GABA turnover rate was measured in terms of GABA accumulation using ethanolamine-O-sulfate (EOS) (2 g/kg, s.c.)., a GABA-T inhibitor, following the method of Leach and Walker (16). Protein was estimated according to the method of Lowry *et al.* (17) using bovine serum albumin as standard.

Measurement of Rectal Temperature : Rectal temperature considered as an index of body temperature (BT) was recorded with a thermistor probe inserted 2 cm into the rectum of normal and HET exposed rats. The rectal temperature of HET-exposed (both control and drug treated) rats was recorded at the end of the 2 h of HET exposure.

Statistical Analysis : The statistical significance between the mean values were assessed by Tukey test for analysis of variance (ANOVA).

RESULTS

Table 1 shows that exposure of rat to HET $(40^{\circ}$C) for 2 h significantly increased hypothalamic GABA level and its turnover with a significant decrease in GABA-T activity.

A single administration of morphine (1 mg/kg, i.p.) to rat either kept at 28°C (room temperature) or exposed to HET significantly reduced the steady-state level of hypothalamic GABA and GAD activity. The hypothalamic GABA-T activity of HET exposed rat treated with morphine was significantly inhibited though the apparent inhibition of GABA-T in hypothalamus of rat tested with morphine at 28°C was observed. No significant inhibition in GABA turnover (EOS-induced GABA accumulation) was observed under these conditions of morphine treatment and/or HET exposure. But the hypothalamic GABA turnover of morphine treated rat exposed to HET was significantly increased (35%) with respect to that observed in morphine treated rat at 28°C. Table 1 further shows that naloxone (1 mg/kg, i.p.) did not produce any significant effect on any of the parameters of hypothalamic GABA excepting the GAD activity and GABA turnover which were increased at room temperature. But treatment of HET exposed rat with naloxone significantly increased hypothalamic GABA level and turnover. The GABA-T activity in hypothalamus was significantly inhibited under similar condition.

Table 1. Effect of Morphine or Naloxone on the Steady-State Level of GABA, its Metabolising Enzymes and Turnover in Hypothalmus of Rat at Room Temperature (28°C) or Exposed to HET (40°C)

Parameters	Control (Saline)		Morphine		Naloxone	
	28°C	40°C	28°C	40°C	28°C	40°C
Steady-state level of GABA	100.0 ± 4.4	144.8[a] ± 5.6	69.8[a] ± 4.7	52.9[a] ± 4.4	117.3 ± 5.7	132.0[a] ± 5.8
GAD Activity	100.0 ± 3.2	101.8 ± 3.2	58.0[a] ± 2.9	44.4[a] ± 2.5	127.4[a] ± 2.8	123.0 ± 3.9
GABA – T Activity	100.0 ± 5.4	70.0[a] ± 3.0	73.3 ± 6.7	58.7[a] ± 3.7	88.7 ± 4.5	67.2[a] ± 4.0
GABA turnover (EOS-induced GABA accumulation).	100.0 ± 4.0	133.0[a] ± 5.3	83.3 ± 2.1	112.5[a] ± 4.1	137.8[a] ± 3.8	141.7[a] ± 4.1

Results (in percent) are expressed as mean ± SEM (n=4-6).

The control values of Steady- State level of hypothalamic GABA (28.22 ± 1.06 n mole/mg protein), GAD activity (363.65 ± 11.92 n mole GABA/mg protein/h), GABA – T activity (381.85 ± 28.41 n mole SSA/mg protein/h) and GABA turnover (2.33 ± 0.16 n mole GABA accumulated/mg protein/h) of rat at 28°C were considered as 100.

Significantly different from control at 28°C or 40°C or Corresponding drug at 28 ° C [a] P<0.01 (using Tukey test for ANOVA).

Table 2 represents that exposure of rats to HET (40°C) for 2 h significantly raised the BT. Treatment with morphine (opioid against; 1 mg/kg, i.p.) significantly increased the BT of both normal (28°C) and HET exposed rats; whereas, naloxone (opioid antagonist; 1 mg/kg, i.p.) significantly reduced the BT of rat at 28°C and attenuated the HET-induced rise in BT. The BT of normal (28°C) and HET exposed rats was significantly increased with bicuculline (1 mg/kg, i.p.), GABA antagonist. But GABA agonist, muscimol (1 mg/kg, i.p.), significantly reduced the BT of both normal and HET exposed rats. Cotreatment of morphine with bicuculline significantly enhanced (a) the BT of bicuculline treated rats kept at 28°C or exposed to 40°C and also (b) the BT of morphine treated rat exposed to HET. The cotratment of naloxone with bicuculline, on the other hand, failed to alter the bicuculline-induced rise in BT of normal (28°C) and HET exposed rats (Table 2).

Table 2. BodyTemperature (Rectal Temperature) of Normal (28°C) and HET (40°C) Exposed Rats Treated with Agonists and Antagonists of Opioidergic and GABAergic Systems.

Treatment	Dose (mg/kg, i.p)	Change in Rectal Temperature (°C) at	
		28°C (Normal)	40°C
Control (Saline)	–	–	+ 2.8 ± 0.06[a]
Morphine	1	+ 1.8 ± 0.11[a]	+ 3.9 ± 0.08[a]
Naloxone	1	– 0.8 ± 0.04[a]	+ 2.0 ± 0.05[a]
Muscimol	1	– 1.0 ± 0.03[a]	+ 2.0 ± 0.04[a]
Bicuculline	1	+ 1.1 ± 0.03a	+ 3.8 ± 0.07[a]
Morphine + Bicuculline	1 + 1	+ 1.7 ± 0.06[a]	+ 4.8 ± 0.08[a]
Naloxone + Bicuculline	1 + 1	+ 1.3 ± 0.03[a]	+ 3.6 ± 0.09[a]

Results are expresed as mean change ± SEM (n = 4-6).

In HET exposed rats rectal temperature was measured at the end of the 2nd hour of HET exposure. Muscimol, bicuculline, naloxone were injected 30 min, morphine 1 h before measurement of rectal temperature in rats exposed to either 28°C or 40°C. Changes were calculated with respect to control rat maintained at 28°C. The basal rectal temperature of control rat was 36.2° ± 0.14°C.

Significantly different from control at 28°C or 40°C [a]$p<0.01$ (using Tukey test for ANOVA).

+ indicates increase and – indicates decrease with respect to control.

DISCUSSION

The BT of mammals has been found to increase with the increase in degree and duration of exposure to ambient temperature (18). The duration of exposure to 40°C in the present study was maximum up to 2 h, because exposure at this temperature for more than two hours causes significant mortality. Our previous studies have shown that the increase in BT due to the exposure to HET is under the regulation of central GABA through the interaction with cholinergic neuron (11). The knowledge of the involvement of opioid in thermoregulation and the interaction of opioid with other neurotransmitter(s) including GABA, dopamine and choline (5- 8, 10) in CNS led authors to investigate biochemically and pharmacologically the possible mechanism of interaction of opioid with hypothalamic GABA under normal and HET-exposed condition.

The HET-induced increase in BT is further enhanced with morphine. Naloxone, on the other hand, attenuates the HET-induced rise on BT (Table 2). It is also noted that morphine at the same dose produces hyperthermia in control rats at 28°C (room temperature); but naloxone is hypothermic under similar condition (Table 2). Thus, the HET-induced increase in release of endogenous opioid/or opioid like substance and the involvement of opioidergic neuron in thermoregulation under normal and HET exposed conditions may be suggested. The increased release of β- endorphin in HET-induced hyperthermia has already been reported (19). Further, (a) bicuculline - induced increase in BT of both normal (28°C) and HET exposed rat, (b) muscimol-induced hypothermia in normal rat and (c) the attenuation of HET-induced increase of BT with muscimol (Table 2), suggest that GABA may be involved in thermoregulation. Biochemical studies show that short-term (2 h) HET exposure of rat significantly decreases the ratio of GABA- T/GAD and increases the EOS-induced accumulation of GABA in hypothalamus in spite of significant increase in hypothalamic GABA level (Table 1). Since hypothalamic GABA-T activity is inhibited without any appreciable change in GAD activity under HET exposed condition, the hypothalamic GABA level with HET exposure may be increased due to an inhibition of GABA utilisation suggesting that hypothalamic GABA system is inhibited under HET exposed condition. Thus, it may be concluded that central GABA is involved in thermoregulation possibly through inverse relation with BT. Now the question is what is its mechanism? How opioid or opioid like substance is involved in thermoregulation under normal and HET-exposed conditions?

The treatment with morphine at $28^{\circ}C$ significantly decreases the GAD activity and reduces the GABA content in hypothalamus with apparent inhibition in its GABA-T activity (Table 1) indicating that morphine at room temperature ($28^{\circ}C$) inhibits the hypothalamic GABA metabolism (both synthesis and utilization of GABA). The exposure of morphine treated rats to HET significantly reduces the hypothalamic GABA content and GAD activity indicating an inhibition in GABA synthesis and its release (20). Moreover, the decrease in hypothalamic GABA-T activity and increase (with respect to the effect of morphine treatment at $28^{\circ}C$) in its GABA turnover (Table 1) under similar condition suggest that morphine under HET-exposed condition reduces the hypothalamic GABA utilisation and hence the GABA system. Further, it seems that morphine produces greater inhibitory effect on hypothalamic GABA ergic activity under HET-exposed condition than that observed with morphine at $28^{\circ}C$ or HET alone. Pharmacologically the potentiating effect of morphine on bicuculline-induced increase in BT of normal and of HET-exposed rats (Table 2) supports the idea that HET may stimulate the release of endogenous morphine-like substance/opioid (19) and morphine inhibits the GABAergic activity and increases the BT. The inhibitory effect of β-endorphin on central GABA has already been shown by other (21). No significant effect of naloxone on bicuculline-induced increase in BT of both normal and HET-exposed rats (Table 2) further supports the above assumption. Biochemical studies show that the treatment of naloxone to normal rats increases the hypothalamic GABAergic activity by increasing its GABA synthesis (because th ratio GABA- T/GAD is significantly reduced without any appreciable effect on GABA-T) and GABA turnover (Table 1). But exposure of naloxone treated rats to HET reduces the naloxone-induced hypothalamic GABA utilisation by inhibiting the GABA-T activity, the ratio GABA-T/GAD and increase of EOS-induced GABA accumulation and GABA content (Table 1) and hence reduces the GABAergic activity. This result may explain the naloxone-induced hypothermia of normal rat and attenuation of HET-induced increase of BT (Table 2).

In recent past we have shown that central GABA has an inbibitory regulation over its cholinergic activity in thermoregulation (8). Thus, it may be concluded, finally, from the present study that central opioidergic neuron regulates the BT under both normal and HET-exposed conditions through the interaction with hypothalamic GABAergic activity.

Acknowledgment : The present work was supported by Indian Council of Medical Research, New Delhi; University Grants Commission, New Delhi and University of Calcutta, India.

REFERENCES

1. Blatteis CM. Hypothalamic substances in the control of body temperature : General Characteristics. Fed. Proc. 1881:40: 2735- 2740

2. Bligh J. The Central neurology of mammalian thermoregulation. Neunroscience 1979 : 4: 1213-1236.

3. Siesjo BK. Hypothermia and Hyperthermia. In Siesjo BK, editor. Brain Energy Metabolism, New York, Wiley, 1978 : 324-344.

4. Lipton JM, Clerk WG. Neurotransmitters in temperature control. Ann. Rev. Physiol. 1986 : 48 : 613-623.

5. Lin MT, Chan FF, Churn YF, Fung JC. The role of the cholinergic system in the central control of thermoregulation in rats. Can. J. Physiol. Pharmacol. 1997 : 57: 1205-1212.

6. Minano FJ, Sancibrian M, Serrano JS.Hypothermic effect of GABA in concious stressed rats : Its modification by cholinergic agonists and antagonists.J.Pharm.Pharmacol.1987: 39 : 721-726.

7. Bligh J, Cottle WH, Maskrey M. Influence of ambient temperature on the thermoregulatory response of 5-hydroxytryptamine, noradrenaline and acetylcholine injected into the lateral cerabral ventricles of sheep, goats and rabbits.J. Physiol.1971 : 212 : 377-392.

8. Adler NW, Geller EB, Rosow CE, Cochin J. The opioid system and temperature regulation. Ann. Rev. Pharmacol. Toxicol. 1988 : 28 : 429-449.

9. Paolino RN, Bernard BK. Environmental temperature effects on the thermoregulatory response to systemic and hypothalamic administration to morphine. Life Sci. 1968 : 7 : 857- 863.

10. Milanes MV, Cremades A,Vargas ML, Arnaldos JD. Effect of selective monoamine oxidase inhibitonrs on the morphine-induced hypothermia in restrained rats. Gen. Pharmacol. 1987 : 18 : 185-188.

11. Biswas S, Poddar MK, Does GABA act through dopaminergic/cholinergic interaction in the regulation of higher environmental temperature-induced change in body temperature. Meth. Find. Exp. Clin. Pharmacol. 1990 : 12 : 303-307.

12. Poddar MK, Dewey WL. Effects of Cannabinoid on Catecholamine uptake and release in hypothalamus and striatal synaptosomes. J. Pharm. Exp. Ther. 1980 : 214 : 63 - 67.

13. Lowe IP, Robins E, Eyerman GS.The fluorometric measurement of glutamate decarboxylase and its distribution in brain. J. Neurochem. 1958 : 3 : 8-18.

14. Mac Donnell P, Greengard O. The distribution of glutamate decarboxylase in rat tissue : Isotopic vs fluorometric assay. J. Nenurochem. 1975 : 24 : 615-618.

15. Sytinsky IA, Guzikov BM, Eremin VP, Konovalona NN. The gamma aminobutyric acid system in brain during acute and chronic ethanol intoxication. J. Neurochem. 1975 : 25: 43-48.

16. Leach MJ, Walker JMG. Effect of ethanolamime-O-sulfate on regional GABA metabolism in mouse brain. Biochem.Pharmacol. 1977 : 26 : 1569-1572.

17. Lowry OH, Rosenbrough NH, Farr AL, Randall RJ. Protein measurement with Folin Phenol reagent. J. Biol. Chem. 1951 : 193 : 265-275.

18. Ghosh S, Poddar MK. Higher environmental temperature- induced increase in body temperature. Neurochem. Res.1993 : (in press).

19. Robins HI, Kalin NH, Shelton SE. Neuroendocrine changes in patients undergoing whole body hyperthermia. Int. J. Hyperthermia. 1987 : 3 : 99-106.

20. Tapia R, Sandoval ME, Contreras P. Evidence for a role of glutamate decarboxylase activity as a regulatory mechanism of cerebral excitabitity. J. Neurochem. 1975 : 24 : 1283-1285.

21. Nicoll RA, Alger BE, Johr CE. Enkephalin blocks inhibitory pathway in the vertebrate CNS. Nature. 1980 : 287 : 22- 25.

Temperature Regulation
Advances in Pharmacological Sciences
© 1994 Birkhäuser Verlag Basel

BILATERAL DIFFERENCE IN TYMPANIC TEMPERATURES REFLECTS THAT IN BRAIN TEMPERATURE

Tokuo Ogawa, Junichi Sugenoya, Norikazu Ohnishi, Kazuno Imai, Takae Umeyama, Motohiko Nishida, Yoshikazu Kandori and Akira Ishizuka. Departments of Physiology and Radiology, Aichi Medical University, Japan.

Summary
Bilateral difference in tympanic temperatures was resulted from tilting the head or body to one side, stellate ganglion block, hemi-face cooling, compression of a unilateral carotid artery or angularis oculi vein, and from brain damage due to unilateral cerebral infarct or hemorrhage. It is inferred that regional or unilateral imbalance in the brain between metabolic heat production and heat loss by circulatory convection and conduction to the surroundings may result in unevenness or asymmetries in the brain temperature which are then reflected in the bilateral difference in tympanic temperatures.

Introduction

Tympanic temperature (Tty) has been widely used as an indicator of brain temperature. Although some investigators dispute its significance (1,2,3), it has been demonstrated by many investigators that Tty correlates better with thermoregulatory responses (4,5,6,7,8) and parallels brain temperature better than other deep body temperatures, such as oesophageal temperature (unpublished data) which has been considered a 'standard' core temperature.

The right and left Tty's are not always equal, but, even with careful measurement, obvious bilateral differences up to 0.05°C are not infrequently recorded, and in some occasions, the right and left Tty's move in mutually opposite directions. Such unparallel changes in the right and left Tty's may likely be caused by bilateral differences in brain temperature.

Brain temperature is determined by the balance between heat production due to cerebral metabolism and heat loss due to circulatory convection and conduction to the surroundings, and regional or unilateral impairment of this balance may result in unevenness or asymmetries in the brain temperature, which may be reflected in the difference between the right and left Tty's. This paper surveys a series of our studies concerning the mechanism and significance of bilateral differences in Tty's.

Bilateral Tty's were measured continuously using a thermistor element with soft spring coil, whose placement on the tympanic membrane was easy and caused little pain (9).

Effect of head and body positions

Long-term simultaneous recording of the right and left Tty's (8-48 hr) showed frequent unparallel changes in them which were closely associated with the posture of the body and/or head. Flexion of the head to a side in an upright sitting or standing position, rotation of the head to a side in a supine position, and lying on a side, all caused a rise in the Tty on the ipsilateral (lower) side and a fall in that on the contralateral (upper) side (Fig. 1). Brain scintigraphy by means of single photon emission computed tomography (SPECT) using 99mTc-hexamethylpropyleneamine oxime (99mTc-HMPAU) showed that rotation of the head to the right caused a slight decrease of blood flow in the right hemisphere and a slight increase in the left hemisphere. Those observations suggest that an increase in blood flow facilitates cooling of the brain and that the distribution of blood flow to the right and left hemispheres is influenced by a reflex involving vestibular stimulation (10).

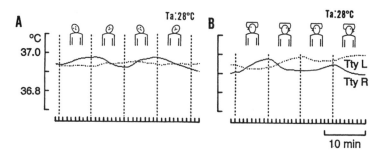

Fig. 1 Changes in the bilateral tympanic temperatures in response to changes in the head position: A, lateral flexions in a sitting position; B, rotations in a supine position.

Effect of stellate ganglion block

Stellate ganglion blocking by local anesthesia was followed by a slight but significant drop in the ipsilateral Tty sustained for about an hour (11). An increase in ocular blood flow was recognized on the block side by oculo-cerebrovasculometry. An increase in blood flow of the ipsilateral hemisphere was evidenced by SPECT with 99mTc-HMPAU. Sympathetic control is generally not considered to play an important role in cerebral blood flow, but the above observations suggest that the release of the sympathetic control of the cerebral vasculature induces a slight but significant increase in blood flow, thus facilitating brain cooling on that side.

Effect of hemi-face cooling

Blowing on one side of the head with cool air or ventilating a circumscribed area of the lateral neck at a high ambient temperature generally induced a fall in the ipsilateral Tty, which was accompanied in many cases by a slight decrease in forearm sweating rate on the same side (12,13). The fall in the ipsilateral Tty is considered to have been resulted from cooling of the brain as well as the tympanic membrane by means of circulatory convection and partly by conduction, as has been suggested by Cabanac et al. (14). The results suggest that the maneuver may cause asymmetry in brain temperature and even in hypothalamic temperature which induces a slight asymmetric change in sweating activity.

Effect of vascular compression in the head and neck

A unilateral rise in Tty was also noted during compression of the ipsilateral carotid artery (Fig. 2A). It was distinct in a warm environment, while the face was cooled by blowing cool air upon it, as long as the Tty was higher than the oesophageal temperature. A slight but definite rise in a unilateral Tty was also induced by compression of the ipsilateral angularis oculi vein at the medial ocular angle, which would block the communication between the cavernous sinus and veins on the face, while the face was blown with cool air in a warm environment (Fig. 2B). Such maneuvers appear to disturb cooling of the arterial blood perfusing the brain by the venous blood from the face and skull by means of the countercurrent exchange mechanism (15).

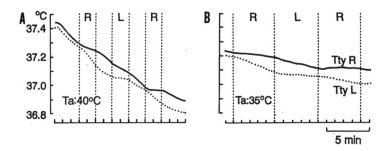

Fig. 2 Changes in the right and left tympanic temperatures during compression of a unilateral carotid artery (A), and of a unilateral angularis oculi vein at the medial ocular angle (B), while the face was blown with cool air in a warm environment.

All the above results reveal that a unilateral increase in cerebral blood flow facilitates and its decrease impedes cooling of the hemisphere.

Effect of brain damage

Furthermore, significant asymmetry in Tty's was observed in patients with unilateral cerebrovascular disorders, Tty being lower on the side where brain was damaged (16). The spread of the brain damage determined by computed tomography correlated well with the degree of the bilateral difference in Tty's (Fig. 3). Those observations suggest that a localized decrease in heat production due to impaired local metabolism results in a lowering of the hemispheric temperature which is reflected in the Tty on the ipsilateral side.

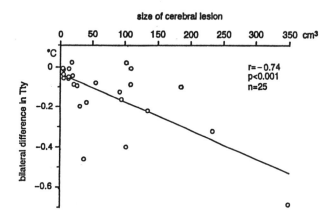

Fig. 3 Relation of the bilateral difference in tympanic temperatures (Tty on the affected side minus that on the intact side) to the size of the lesion in patients with cerebral infarct, estimated from computed tomography images.

In conclusion, asymmetry in either metabolic rate or in cerebral blood flow, or localized imbalance between them elicit an asymmetric change in hemispheric temperatures which is reflected in bilateral Tty's.

REFERENCES

1 Shiraki K, Konda N, Sagawa S. Esophageal and tympanic temperature responses to core blood temperature changes during hyperthermia. J Appl Physiol 1986; 61:98-102.

2 Brengelmann GL. Dilemma of body temperature measurement. In: Shiraki K, Yousef MK, editors. Man in stressful environments, thermal and work physiology. Springfield: Thomsa, 1987; 5-22.

3 Shiraki K, Sagawa S, Tajima F, Yokota A, Hashimoto M, Brengelmann GL. Independence of brain and tympanic temperatures in an unanesthetized human. J Appl Physiol 1988;65:482-6.

4 Cabanac M, Caputa M. Open loop increase in trunk temperature produced by face cooling in working humans. J Physiol (Lond) 1979;289:163-74.

5 Caputa M, Cabanac M. Muscular work as thermal behavior in humans. J Appl Physiol 1980;48:1020-3.

6 Hirata K, Nagasaka T, Noda Y, Nunomura T. Finger vasodilation correlates better with tympanic than esophageal temperature. Eur J Appl Physiol 1988;57:735-9.

7 Ogawa T, Ohnishi N, Yamashita Y, Sugenoya J, Asayama M. Effect of facial cooling during heat acclimation process on adaptive changes in sweating activity. Jpn J Physiol 1988;38:479-90.

8 Brinnel H, Cabanac M. Tympanic temperature is a core temperature in humans. J Therm Biol 1989;14:47-53.

9 Masuda M, Uchino K. A device for measuring tympanic membrane temperature in man. Jikeikai Med J 1978;25:95-9.

10 Ogawa T, Sugenoya J, Ohnishi N, Natsume K, Imai K, Kandori Y, et al. Effects of body and head positions on bilateral differences in tympanic temperatures. Eur J Appl Physiol. In press.

11 Umeyama T, Kugimiya T, Ogawa T, Ichiishi N, Hanaoka K, Wakasugi B. Unilateral drop of tympanic temperature caused by stellate ganglion block (in Japanese). Pain Clinic 1992;13:59-61.

12 Ogawa T, Yamashita Y, Ohnishi N, Natsume K, Sugenoya J, Imamura R. Significance of bilateral differences in tympanic temperature. In: Mercer JB, editor. Thermal physiology 1989, Amsterdam: Elsevier, 1989, 217-22.

13 Ogawa T, Yamashita Y, Sugenoya J, Ohnishi N, Natsume K, Imamura R, et al. Evaluation of bilateral difference in tympanic temperature (in Japanese). Jpn J Biometeor 989;26:161-7.

14 Cabanac M, Germain M, Brinnel H. Tympanic temperature during hemiface cooling. Eur J Appl Physiol 1987;56:534-9.

15 Ogawa T, Sugenoya J, Ohnishi N, Natsume K, Imai K. Bilateral differences in tympanic temperatures induced by vascular compression in the head and neck. Jpn J Physiol 43 Suppl. In press.

16 Nishida M. Cerebrovascular disease and tympanic temperature (in Japanese). J Aichi Univ Med Assoc. In press.

Temperature Regulation
Advances in Pharmacological Sciences

TEMPERATURE INTERHEMISPHERIC BRAIN ASYMMETRY AS A SIGN OF FUNCTIONAL ACTIVITY

I.K. Yaitchnikov and V.S. Gurevitch

Laboratory of Biochemistry, Institute of Cardiology, 15 Parkchomenko Str., St. Petersburg, 194018, Russia.

Summary

Dubl-Coupl device for precise multichannel brain temperature monitoring has been created. It was shown, that the value of interhemispheric temperature asymmetry of limbic structures may be influenced by various factors as temperature, chemical substances, behavioural reactions. Large asimmetry under thermoneutral environments seems to predict more effective resistance of animal to hot load, but small asymmetry before animal heating predict quick rise of core temperature.

Introduction

We have began our study with attempt to estimate in conscious rabbits whether heating of the preoptic anterior hypothalamic area (PAHA) induces synchronization of the electroencephalogram and PAHA overheating includes arousal of it according to C. Von Euler data obtained under anaesthesia [1]. The first surprise was the obvious absence of any electroecephalogram changes under PAHA heating within normal limits of core temperature [2]. The second one was the presence of electroencephalogram changes only as a result of interaction of local PAHA heating and temperature characteristics of core, but not as a result of current or evoked thermoregulatory reactions [2, 3]. Therefore, we were to decide what is brain temperature map in conjuction with core temperature with main program of body life. Our literature findings were not very large [4, 5, 6]. There is a small amount of individual works for the many years but they are not enough to be summarized in a number of reviews.

Thus, the present experimental study consists of several stages : a) To measure various brain structures temperature simultaneously during long period of time with the highest precision; b) To compare brain and body temperature changes in the dynamics *) under various temperature environments, **) under electrical stimulation of hypothalamic structures and ***) under pharmacological influences.

Materials and Methods

The experiments were carried out in 25 male rabbits after stereotaxic implantation of 12 thermocouples and 4 bipolar nichrome electrodes symmetrically in right and left brain hemispheres. An experiment lasted 4-6 hours, next one began 7-10 days, study programme lasted 1-1.5 year in one animal. Neutral (18-24°C), warm (25-30°C) and hot (42-45°C) environments were created by hot wall chamber. Temperature of ears (right and left) and rectum, heart and respiratory rate, and muscles electrical activity were recorded simultaneously. For thermometry of one point of the nervous tissue we used two identical copper-constantan couples (80mkM diameter). One of the couples was implanted (Fig. 1) with it "hot" junction into the brain and was inserted with it referent junction into thermal gradient neutralizer tube which was maunted on the head of the rabbit. The second couple was inserted into accurate temperature constant device with it referent junction and into thermal gradient neutralizer with it "hot" junction. In this way good thermal contact between both thermocouples was obtained and it resulted in good compensation of outside temperature signals from animal environments. Electric circuit for measuring was formed of both thermocouples (by means of their electric contact with opposite electric polarity), amplifier and recorder. The precision of measurement depends on properties of each element of the circuit and really was performed with an accuracy of ±0.01°C for absolute significance within the range of 15-45°C and of ±0.001°C for relative significance within the range of 0.2°C between channels. However, the precision at special cases could be provide higher [7]. This device was named "Dubl-Coupl".

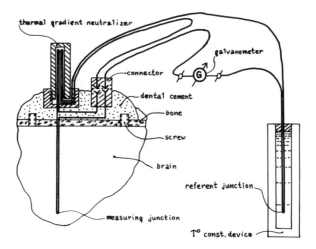

Fig.1. Flow diagram of the Dubl-Coupl device for contact local precise temperature measurements in vital subjects of inquiry (See explanations in the text).

Electric stimulation of the nuclei of medial forebrain bundle or pareventricular systems inside the hypothalamus was performed by rectangular impulses (ims, 150Hz) with step by step (during 3-5s) rised amplitude up to appearance of behavioural reactions (50-80mkA). Aqueous solutions of drugs (chlorpromazine, haloperidol, aethimizol, caffeine, amphetamine) and chemical substances (derivates of imidazoldicarbonic acid, taurine and glutamate) were injected intravenousely equimollary according to aethimizol effective dose of 0.1mMol. All stereootaxic coordinates of investigated "points" of the brain tissue were histologically verifyed. Data values are expressed as the mean±S.E. Comparisons were made using the Student's "t" test for unpaired data. The level of statistically significant difference was P<0.05.

Results and Discussion

Under rest and thermoneutral condition rabbit's brain between A3 amd P5 frontal sections [8] may be characterized with several attributes. Each hemisphere is warmer at basal level relative to dorsal one, at medial part relative to lateral one and at basal level caudal part relative to rostral one. As a whole, right hemisphere is warmer, than left one. The rate of temperature variations of the various nuclei relative to each other inside each hemisphere is 0.351±0.044°C and between homotopic nuclei is 0.420±0.039°C (P<0.05). Both mono- and bilateral temperature amplituide variations are the largest in neocortex and in amigdala, while they are moderate in dorsal hippocampus. In the hypothalamus temperature fluctuations are the smallest. Other data [9] makes it possible to say, that the hypothalamus is thertmostabilized. In the picture of 6 simultaneous thermogram pairs we must note prominent spontaneous variability of amigdala temperature and interhemispheric temperature gradient (ITG) between right and left dorsal temperature, therefore speaking about brain temperature we mean right hypothalamic temperature (right medial preoptic area is preferable). For example, if the hypothalamus does not change, we say that temperature of amigdala decreases. But, if both amigdala and hypothalamic temperature change down, we say that temperature of the brain decreases. In comparison with rectal temperature hypothalamic temperature according to its dynamics corresponds more than other brain nuclei temperature. Both these temperatures under its above mentioned conditions do not differ from each of other more than 0.31±0.09°C. Therefore, to be correct, we must say that core temperature increases if rectal temperature increases and brain temperature keeps within above noticed limits. In the opposite case we say that rectal temperature increases and brain temperature stays at different temperature level. Figures 2 and 3 present the examples of multichannel recording of vegetative reactions and temperature characteristics of animal.

Under warm environments (when rabbits may compensate thermal load) we have observed: *) stabilization (P<0.05) of temperature interrelationships among all channels of brain temperature recordings during beginning, steady state and postwarming periods, **) large decrease's (even 1.12±0.21°C) of brain temperature (P<0.005) while rectum temperature was within the normal range (38.73±0.13°C - 39.36±0.12°C). If preheating period is characterized with the absence of dorsal hippocampus ITG, animals, as a rule, were not able to compensate warming. Under overheating (when rabbits cannot overcome thermal load) the time of survival, or helpful redistribution of core temperature are signs of successful fight of animal. Overheating induced: *) decrease of dorsal hippocampus ITG too (P<0.05), **) cooling brain in comparison to rectal one. The later was estimated under step-regression anaylsis (rectal temperature was taken as primary, brain temperature was taken as secondary).

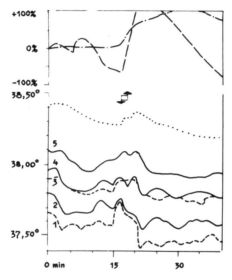

Fig.2. Interhemispheric temperature gradient distributions in rabbit brain induced by electric stimulation of n.ventromedialis hypothalami.

Thermogramms: hippocampus dorsalis sinister - 1 and - dexter 2, n. ventromedialis hypothalami sinister - 3 and - 4 dexter, formatio reticularis dexter - 5, "..." - respiratory rate. Arrows indicate start and finish of electric stimulation of the n.ventromedialis hypothalami dexter.

Time scale is given as minutes and temperature scale is given as degrees Celsius, "%" denotes magnitude relative to background value.

Of course, after learning these findings, we were occupied with the idea to see the work of the old theory about synnergic function of, so called, "cold" and "hot" centers in the hypothalamus by means of their stimulation with electric impulses. We were very suprised to obtain the decrease of ITG in dorsal hippocampus in response to medial preoptic area stimulation early before the appearance of the thermoregulatory reactions and just before the appearance of behavioural (agression) reaction. Enlargement of the dorsal hippocampus ITG was seen with hypothalamic supramammilary area stimulation at the moment before switching behavioural (orientation) reaction. These effects are very raliable (P<0.05). After many trials we have concluded, that all active hypothalamic points are connected by two systems: medial forebrain bundle and periventricular. As we are able to suggest, we have obtained by experimental way the validation of reduction of the rabbit's resistance to heat load if the stimulation diminishes ITG at least in the dorsal hippocampus.

The reasons of our drugs choice are obvious from the behavioural point of view because some of them depress and other stimulate brain functions. These drugs are well known among the thermophysiologists too. The data have shown, that neuroleptics under thermoneutral environments decrease core temperature, do not change and stabilize ITG, remove brain temperature waves. Under hot load neuroleptics stabilize ITG, decrease animal resistance. Psychostimulants increase core temperature, enhance temperature rythmic activity of the brain structures, sometime enlarge ITG under thermoneutral environments and stabilize ITG under hot load.

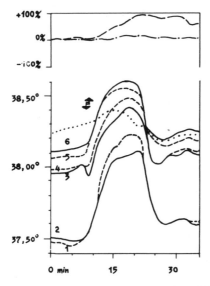

Fig.3. Interhemispheric temperature gradient distributions in rabbit brain induced by electric stimulation of n.lateralis hypothalami dexter.

Thermogramms: hippocampus dorsalis - 1 sinister and - 2 dexter, n.lateralis hypothalami - 3 dexter and 4 sinister, n. ventromedialis hypothalami - 5 sinister and 6 dexter. Others signs are as of Fig.2.

However, aethimizol (derivate of caffeine [10]) decreases brain temperature much more, than rectal one, enlarges ITG under thermoneutral environments. Under overheating aethimizol increases survive time and stabilise ITG. We have found several substances [11] after brain ventriculum injection and imidazoldicarbonic acid derivates which joint useful properties as described above.

If the investigators pay attention to the revealed significance of temperature asymmetry of brain structures in thermal physiology we believe that it is the main outcome of presented paper. Homeostasis of inner medium of the organism under rest [12] and prior conjuction under harmful temperature conditions are two sides of one regulatory system, which is named thermoregulatory.

References

1. EULER, C., VON, SODERBERG, U. (1957). The influence of hypothalamic thermoceptive structure on the electroencephalogramm and gamma- motor activity. *EEG Clin. Neurophysiol.* **9**, 391-408.

2. YAITCHNIKOV, I.K. (1972). Bioelectrical activity of the various parts of the rabbit's brain in response to local change of the central thermosensor temperature. *Phiziol. Jurnal of the USSR.* **58**, 350-355

3. YAITCHNIKOV, I.K. (1979). Pecuillarity of the tempreature influences on central thermocentre). *Phiziol. Jurnal of the USSR* . **65**, 224-231.

4. DONHOFFER, S.Z. (1980). Homeothermia of the brain. *Budapest..*

5. IVANOV, K.P. (1972). Bioenergetics and temperature homeostasis. *Leningrad.*

6. SZELENY, Z. (1986). Lower brainstem mechanisms of body temperature regulation in the guinea-pig. *J. therm. Biol.* 85-86.

7. YAITCHNIKOV, I.K. (1984). A technique for continuous precise thermometry of cerebral structures in chronic animal experiment. *Phiziol. Jurnal of the USSR,* **70**, 91-93.

8. SAWYER, C.H., EVERETT, J.W., GREEN, J.D. (1954). The rabbit diencethallon in stereotaxic coordinates. *J. Compar. Neurol,* **101**, 801-832.

9. YAITCHNIKOV, I.K. (1977). Influences of induced temperature changes of anterior-, posterior hypothalamus and reticular formation on brain temperature mapping. *Patol. Phiziol.* **4**, 76-77.

10. BORODKIN, YU.S., YAITCHNIKOV, I.K. (1988). Analysis of hypothermic properties of aethimizol. *Pharmacol. i Tokssikol.* **4**, 34-35.

11. GUREVITCH, V.S. (1986). Taurines. *Leningrad.*

12. WERNER, J. (1990). Functional mechanisms of temperature regulation, adaption and fever. In: *Schonbaum, E., Lomax, P., editors. Thermoregulation: Physiology and Biochemistry, Pergamon Press, Ink. (New York).* 185-208.

Temperature Regulation
Advances in Pharmacological Sciences
© 1994 Birkhäuser Verlag Basel

CEREBRAL AND RELATED TEMPERATURES IN NORMOTHERMIC SUBJECTS

Zenon Mariak, Janusz Lewko, Marek Jadeszko, Henryk Dudek.
Department of Neurosurgery, University Medical School Bialystok
15-276 Bialystok, M.C. Sklodowskiej 24a, Poland.

Summary

Basing on the material from direct measurements of cerebral temperatures in 18 neurosurgical patients the authors discuss thermal gradients between cerebral and related temperatures in normothermic subjects. Significant correlation was shown to exist between the subdural, epidural, and tympanic temperatures with small and rather stable offsets between them during continuous overnight recordings. Tty most closely approaches the subdural temperature in conscious subjects, but significant, temporal split between these two temperatures was noted in a patient with cerebral pathology.

Introduction

The information regarding the human brain temperature is still very sparse. Up to date only few authors reported direct measurements of cerebral temperature in human subjects - usually neurosurgical patients (1,2,3,4). The results are still preliminary and not always consistent, since the authors discuss together patients who were in different clinical states at the time of investigation.

We have also conducted measurements of brain temperatures in neurosurgical patients. This study is a retrospective analysis of selected subjects who were normothermic, conscious, and did not show clinical signs of focal cerebral pathology at the time of examination. The results could be related to physiological situation in which obtaining of such information is hardly possible.

Materials and methods

Group 1. Continuous recordings of subdural and epidural temperatures were carried out in 11 patients, aged 45-68 years, whose operations were due to subacute subdural haematomas. All patients were conscious and in good general condition. They were informed of the protocol and gave their consent. Miniature, teflon - coated thermocouples were introduced into the subdural and epidural spaces together with a draining latex strip which is always used in surgery of subdural haematomas. Tty was measured with a 0.2 mm thermocouple placed on the anterior lower quarter of the tympanic membrane. Tes was obtained from a probe placed in the oesophagus, 0.42 m below the nares. The temperatures were recorded on a computer-backed registrator and averaged hourly.

Group II. Single measurements of tympanic, rectal, oesophageal and carotid blood temperatures (TWillis) were carried out in 7 conscious patients who had carotid arteriography due to minor subarachnoid haemorrhages. The artery was entered with a Venflon needle and a teflon-sealed, flexible thermacouple (diameter 0.6 mm) was introduced and pushed in a position compatible with that of the circle of Willis.

The temperatures were read with a BAT 10 thermometer (Physitemp-USA) with an accuracy of $0.1^{o}C$ in absolute temperature reading and $0.01^{o}C$ in differential mode.

Results

Fig. 1 shows some examples of typical 24-hours courses of Tre, Tsubdur, Tepidur, Tes and the temperature of the skin on forehead (Tskin) obtained in patients from group I. These data emphasize the parallel behaviour of the temperatures and rather consistent offsets between them obtained in a majority of measurements. The mutual offsets between the recorded temperatures were calculated every hour by subtraction and these values were averaged for all subjects investigated (Fig.2). The following mean differences were found: Tre - T subdur = 0.23 $0.18^{o}C$. Tsubdur - Tty = 0.03 $0.12^{o}C$. Tepidur - Tty = 0.18 $0.07^{o}C$.Tsubdur - Tepidur = 0.2 $0.12^{o}C$. A transient split between Tsubdur and Tty was noted in one subject (Fig.3). The difference between both temperatures, which exceeded $1^{o}C$, persisted over 10 hours. This patient deteriorated clinically and multilocal infarctions within his brain were found in subsequent CT examination.

The ventricular temperature was recorded in two patients and a gradient of 0.2 and $0.3^{o}C$ respectively, was observed with the epidural temperature.

In the group II we found the following mean differences between the temperatures: Tty - TWillis - Tes = 0.25 $0.13^{o}C$ (Fig.2).

Discussion

Despite of increasing interest of neurosurgeons in cerebral temperature many basic questions are still unanswered. Among the problems which need clarification in relation of tympanic temperature to the temperature of the brain (5,6,7) and mutual relation of tympanic temperature to the temperature of the brain (5,6,7) and mutual relations between particular cerebral and related extracerebral temperatures (8). Our data indicate that in normothermic subjects Tsubdur, Tepidur and Tty follow similar time courses during the period of recording. The average offsets

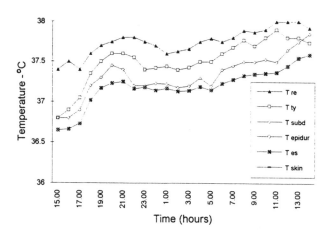

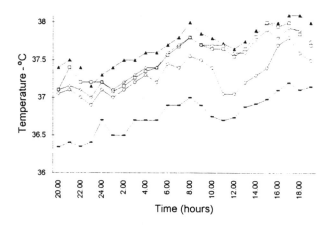

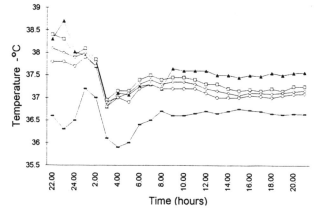

Fig. 1. Examples of 24 hour records of cerebral, tympanic, oesophageal and frontal skin temperatures.

between these temperatures have not exceeded $0.2^{\circ}C$, being close to $0^{\circ}C$ in the case of Tty - Tsubdur. Fig. 4 shows very good correlations between Tty, Tsubdur and Tepidur. Tty could therefore be regarded as an ideal approximation of Tsubdur in most normothermic subjects, but more caution may be suggested in this matter when dealing with patients with cerebral

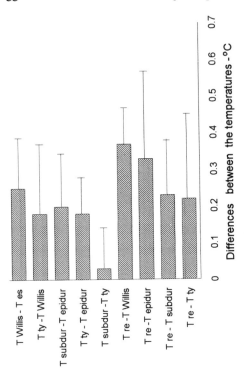

Fig.2. Mean offsets between particular cerebral and related temperatures.

pathology (Fig.2). Intracranial measurement of cerebral temperature seems unavoidable in clinical practice (2). The epidural space seems to be a site of choice for such measurement owing to a small and relatively stable offset between these two temperatures. Epidural measurement of the brain temperature is relatively easy and safe for the patient, whereas direct measurements of Tventr or Tbr will always remain uncommon. From our preliminary observations and from the literature one can suggest that centripetal thermal gradients do exist in human brain, Tventr being $0.2 - 0.5^{\circ}C$ higher than Tepidur and Tsubd (1,2). This is also suggested indirectly by our results of measurements of the temperature within the carotid artery. It has been accepted that during normothermia a main bulk of cerebral metabolic heat is eliminated by circulating blood. From the rates of cerebral metabolism and blood flow one can calculate $0.35 - 0.5^{\circ}C$ as a theoretical difference between the average temperature of the brain and that of incoming arterial blood. In our material a mean difference between Tty and TWillis amounted only to $0.18+/- 0.07^{\circ}C$ which means that the temperature corresponding to the

balanced heat exchange prevails somewhere between the ventricles and the surface of the brain
(9). Further exploration of this problem needs direct recording of all temperatures involved.

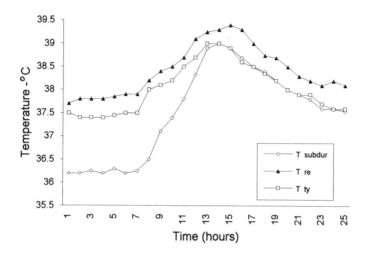

Fig. 3. The time course of the temparatures in one of the examined subjects. Note significant
transient split between Tty and Tsubdur.

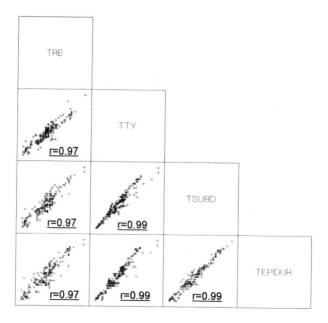

Fig. 4. The plot of mutual correlations between some temperatures examined.

Acknowledgements: This work was supported by the Polish Committee for Scientific Investigation. Grant No 402249101

References

(1)   Shiraki K, Sagawa S, Tajima F, Yokota A, Hashimoto M, and Brengelmann L. Independence of brain and tympanic temperatures in an unanaesthetized human. J Appl Physiol 1988; 65: 482-486.

(2)   Mellergard P and Nordstrom C C. Epidural temperature and possible intracerebral temperature gradients in man. Brit J Neurosurg 1990; 4: 31-38.

(3)   Mellergard P, Nordstrom Ch. Intracerebral temperature in neurosurgical patients. Neurosurgery 1991; 28: 709-13.

(4)   Sternau L, Thompson C, Dietrich W D, Busto R, Globus M T, Ginsberg M D. Intracranial temperature - observation in the human brain (Abstract). J Cereb Blood Flow Metab 1991; 11 (suppl 2): S 123.

(5)   Brinnel H and Cabanac M. Tympanic temperature is a core temperature in humans. J Therm Biol 1989; 14: 47-53.

(6) Brengelmann G L. Dilemma of body temperature measurement. In: Shiraki K and Yousef M K, editors. Man in Stressfull Environments _ Thermal and Work Physiology. Springfirld Il: Thomas, 1987: 5 - 22.

(7)   Nielsen B. Natural cooling of the brain during outdoor bicycling? Pfluegers Arch 1988; 411: 456-461.

(8)   Narebski J. Human brain homeothermy during sleep and wakefulness: an experimental and comparitive approach. Acta Neurobiol Exp 1985; 45: 63 - 75.

(9)   Mariak Z, Bondyra Z, Piekarska M. The temperature within the circle of Willis versus tympanic temperature in resting normothermic humans. Eur J Appl Physiol 1993 (in press)

Temperature Regulation
Advances in Pharmacological Sciences
© 1994 Birkhäuser Verlag Basel

NON-THERMOMETRIC MEANS OF ASSESSING CHANGES OF BRAINSTEM TEMPERATURE: THE QUESTION OF SELECTIVE BRAIN COOLING IN HUMANS

Torsten Langer[1], Bodil Nielsen[2] and Claus Jessen[1]
Physiologisches Institut der Universitat Giessen, Germany (1)
August Krogh Institute, University of Copenhagen, Denmark (2)

Summary
The latencies of the acoustically evoked brainstem potentials depend on brainstem temperature. The sensitivity of this method in detecting changes of temperature is 0.4°C. It was used to see whether face fanning or scalp cooling in hyperthermic human subjects dissociate brainstem temperature from general core temperature. No such evidence was found. It is concluded that external cooling does not decrease brainstem temperature significantly below that of the rest of the body core.

Introduction

Many mammalian species can maintain, during heat stress, the temperature of the brain below the temperature of the rest of the body core. The process is termed selective brain cooling (SBC) and is most evident in artiodactyls, in which the carotid rete enables a heat exchange between the arterial route to the brain and the venous return from the evaporating surfaces of the upper respiratory tract. Although humans do not possess a carotid rete, it has been claimed that SBC occurs also in hyperthermic human subjects when cool venous blood is provided by the sweating face or other surfaces of the head to flow to the inner of the skull (1). The hypothesis relies entirely on the questionable (2) assumption that tympanic temperature is a reliable index of brain temperature and has been difficult to test since direct measurements of brain temperature in substantially heat-stressed humans are precluded. More recently, a non-thermometric approach was introduced. The interpeak latencies (IPLs) of the acoustically evoked brainstem potentials depend on brainstem temperature, and their changes are proportional to changes in local temperature.

This property was used to see whether brainstem temperature can be uncoupled from general body core temperature (esophageal temperature, T_{es}) by face fanning. No such effects were observed, and it was concluded that there was no evidence for brainstem cooling in hyperthermic humans (3, 4). The present study was designed to test the possibility that cooling the scalp would affect brainstem temperature by a heat transfer via the emissary veins of the parietal and mastoid bones (5).

Methods

Clicks of 70 dB SL were delivered monaurally at a rate of 30 Hz. The resulting EEG activity was averaged over 1500 stimuli, and the IPL I-V, which reflects the conduction velocity and synpatic delays between the acoustic nerve and the colliculus inferior, was determined to the nearest 20 µs (Fig. 1). On 11 healthy volunteers, 10 such measurements were made in each of three conditions. First, the subjects were resting in a thermoneutral water bath until T_{es} became stable at a normal level (condition A: normothermia). T_{es} was then increased by a hot water bath. When T_{es} had stabilized at a hyperthermic level, another 10 determinations of the IPLs were made while the head of the subject was enclosed in a hood containing warm and water-saturated air (condition B: hyperthermia, no direct heat loss from the head). The hood was then removed and the head was locally cooled by a helmet. It consisted of a large number of small water-perfused plates, which covered the head from the forehead to the neck and withdrew heat at a rate of 40-50 W. When T_{es} had stabilized at the same level as in condition B, a final set of determinations was done (condition C: hyperthermia, facilitated heat loss from the head).

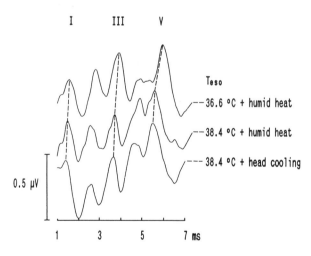

Fig. 1: Acoustically evoked brainstem potentials at conditions A, B and C. The latency between peaks I and V depends on local brainstem temperature.

Results

In order to determine the temperature dependence of the IPL I-V, condition A (normothermia) and condition B (hyperthermia, no direct heat loss from the head) were compared. In both conditions, any local heat loss from the head was blocked so that a fixed correlation between brainstem temperature and general body core temperature (T_{es}) could be assumed. Fig. 2 comprises the results of this and the two preceding series (3, 4), which were done under similar conditions. In all subjects IPLs I-V were shorter at higher T_{es}. IPLs I-V decreased from 4.04 ± 0.22 ms at 37.0 ± 0.3°C T_{es} to 3.80 ± 0.20 ms at 38.6 ± 3°C T_{es}. In each subject the differences were statistically significant (p < .01). The average change of IPL I-V from condition A to condition B amounted to -0.149 ± 0.036 ms/°C increase of T_{es}. Due to the scatter of IPLs I-V in repetitive trials at thermally identical conditions, the minimum change of brainstem temperature detectable by a change of IPLs I-V was found to be 0.30 ± 0.20°C. For the second part of an experiment (condition C) the hood was removed and the cooling helmet was put on to promote the local heat loss from the head.

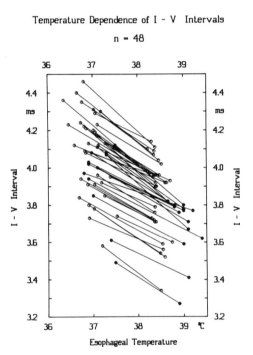

Temperature Dependence of I - V Intervals

n = 48

Fig. 2: A single symbol shows the mean value of 5-10 determinations of IPL I-V at constant T_{es}. Lines join data from the same subject obtained at low and high T_{es}. Open symbols: T_{es} of the resting subjects was raised by a water bath. Closed symbols: T_{es} was raised by exercise.

Heat was removed at a rate of 40-50 W, and scalp skin temperature decreased from 39 to 24°C. By adjusting the temperature of the water bath, T_{es} was maintained at the same level as before (condition B: 38.47 ± 0.13°C, condition C: 38.48 ± 0.12°C). 15 min after head cooling had commenced, another 10 determinations of the IPL I-V were done.

Because conditions B and C were equal with regard to T_{es}, but different with regard to potential SBC, change or constancy of the IPLs would decide on the occurrence or absence of brainstem cooling in response to head cooling. Paired tests showed no differences in 7 experiments. In 3 subjects IPL I-V decreased further in condition B (indicating rising temperature of the brainstem in spite of external cooling), while it increased in another. The mean values of all 11 subjects were 3.82 ± 0.23 ms (condition B), and 3.82 ± 0.20 ms (condition C).

In Fig. 3, the individual means of IPL I-V during head cooling are plotted versus those data which were obtained when the hood had blocked any local heat loss from the head. The figure includes data from two previous studies, in which face fanning was used to promote heat loss from the head, while all other conditions were similar (3, 4). Clearly, the symbols do not deviate systematically from the line of identity, indicating that brainstem temperature had not changed when the local conditions for direct heat loss from the head were greatly improved.

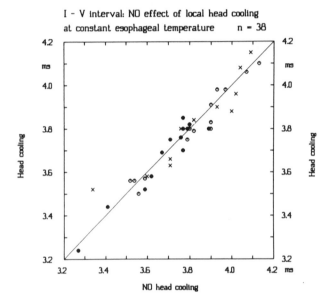

Fig. 3: A single symbol shows the individual mean of IPL I-V during head cooling plotted versus the individual mean of IPL I-V when any local heat loss from the head was blocked. Open circles: face fanning during rest. Closed circles: face fanning following exercise. Crosses: cooling helmet during rest.

Discussion

This study has shown that neither face fanning nor scalp cooling were able to lower brainstem temperature significantly below that of the rest of the body core. The sensitivity of the method in detecting changes of brainstem temperature is approximately 0.4°C. Insofar, the results contradict previous reports (1, 5-7), in which the means of local head cooling were similar to those of our study while their alleged cooling effects on the brain were assessed from changes in tympanic temperature. Downward deviations of tympanic temperature of 1.5°C were reported and taken to suggest equally large changes of brain temperature. In a treatise on heat stroke, face fanning was recommended as a secondary therapy, since "it would keep the brain of a patient near 38.5°C when his rectal temperature reaches 40°C" (6). Thus, our data confirm previous reservations about the reliability of tympanic temperature as an index of brain temperature (2). The absence of cooling effects on the brainstem does not preclude the possibility that other regions of the brain were cooled by face fanning. It should be noted, however, that the proponents of the hypothesis of SBC in human have never considered the possibility that some regions of the brain could be cooled and others not. On the contrary, downward displacements of tympanic temperature were always implicitly and sometimes explicitly (7), taken to indicate selective cooling of the whole intracranial activity. No such evidence was found.

References

1.   CABANAC, M., CAPUTA, M. (1979). Natural selective cooling of the human brain: Evidence of its occurrence and magnitude. *J. Physiol. (Lond)* **286**, 255-264.

2.   BRENGELMANN, G.L. (1987). Dilemma of body temperature measurement. In: *Shiraki, K., Yousef, M.K. eds. Man in Stressful Environments - Thermal and Work Physiology. Springfield: Thomas* 5-22.

3.   JESSEN, C., KUHNEN, G. (1992). No evidence for brainstem cooling during face fanning in humans. *J. Appl. Physiol.* **72**, 664-669.

4.   NIELSEN, B., JESSEN, C. (1992). Evidence against brain stem cooling by face fanning in severely hyperthermic human subjects. *Pfluegers Arch.* **422**, 168-172.

5.   CABANAC, M., BRINNEL, H. (1985). Blood flow in the emissary veins of the human head during hyperthermia. *Eur. J. Appl. Physiol.* **54**, 172-176.

6.   CABANAC, M. (1983). Face fanning: a possible way to prevent or cure brain hyperthermia. In: *Khogali, M., Hales, J.R.S., eds. Heat Stroke and Temperature Regulation. Sydney: Academic Press* 213-221.

7.   CABANAC, M., GERMAIN, M., BRINNEL, H. (1987). Tympanic temperatures during hemiface cooling. *Eur. J. Appl. Physiol.* **56**, 534-539.

Temperature Regulation
Advances in Pharmacological Sciences
© 1994 Birkhäuser Verlag Basel

EFFECTS OF SELECTIVE α-ADRENERGIC BLOCKADE ON CONTROL OF HUMAN SKIN BLOOD FLOW DURING EXERCISE

W. Larry Kenney, The Pennsylvania State University,
102 Noll Laboratory, University Park, PA 16802 USA

SUMMARY

Nonacral human skin blood flow (SkBF) is determined by the balance between α-adrenergic vasoconstriction and active vasodilation. In separate studies, we used pharmacological agents to selectively block either α_1- or α_2-adrenoceptors to study the potential role of each receptor subtype in the control of SkBF during exercise. α_1-adrenergic blockade (prazosin) increased resting SkBF but did not alter its control during exercise. Results from a subsequent α_2-blockade (yohimbine) study suggest that α_2-receptors mediate the relative vasoconstriction in the skin during the sympathetic activation accompanying the first few minutes of exercise. However, this vasoconstriction is overridden by active vasodilation as core temperature increases further. Both studies confirm the idea that the functional limit to increasing SkBF during prolonged exercise is due to an alteration in active vasodilation rather than vasoconstriction.

INTRODUCTION

At rest in thermoneutral conditions, human skin blood flow (SkBF) is determined primarily by tonic sympathetic adrenergic activity. In acral areas, SkBF appears to be controlled solely by reflex changes in adrenergic vasoconstriction (VC); over the rest of the body surface, however, SkBF is controlled by two branches of the sympathetic nervous system -- the VC system and an active vasodilator system for which the neurotransmitter is unknown. Adrenergic VC in human skin is mediated through postjunctional α-adrenoceptors; β receptors are sparse or absent [1]. Based primarily on the development of receptor-specific agonists and antagonists, α-adrenoceptors are pharmacologically classified as either α_1 (actions inhibited by prazosin, doxazosin, terazosin, etc.) or α_2 (actions inhibited by rauwolscine, yohimbine, etc.).

During dynamic exercise, SkBF increases to facilitate heat loss. The level of SkBF is determined by a balance between adrenergic VC and active vasodilation, and each effector branch appears to have distinct roles in its control [2]. At the onset of exercise there is a VC of forearm skin which results in a relative reduction in SkBF. This effect requires an intact VC system, since blocking VC via local iontophoresis of bretylium eliminates this response [3]. On the other hand, the elevation of the core temperature (T_c) threshold for vasodilation during exercise is due to a delay in the activation of the active vasodilator system [4]. Once this threshold is reached, as T_c rises further there is an accompanying steep rise in SkBF. Finally, the rise in SkBF becomes attenuated, i.e., it reaches a functional plateau at higher T_c's [5]. The goal of the two studies described in this paper was to delineate the potentially separate roles of α_1- vs α_2-mediated VC in the control of SkBF during dynamic exercise. In the first study [6], α_1-receptors were blocked systemically with oral prazosin; in the second, α_2-blockade was affected using oral yohimbine. We hypothesized that α-blockade would exert an effect during the initiation phase of exercise, but that this effect would disappear as active vasodilator activity increased and that α-blockade would not affect the attenuation in SkBF at higher core temperatures.

METHODS

Subjects. All procedures utilized in this study were approved in advance by the Biomedical Committee of the Institutional Review Board of the Pennsylvania State University. A total of 14 healthy, active men volunteered to participate in the studies -- 9 in Study 1 and 5 in Study 2. Preliminary screening included a graded exercise test to volitional fatigue performed on a motor-driven treadmill for the assessment of maximal oxygen consumption (VO_2max). All subjects were normotensive, non-smokers, and none was taking other medications of any kind.

Drug regimens. Study 1 was a randomized double-blind trial with subjects taking either prazosin (P; Minipress, Pfizer) or a placebo. A total of 3 mg P was given orally over the 4-h period prior to testing. In Study 2, subjects began oral ingestion of yohimbine (Y; Yohimex, Kramer) 3-4 d prior to testing. The yohimbine dosage was 16.2 mg$\cdot$d^{-1} plus an additional 5.4 mg taken immediately prior to the exercise test. Two separate techniques -- face cooling and graded local heating -- were used to verify drug efficacy. Those results are not described here but can be found in references 6 and 7.

SkBF measurement. Forearm blood flow (FBF) was measured by venous occlusion plethysmography using a mercury-in-Silastic strain gauge. During whole-body heating [8] and dynamic leg exercise [9] increases in FBF are commonly used as an index of increases in SkBF, since forearm muscle blood flow is not elevated during these procedures. FBF was measured in triplicate every minute, and forearm vascular conductance (FVC; 1 unit = ml$\cdot$100ml^{-1}$\cdot$min^{-1}$\cdot$mmHg^{-1}) was calculated from FBF and mean arterial pressure (MAP).

Exercise protocol. Each study began with 30 min of baseline measurement, during which time the ambient temperature was either held at 36 °C (Study 1) or raised from 23 °C to 36 °C over a 30-min period (Study 2). After this period, with ambient temperature fixed at 36 °C, each subject (clad in shorts and a tee-shirt) began exercising on a cycle ergometer positioned with the pedals horizontally in front of the hips at extension. Exercise intensity was preset to approximate 40-50% of each subject's VO_2max. Exercise was continued for 45-60 minutes until each subject's esophageal temperature (T_{es}) exceeded 38.3 °C.

Measurements. T_{es} was measured with a calibrated Type T thermocouple inside a sealed infant feeding tube fed through the nose and into the esophagus to the approximate level of the right atrium. Skin temperature was measured at four sites (chest, upper arm, thigh, and calf) using uncovered Type T thermocouples and mean skin temperature (T_{sk}) was calculated as the weighted average of the four sites. Heart rate (HR) was monitored from a single-lead electrocardiogram. T_{es}, T_{sk}, HR, and cycle rpms during exercise were recorded and stored on a dedicated computer via a Keithley Series 500 data collection system and displayed every 2.5 min during the experiment. Oxygen uptake was measured in duplicate during exercise (at about the 30-min mark) from expired air collected in a Douglas bag. Blood pressure was measured by brachial auscultation every 5 min.

Statistical analyses. Analysis of variance with repeated measures was used to examine potential drug effects on T_{es}, T_{sk}, HR, MAP, and FVC during exercise. FVC was plotted against T_{es} for each subject and various phases of the FVC:T_{es} curve were compared. In all cases, a significance level of 0.05 was used.

RESULTS

Study 1. Resting baseline T_{es} was significantly lower ($p<0.01$) after prazosin administration; however, the T_{es} response to exercise was unaltered by the drug. There was no drug effect on T_{sk} at rest or during exercise. As expected, MAP was significantly lower at rest and during exercise ($p<0.03$) with prazosin and exercise HR was significantly higher ($p<0.02$). Prazosin increased baseline FBF and FVC ($p=0.05$). When exercise FVC was plotted against T_{es} (Fig. 1), the response patterns seen with prazosin and placebo were coincident.

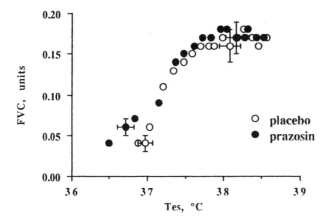

Figure 1. Forearm vascular conductance (FVC) vs esophageal temperature (T_{es}). Each point represents the mean for 9 subjects exercising at ~40% VO2max at an ambient temperature of 36 °C. 1 unit = 1 mmHg·ml^{-1}·100ml·min; bars represent ± 1 SEM.

Study 2. The effects of drug treatment with yohimbine on resting cardiovascular variables are shown in Table 1. Each value presented represents the average value for 5 subjects measured from min 5 through min 30 of the resting period prior to exercise (seated rest at 23 °C ambient temperature). Y significantly increased resting MAP by an average of 11 mm Hg ($p <$

0.05) and HR by 13 bpm ($p < 0.05$). There was no significant drug effect on resting FVC, as the pressor effect of the Y was offset by a slight increase (not statistically significant) in FBF.

Table 1. Resting and exercise (at 15, 30, and 45 min) responses of subjects in the control (C) and α_2-blockade (Y) conditions. Data represent mean values $\pm$ SEM for 5 subjects. Environmental conditions were 36 °C dry-bulb temperature and 30 °C wet-bulb temperature. An asterisk denotes a significant drug effect at $p < 0.05$.

min		rest	15 min	30 min	45 min
MAP, mm Hg	C	88±2	89±5	87±4	85±4
	Y	99±2*	98±4*	96±4*	93±5*
HR, bpm	C	58±3	123±4	135±6	141±5
	Y	71±6*	130±5	148±5	158±3
VO$_2$, l·min^{-1}	C			2.0±0.1	
	Y			2.2±0.2	

Raising ambient temperature to 36 °C over the 30-minute period prior to beginning exercise elevated T_{sk} by 2.5 ± 0.1 (C) and 2.6 ± 0.2 (Y) °C, respectively and during the final 30 min of exercise, T_{sk} averaged 36.3 ± 0.1 °C across all tests with no drug effect. During exercise in the heat, Y had no significant effect on VO$_2$ (Table 1) or the increase in T_{es} (which rose linearly from 36.5 °C to about 38.4 °C in each case). Y resulted in a slight tachycardia during exercise (N.S.) and a significantly elevated MAP ($p < 0.05$) throughout the exercise bout (Table 1). Figure 2 shows the FVC:T_{es} response for one representative subject. At the onset of exercise, there was a steep increase in FVC over the initial 5-8 min with Y, but the steepest portion of the FVC:T_{es} was delayed for several minutes in the C tests. That is, Y treatment caused FVC to increase linearly with increasing T_{es} from the onset of exercise, whereas this steep rise was preceded by a more gradual increase in the C condition, yielding more of a sigmoid curve. After T_{es} had increased by 0.3 °C, the corresponding increase in FVC was 7.2 ± 1.8 units for the Y trials, but only 3.4 ± 0.6 units for the C tests ($p < 0.05$). There were no further drug effects on FVC after this initial phase.

DISCUSSION

In Study 1, prazosin was selected as the α antagonist for the following reasons [6]: (a) the main action of prazosin is at the arteriole but it is devoid of direct dilator activity, (b) with prazosin, prejunctional alpha₂ receptors remain functional, preventing the disproportionate increase in norepinephrine release, renin release, and cardiac output seen with nonselective antagonists, (c) prazosin markedly reduces the skin vasoconstrictor response to such stimuli as deep inspiration and contralateral hand cooling, and (d) baroreceptor control of arterial pressure homeostasis is maintained with prazosin.

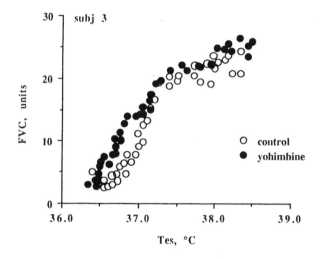

Figure 2. Forearm vascular conductance (FVC, units of ml·100ml^{-1}·min^{-1}·mmHg^{-1}·100) vs esophageal temperature (T_{es}) for one representative subject. These data were collected during separate bouts of dynamic leg exercise, one with and one without yohimbine treatment.

The two main results of Study 1 are: (a) α_1 adrenergic receptors mediate resting vasoconstrictor tone in forearm skin, and (b) the limit to the increase in skin blood flow during prolonged exercise does not appear to be the result of the imposition of vasoconstrictor drive acting through α_1 receptors. These conclusions are based on the finding that blocking α-induced VC in the forearm results in an increase in resting baseline FVC but affects neither the time course nor the magnitude of FVC during exercise.

Prazosin could potentially manifest an effect on skin blood flow systemically, as opposed to via peripheral action. Since systemic blood pressure is reduced, skin blood flow could be decreased purely due to a fall in upstream perfusion pressure. This problem illustrates the need to monitor arterial pressure in FBF studies in which there is reason to believe that MAP may be altered. To better elucidate peripheral effects, we reported FVC, correcting our FBF values for the decreased driving pressure caused by acute administration of prazosin.

If the attenuated rise in FBF at high internal temperatures were due to reflex VC acting through alpha₁-adrenergic receptors, we would have expected prazosin treatment to alter either the time course or the magnitude of the FVC:T_{es} response. Since it did neither (Fig. 1), two possibilities exist. First, the plateau phase of the FVC response could indeed be due to a cardiopulmonary baroreflex which acts primarily through alpha$_2$ (or some other) receptor system. A second possibility is that, rather than imposing VC drive, active vasodilation may be decreased at high internal temperatures. This hypothesis depends on active vasodilation being on the efferent arm of blood pressure regulation as well as subserving temperature regulation. Kellogg et al [10] demonstrated that application of lower-body negative pressure during hyperthermia elicited a withdrawal of active dilator tone, implying that this system is, indeed, under baroreflex control. Clearly, more work is necessary to elucidate control characteristics of the active vasodilator system, a system for which the neurotransmitter is not yet known.

Study 1 did not exclude the possibility that alpha$_2$ adrenergic receptors are integrally involved in the attenuated rise in FVC at high core temperatures. Therefore, study 2 was designed to assess the effects of an α_2-adrenergic antagonist, yohimbine, on forearm vascular conductance (FVC) during exercise. The drug was given orally, and its efficacy in blocking α_2 receptors systemically is demonstrated by its effect on resting HR (tachycardia) and MAP (increased blood pressure). The lack of an effect of yohimbine on resting FVC is puzzling, as it suggests that either no α_2-activation is present at rest or that blockade was insufficient to cause any baseline effect. The central effects of the drug (pressor and heart rate effects) argue against the latter. With respect to the former, the effect of the Y in increasing MAP was offset by a slight increase (not statistically significant) in FBF. Since our baseline FBF measurements were in the range of 2.0 - 2.5 ml·100ml^{-1}·min^{-1}, the magnitude of the increase in FBF needed to offset a 10-12% increase in MAP is within the margin of error of the measurement. We are therefore reluctant to speculate about a lack of effect of Y on baseline FVC. There are reports of both increased [11] and decreased [12] FVC with yohimbine treatment.

The results of Study 2, coupled with the lack of effect of α_1-blockade on forearm SkBF at the initiation of exercise [6] suggests that any relative VC is mediated through α_2-receptors. While not influencing the T_c at which SkBF begins to increase above control values, α_2 blockade with Y significantly affected the rate of rise of SkBF during the initial stages, i.e. the first 5-8 min (first 0.2-0.4 °C rise in T_c). Without Y, FVC increases more slowly at first, essentially doubling after a ΔT_{es} of 0.3 °C. With Y, FVC increased twice as much during this initial phase. Our data cannot clearly distinguish between the effects of α_2-blockade on the exercise induced VC at the onset of exercise and possible effects on the T_c threshold. However, our data are consistent with the idea that the active dilator system determines the threshold T_c for skin vasodilation [4], but that α_2-mediated VC is still present during the early stages of vasodilation. Secondly, the present finding support the hypothesis that the attenuated rise in SkBF at higher T_c's [5] is due to a limit to further increases in vasodilator activity, since neither α_1- (Study 1) nor α_2-adrenergic blockade (Study 2) has a significant influence on this phase of the response.

REFERENCES

[1] Rowell, L.B. 1986. Human Circulation: Regulation during Physical Stress. Oxford University Press, New York.

[2] Kenney, W.L. and J.M. Johnson. 1992. Control of skin blood flow during exercise. Med. Sci. Sports Exerc. 24:303-312.

[3] Kellogg, D.L., J.M. Johnson, and W.A. Kosiba. 1991. Competition between cutaneous active vasoconstriction and active vasodilation during exercise in humans. Am. J. Physiol. 261:H1184-H1189.

[4] Kellogg, D.L., J.M. Johnson, and W.A. Kosiba. 1991. Control of internal temperature threshold for active cutaneous vasodilation by dynamic exercise. J. Appl. Physiol. 71:2476-2482, 1991.

[5] Brengelmann, G.L., J.M. Johnson, L. Hermansen, and L.B. Rowell. 1977. Altered control of skin blood flow during exercise at high internal temperatures. J. Appl. Physiol. 43:790-794.

[6] Kenney, W.L., C.G. Tankersley, D.L. Newswanger, and S.M. Puhl. 1991. Alpha$_1$ adrenergic blockade does not alter control of skin blood flow during exercise. Am. J. Physiol. 260:H855-H861.

[7] Kenney, W.L., D.H. Zappe, C.G. Tankersley, and J.A. Derr. 1993. Effect of systemic yohimbine on the control of skin blood flow during local heating and dynamic exercise. Am. J. Physiol. in press.

[8] Detry, J.M., G.L. Brengelmann, L.B. Rowell, and C. Wyss. 1972. Skin and muscle components of forearm blood flow in directly heated resting man. J. Appl. Physiol. 32:506-511.

[9] Johnson, J.M. and L.B. Rowell. 1975. Forearm skin and muscle vascular responses to prolonged leg exercise in man. J. Appl. Physiol. 39:920-924.

[10] Kellogg, D.L., J.M. Johnson, and W.A. Kosiba. 1990. Baroreflex control of the cutaneous active vasodilator system in humans. Circ. Res. 66:1420-1426.

[11] Bolli, P., B.H. Ji, L.H. Block, W. Kiowski, and F.R. Buhler. 1984. Adrenaline induces vasoconstriction through postjunctional alpha$_2$-adrenoceptors and this response is enhanced in patients with essential hypertension. J. Hypertension 2 (suppl 3):115-118.

[12] Kubo, S.H., T.S. Rector, S.M. Heifetz, and J.N. Cohn. 1989. α_2-receptor-mediated vasoconstriction in patients with congestive heart failure. Circ. 80:1660-1667.

ACCLIMATION TO 3 DIFFERENT CLIMATES WITH THE SAME WET BULB GLOBE
TEMPERATURE

Barbara Griefahn and Paul Schwarzenau
Institute for Occupational Physiology, Department of Environmental Physiology and
Occupational Medicine, Ardeystrasse 67, D-4600 Dortmund 1, Fed. Rep. Germany

Summary

It was hypothesized that acclimation is similar for any climate with the same value of a given thermal index. Therefore, 8 subjects each were exposed during 15 consecutive days to either a warm-humid, a hot-dry, or a climate with radiant load with ever the same wet bulb globe temperature (WBGT). As expected, heart rates and core temperatures decreased gradually, and sweat rates increased. Heart rates level off first, followed by core temperatures, and finally by sweat loss. The courses of acclimation are similar for the 3 climates, but this process is most rapid during the warm-humid condition and it needs the longest time for radiant load.. The hot-dry condition is the most strainous.

1 Introduction and hypothesis

The process of acclimation is characterized by gradual physiologic adaptations to repeated and/or continued heat exposure and finished if the physiologic functions level off to a new plateau, at best to the baseline in temperate climates.

Numerous studies on acclimation revealed the principal physiologic alterations and their mechanisms, but several problems are still unresolved. For instance, contradictory reports concern acclimation to various thermal conditions. More efficient and faster adaptation was reported for hot-dry but also for warm-humid climates. These discrepancies probably result from the fact that the climates compared were not equivalent.

Wet bulb globe temperature (WBGT) which integrates air temperature, humidity, velocity, and radiation is a most simple and widely accepted index for heat. According to the theory of thermal indices the process of acclimation must be the same for any climate with the same WBGT, regardless of the actual variation of the physical parameters. Similar physiologic responses to repeated exposure to various but equivalent climates are expected. The present study was designed to proof this hypothesis for 3 equivalent but differently composed climates with the same WBGT.

2 Material and methods

3 experimental series with 15 consecutive days each were executed in the lab. To avoid natural (partial) acclimation the experiments took place during the winter. Throughout the entire series the subjects were exposed to either of the 3 following climates which are equivalent in terms of the WBGT but differ considerably with regard to the thermal parameters:
- warm-humid: t_a=37 °C, P=4.40 kPa, v_a=0.3 m/s, t_R= t_a, WBGT=33.5 °C,
- hot-dry: t_a=50 °C, P=1.90 kPa, v_a=0.3 m/s, t_R= t_a, WBGT=33.6 °C,
- radiant load: t_a=25 °C, P=1.26 kPa, v_a=0.5 m/s, t_R=90.8°C, t_g=58.2 °C, WBGT=33.4 °C.

Experimental procedure: The experimental procedure was invariably maintained during the 15 days of each series. First, the subjects rested 10 minutes on a chair in a neutral climate (t_a = 22 °C, RH: 40-60 %). Thereafter they spent another 10 minutes sitting on a chair in the climatic chamber and then walked on a treadmill during 4 successive 25-minutes periods (4 km/h, inclination = 0°). They finally rested 15 minutes in the neutral climate. Each of these periods were followed by a 3-minutes break. The first interval just before moving into the climatic chamber endured 5 minutes.

Subjects: Each series was completed by 8 healthy subjects (2 females, 6 males, 21-32 yrs). As some subjects participated in 2 or even 3 series, a complete loss of acclimation was guaranteed by an interval of at least 52 days between consecutive series.

Clothing and fluid supply: Male subjects wore cotton shorts, socks and gym shoes (0.1 clo), the females wore additionally a T-shirt (0.2 clo). The subjects received herb tea ad libitum.

Physiologic parameters and mood: During the entire experiments heart rates (ECG), rectal temperatures (thermistors, 10 cm above the anus), and skin temperatures (forehead, chest, leg) were continuously recorded throughout. During each break sweat loss was determined by weighing the subjects with an accuracy of 5 grams and actual mood was assessed using suitable scales.

If rectal temperature exceeded 38.7 °C the treadmill walk was terminated but the subjects stayed in the climatic chamber until the regular end of the last working period.

Treatment of the data, statistics: Heart rates and core temperatures were averaged for every consecutive minute and missing values were estimated by a logistic growth function. Further statistics base on total sweat loss during the entire experiment and on core temperatures as averaged over the 25 data of the 4th working period. The cardiac strain is indicated by the difference between the average heart rates of the 4th working period minus the heart rates recorded during the initial rest in the neutral climate (stress-related increase of the heart rates). These data were then averaged over the beginning and over the end of the experiments (2 initial resp. 2 final days).

First of all the physiologic data during the initial rest in the neutral climate were proofed to be equal. This allows to relate differences between climates to the particular thermal stress.

3 Results and discussion

3.1 The general course of acclimation

The physiologic variables revealed characteristic signs of acclimation, which are supported by many studies (1): Core temperature and cardiac strain decreased whereas sweat loss increased (fig. 1,2). The alterations between the beginning and the end of the 15 experiments are significant on the1%-level for each variable and for each climate (table 1, t-test for dependent samples). A single exception concerns sweat loss during repeated exposure to radiant load: where heart rates and core temperatures decrease considerably, whereas sweat loss remains constant throughout. The point of acclimation, whereafter no further significant alterations occur was determined by means of a logistic function and defined as the calculated final size minus 0.1 °C for core temperatures, and minus 5 bpm for heart rates. This point varies with the physiologic variables: heart rates level off first, after 3.6 to 5.2 days, core temperatures need 4.1 to 6.9 days, and sweat loss 5.4 to 9.5 days. This succession is reported by other experimenters as well. Thus, full acclimation was achieved and statistically confirmed for the 3 variables in each climate. But the final values shift to a higher plateau as compared to the same physical load executed in a neutral climate. Heart rates and core temperatures are then 20 - 30 bpm resp. 0.1 - 0.4 °C less than in the climates applied here.

Parameter	warm-humid (w)				acclimation	
	days 1/2		days 14/15		t-test	point
	AM	SD	AM	SD	p	day
increase of HR	44.4	10.4	34.1	8.7	0.006	3.6
core temperature	37.9	0.27	37.6	0.18	0.003	4.1
sweat loss	917	226	1185	254	0.020	5.4
	hot-dry (h)				p	day
increase of HR	56.2	13.4	41.9	7.8	0.003	4.4
core temperature	38.1	0.16	37.7	0.23	0.001	5.9
sweat loss	1524	161	1705	143	0.005	9.5
	radiant load (r)				p	day
increase of HR	47.3	11.5	32.3	5.8	0.001	5.2
core temperature	38.1	0.38	37.6	0.20	0.004	6.9
sweat loss	1644	176	1652	161	0.745	-
t-test between	days 1&2			days 14&15		
climates (p)	w:h	w:r	h:r	w:h	w:r	h:r
increase of HR	.086	.719	.246	.104	.572	.048
core temperature	.086	.572	.443	.334	.443	.288
sweat loss	.007	.005	.359	.007	.007	.572

Table 1

3.2 The influence of the actual thermal condition on acclimation

According to the hypothesis each of the 3 climates is supposed to induce an almost equal strain. If this is true, the physiologic parameters must not differ between the 3 thermal conditions at any time. This assumption was verified: Core temperatures and cardiac strains were statistically equal at the beginning and at the end of each series regardless of the composition of the thermal parameters (table 1, t-test for independent samples).

However, as compared to the warm-humid condition sweat loss is about one half larger in the hot-dry climate or during radiant load (figures 1,2). After transforming sweat loss into percentages (related to the average of each series) no significant alterations exist anymore.

Despite this general correspondence visual inspection suggests that the courses of the physiologic variables during acclimation are nevertheless influenced by the particular thermal stress (figs 1&2). Indeed, the more powerful Wald's test, which considers the data of each single experiment and their specific temporal development, reveals significant differences between the 3 climates.

The **warm-humid** condition might be the least strainous. The 3 physiologic variables acclimate most rapidly and, compared to both the other thermal conditions, the first exposure causes the least elevations of core temperatures and heart rates which then decline less than during the other climates. Due to the high water vapor pressure sweat loss is about one third less than in hot-dry or in radiant load throughout the 15 experiments, the increase is steeper and the augmentation reaches 19.2 % (268 g, first 2 days vs last 2 days). The average gradient between core and skin temperature is greatest in this climate.

The **hot-dry** condition is apparently the most strainous: Acclimation needs more time than in warm-humid but less than during radiant load. Core temperatures and cardiac strain are throughout the entire series above those

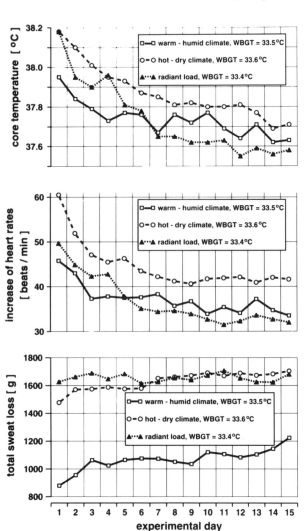

Figure 1

of the other climates. Sweat loss is already high at the beginning and increases moderately (11.9%, 181 g) to the same level as registered during radiant load. The average difference between core and skin temperatures is less than during humid heat but exceeds the gradient recorded during radiant load.

During **radiant load** sweat loss was already largest at the beginning and failed to increase thereafter (0.9 %, 8 g). Core temperature and cardiac strain started at high levels, declined steeper and reached the point of acclimation later than in the other thermal conditions. The gradient between core and skin temperatures is at minimum in this climate.

Cardiac strain and core temperatures are significantly correlated to each other in each of the 3 climates (p < 0.01). The appropriate regressions agree highly (Akaike's information criterion) and the relation can be expressed by a single equation. Both these variables correlate again on the 1%-level with total sweat loss during warm-humid or during hot-dry but the appropriate ascents differ considerably, preventing the calculation of a common equation. Additionally, neither cardiac strain nor core temperatures are associated to sweat loss if radiant temperatures exceed air temperatures. These results suggest that the various mechanisms which contribute to thermoregulation are to some extent independent of each other. This is e.g. supported by Wyndham et al. (2) who conclude that central circulatory and thermoregulatory processes are dissociated in acclimation.

In any condition acclimation is completed first for cardiac strain, second for core temperatures, and third for sweat loss. The points of acclimation depend on the particular thermal condition, they occur first in the warm-humid, thereafter in the hot-dry, and finally in the climate with radiant

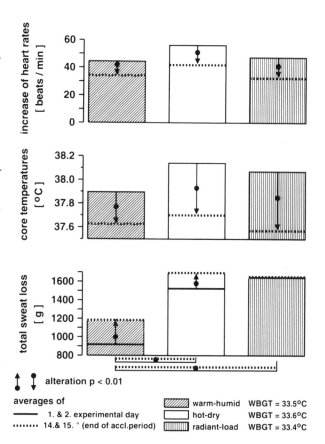

Figure 2

load (where no point of acclimation could be determined for sweat loss which remains almost constant throughout the experiments).

A very few authors compared acclimation to various climates, mostly to hot-dry and to warm-humid (acclimation to radiant load, however, was not yet studied in the lab). The results and the conclusions are contradictory. Among others, Rowell (3) assumes a faster acclimation to hot-dry than to warm-humid. Fox et al. (4) interprete higher sweat rates in the dry condition as a sign for better acclimation. Henane (5), on the contrary concludes from a literature review, that acclimation might be superior in warm-humid as sweat rates in this conditions increase far more than during repeated exposures to hot-dry environments.

Only Shvartz et al. (6) applied two considerably different climates which were equivalent in terms of the WBGT (35 °C). They registered higher sweat rates in the hot-dry than in the warm-humid climate and concluded that acclimation is better in hot-dry.

Both, an overall larger sweat loss in the hot-dry and a larger increase in the humid condition were again found in the present study: But it is certainly insufficient to judge the process of acclimation only on the basis of sweat loss. If the points of acclimation are regarded as a reasonable criterion, acclimation in the warm-humid condition is somewhat superior to both the other equivalent climates and radiant load is the least.

4 Conclusions

Though equivalent in terms of the WBGT the 3 climates are differently strainous; they cause various courses of acclimation which indicate that the mechanisms of thermoregulation during acclimation are dissociated. But, as the differences between the 3 thermal conditions are biologically less important, the WBGT is a useful measure for practical purposes, e.g. for industrial situations. For scientific research, however, another - if any - thermal index is certainly preferable.

5 References

1 Strydom NB, et al.: Acclimatization to humid heat and the role of physical conditioning. J Appl Physiol (1966) 21:636-642
2 Wyndham CH, Rogers GG, Senay LC, Mitchell D: Acclimatization in a hot, humid environment: cardiovascular adjustments. J Appl Physiol (1976) 40:779-785
3 Rowell LB: Human circulation regulation during physical stress. Oxford University Press, New York 1986
4 Fox RH, Goldsmith R, Hampton IFG, Hunt TH: Heat acclimatization by controlled hyperthermia in hot-dry and hot-wet climates. J Appl Physiol (1967) 22:39-46.
5 Henane R: Acclimatization in man: giant or windmill: In: Szelényi Z, Székely M (eds): Contributions to thermal physiology. Pergamon Press, Oxford 1981, pp 275-284.
6 Shvartz E, Saar E, Meyerstein N, Benor D: A comparison of three methods of acclimatization to dry heat. J Appl Physiol (1973) 34:214-219

SENSIBLE HEAT LOSS AFTER SYSTEMIC ANTICHOLINERGIC TREATMENT

M.A. Kolka, L.A. Stephenson and R.R. Gonzalez
US Army Research Institute of Environmental Medicine,
Natick, MA, 01760-5007, USA

Summary

Systemic atropine administration decreases eccrine sweat gland release via competitive inhibition at cholinergic receptors and augments skin blood flow by a yet unknown mechanism. Increased skin blood flow promotes auxiliary radiative and convective heat flux by 20-100% in both warm and cold environments. These studies showed that the systemic dose of atropine (2 mg, im) was sufficient to block cholinergic sweat gland activity and core and skin temperatures were higher after atropine compared to control experiments. There was an enhanced skin blood flow thermosensitivity to esophageal temperature drive following atropine administration.

Introduction

Atropine or atropine-like antimuscarinic drugs block the action of acetylcholine at post-ganglionic cholinergic nerves, neuronal and ganglionic muscarinic receptors and smooth muscle cells that lack cholinergic innervation [1]. The most common cardiovascular effect of antimuscarinic drugs is an increased heart rate via blocking vagal stimulation of M_2 muscarinic receptors [1]. Since most vascular beds lack significant cholinergic innervation, the effects on blood vessels and blood pressure are not readily apparent [1]. However, during exercise in either hot or cold environmental conditions, systemic atropine administration dilates cutaneous vessels by a yet unknown mechanism which may be a compensatory vasodilator response to offset the rise in body temperature, or may be unrelated to cholinergic blockade [1]. This increase in skin blood flow after system atropine administration [2] increases radiative and convective heat flux in subjects whose skin temperature is warmer than the environment, and decreases heat gain in subjects whose skin temperature is lower than the ambient temperature. The increase in skin blood flow seen after atropine injection is greater than that occurring after the release of vasoconstrictor activity [3], and occurs even in cooler environments where the mean weighted skin temperature (as well as the local skin temperature) is below 33°C.

Materials and Methods

The effects of anticholinergic therapy, in this case intramuscular atropine sulfate (2 mg), on dry (sensible) and wet (insensible) heat exchange was studied in 22 healthy, young men in a series of studies. Their average (±SD) age was 21±3 yr; mass, 77.9±8.6 kg; surface area, 1.98±0.12 m², height 1.80±0.07 cm; body fat, 15.0±3.7%; and maximal aerobic power, 3.79±0.38 L•min⁻¹. Experiments were done across a range of environmental conditions with dry bulb temperatures from 22 to 48°C and dew point temperatures from 7 to 24°C. Exercise intensity ranged from 30 to 55% of the measured maximal or peak aerobic power ($\dot{V}O_2$). All subjects were made familiar with all aspects of the testing and measurement procedures before data collection began. Subjects were tested once after the intramuscular injection of atropine sulfate (0.025 mg•kg⁻¹; Elkin-Sinn, Cherry Hill, NJ, USA) and once after the injection of an equal volume of sterile saline. Test days were separated by a minimum of 48 h, and the order of drug presentation was counterbalanced. All experiments were single-blind.

Insensible, or *wet* heat exchange was calculated using partitional calorimetry as:
$E_{sk} = M - (\pm W_k) - (R + C + K)$, in W•m⁻². Specifically, evaporative heat loss from the skin surface was calculated as: $E_{sk} = (g•min^{-1})(0.68 \ W•h•g^{-1})(60 \ min)•m^{-2}$ in W•m⁻². The maximal evaporative power of the environment was calculated as: $E_{max} = h_e(P_{s,sk} - P_{s,dp})$ in W•m⁻². Whole body skin wettedness (w) was calculated from the ratio of E_{sk} to E_{max}. *Sensible* or *dry* heat exchange was calculated from the heat balance as:
$(R+C) = M - (\pm W_k) \pm K - S$, in W•m⁻². Specifically, radiative and convective heat exchange $(R+C)$ was determined as: $R + C = (h_r + h_c)(T_s - T_a)$, in W•m⁻².

Results and Discussion

The expected decrease in evaporative heat loss from the skin (E_{sk}) after atropine administration occurred in all environmental and exercise combinations and averaged -50%. This decrease in sweat secretion and evaporation due to antimuscarinic blockade at the sweat glands, decreased calculated whole body skin wettedness (w) an average of -60%. In contrast to the effect seen on insensible or wet heat loss, radiative and convective heat loss (sensible heat flux) increased after atropine treatment in all conditions studied. This increase ranged from 20 to 100% (20-40

W•m^{-2}) depending on the specific conditions of the individual experiment. These results are summarized in Table 1.

Table 1. Insensible and sensible heat loss in multiple environments.

T$_a$/P$_w$	CONTROL			ATROPINE		
	E$_{sk}$	R+C	w	E$_{sk}$	R+C	w
22°C/0.8kPa	198	96	.56	84	113	.21
30°C/1.0kPa	234	32	.62	83	50	.19
42°C/1.6kPa	257	-85^a	.72	139	-43	.29
48°C/2.4kPa	187	-150	.52	86	-116	.28
30°C/3.0kPa	145	27	.88	74	53	.27

^{a}negative value indicates heat gain

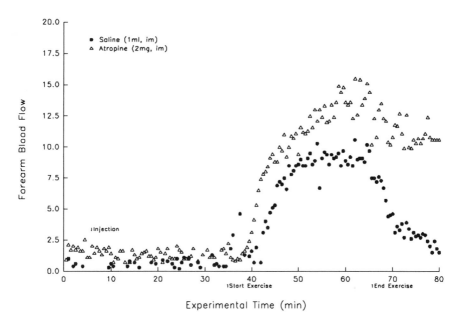

Figure 1. Forearm blood flow (ml•100ml^{-1}•min^{-1}) during saline (control) and atropine experiments for a single subject.

In these studies, the systemic dose of atropine was sufficient to block cholinergic sweat gland activity and increase heart rate (~40 b•min^{-1}) during exercise compared to control experiments. Reports of enhanced cutaneous vasodilation, whether measured directly or calculated from the heat balance, were confirmed by these studies. An example of increased skin blood flow during exercise after the systemic administration of atropine sulfate is shown in Figure 1. In this example (T$_a$ = 30°C, 55% peak V̇O$_2$) atropine or a saline placebo was injected at 5 min and exercise began 30 min after the injection and ended at 65 min. Skin blood flow was 60% higher during exercise after atropine treatment. Both esophageal and mean weighted skin temperature were higher by the end of exercise but not at the time of increased skin blood flow after atropine treatment compared to saline. The data from Figure 1 are presented in Figure 2 as the forearm blood flow plotted against the esophageal temperature. These data indicate the slope of this relationship is increased with atropine treatment.

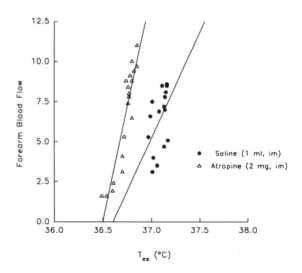

Figure 2. Forearm blood flow (ml•100ml^{-1}•min^{-1}) plotted against esophageal temperature during saline (control) and atropine experiments for a single subject. Note the increased thermosensitivity in the atropine experiment [4].

The observed changes in sensible and insensible heat loss after systemic atropine treatment are in opposite directions, although the increase in dry heat flux does not fully compensate for the decrease in evaporative heat loss. The mechanism(s) responsible for increased skin blood flow during exercise after systemic atropine treatment may include: 1) slightly higher local skin temperature [3], although this appears to be a passive function of the increased skin blood flow; 2) changing baroreflex activity [1,3,5,6], although cardiac output and mean arterial pressure are unchanged; 3) release of sympathetic vasoconstrictor activity [1,3,5], although increases in skin blood flow would be smaller than we observed; 4) the increase in skin blood flow, if neurogenic in nature, is non-cholinergic; 5) the chronotropic effect on the heart such that pulsatile flow or increase shear stress is transduced at the endothelium to increase endothelium-derived relaxing factors [7,8], such as nitric oxide; 6) the presence of other vasodilatory substances, perhaps related to the sweat gland [9]; or 7) a combination of any of the above.

Acknowledgements

The views, opinions and/or findings contained in this report are those of the authors and should not be construed as official Department of the Army position, policy or decision, unless so designated by other Official documentation. Human subjects participated in these studies after giving their free and informed consent. Investigators adhered to Army Regulation 70-25 and United States Army Medical Research and Development Command Regulation 70-25 on the Use of Volunteers in Research. Approved for public release; distribution unlimited.

References

[1] Brown JH. Atropine, scopolamine, and related antimuscarinic drugs. In: Gillman LS, Rall TW, Nies AS, Taylor P, eds. The Pharmacological Basis of Therapeutics, 8th ed. New York: Pergamon Press, 1990:150-65.

[2] Kolka MA, Stephenson LA. Temperature regulation following systemic anticholinergic or anticholinesterase therapy. In: Mercer JB, ed. Thermal Physiology 1989. Proceedings of Satellite Thermal Physiology Symposium, International Congress of Physiological Sciences. Amsterdam: Elsevier Science, 1989:259-64.

[3] Johnson JM. Exercise and the cutaneous circulation. In: Hollozsy JO, ed. Exercise and Sport Sciences Reviews, Vol. 20. Baltimore: Williams and Wilkins, 1992:59-97.

[4] Boulant JA. Hypothalamic control of thermoregulation: Neurophysiological basis. In: Morgane PJ, Panksepp J, eds. Handbook of the Hypothalamus, Vol. 3, Behavioral studies of the hypothalamus, Part A. New York: Marcel Dekker Inc., 1980:1-82.

[5] Taylor P. Cholinergic agonists. In: Gillman LS, Rall TW, Nies AS, Taylor P, eds. The Pharmacological Basis of Therapeutics, 8th ed. New York: Pergamon Press, 1990:122-30.

[6] Lefkowitz RJ, Hoffman BJ, Taylor P. Neurohumoral transmission: The autonomic and somatic motor nervous systems. In: Gillman LS, Rall TW, Nies AS, Taylor P, eds. The Pharmacological Basis of Therapeutics, 8th ed. New York: Pergamon Press, 1990:84-121.

[7] Furchgott RF. The role of the endothelium in the responses of vascular smooth muscle to drugs. Ann. Rev. Pharmacol. Toxicol. 1984;24:175-97.

[8] Furchgott RF, Vanhoutte PM. Endothelium-derived relaxing and contracting factors. FASEB J. 1989;3:2007-18.

[9] Burnstock G. Nervous control of smooth muscle by transmitter, cotransmitters and modulators. Experientia 1985;41:869-74.

Temperature Regulation
Advances in Pharmacological Sciences
© 1994 Birkhäuser Verlag Basel

NEUROPEPTIDE-Y (NPY) REDUCES CUTANEOUS MICROCIRCULATORY BLOOD FLOW AND INCREASES TOTAL BLOOD FLOW IN THE RAT TAIL

Martha E. Heath and John R. Thomas
Naval Medical Research Institute, Bethesda, Maryland, USA 20889-5607

Summary
The in vivo effects of neuropeptide-Y (NPY) on microvascular cutaneous blood flow are described in the tail and foot of the rat and on total blood flow in the tail. NPY (16, 32, and 128 µg/kg, i.v.) induced immediate, marked, dose-related decreases in both tail and foot cutaneous blood flow and significant increases in tail total blood flow, tail blood volume, and tail skin temperature. Findings suggest some tail vessels are dilating in response to NPY and are most likely arteriovenous anastomoses.

Introduction

Neuropeptide-Y (NPY), a 36 amino acid peptide with a wide distribution in the peripheral nervous system, is one of the most profound vasocative substances known [1-5]. The actions of NPY at sympathetic neuroeffector junctions have occasioned research interest because of the prevalence of the peptide in sympathetic nerves and its coexistence with norepinephrine in these nerves. Research studies in man [4] and other mammals have demonstrated marked reductions in blood flow in response to exogenous NPY [6,7]. Studies of isolated blood vessels have shown that NPY induces contraction in some tissues, e.g., the cerebral, skeletal muscle, and coronary, but not all vessels [3,4,8]. The vascular constriction induced by NPY is often slow in onset, but prolonged, compared to that of norepinephrine [9]. NPY produces vasoconstriction by its effect on its own non-adrenergic vascular receptors (three NPY receptors are presently recognized). Additionally, it both potentiates the effect of other vasoconstrictors and modulates vasodilatory effects of acetycholine, adenosine, and norepinephrine [5,6].

NPY has been implicated in development of cardiovascular disease [9] and is released during periods of elevated sympathetic activity and stimulation [9]. Its pronounced and prolonged vasoconstrictive effects in some peripheral vascular beds may precipitate tissue ischemia [10]. Because of these characteristics, NPY may play a major role in many peripheral vascular disorders. It is also reasonable to suggest that NPY is involved in peripheral nerve and tissue injury associated with severe insults such as prolonged thermal stress. There is a clear need for further characterization of the *in vivo* effects of NPY on peripheral blood flow in a whole animal model and to establish its medical relevance to peripheral vascular disorders. The purpose of the present research was to describe in vivo effects of NPY on total blood flow (BF_T) and on the microvascular skin blood flow (BF_{sk}) in appendages of a whole animal model. The hind foot and tail of the rat was used as the model.

Materials and Methods

At least one week prior to experiments, male 300-g Long-Evans rats were anesthetized with ~50 mg/kg sodium pentobarbital and were prepared with a cannula in a jugular vein. On experimental days rats were anesthetized with ~50 mg/kg sodium pentobarbital and gently introduced into a cylindrical plexiglas rat restrainer. Skin temperatures (T_{sk}) on the tail and foot were measured with thermocouples. Tail BF_T was measured with venous occlusion plethysmography by use of mercury-in-silastic strain gauges wrapped twice around the tail connected to an electronic plethysmograph [11]. BF_{sk} of the tail and foot was monitored with two laser Doppler flowmeters [12]. A laser Doppler flow probe and a thermocouple were positioned on the skin adjacent to each other mid-way along the tail. A pneumatic cuff at the base of the tail and a strain gauge ~2 cm distal to the laser Doppler probe were used to measure tail BF_T. A second laser Doppler flow probe was placed on the right foot, and a thermocouple was taped in the same position on the left foot.

   All instrumentation was connected to a computer via an A/D converter. Thermocouple and laser Doppler channels were sampled at 1-sec intervals and averaged for 20 sec. The venous occlusion plethysmographic measure of tail BF_T was accomplished at 20-sec intervals. The occlusion cuff was inflated to 55 mmHg for 5 sec, and tail BF_T was assessed between the second and fourth sec of the cuff inflation.

   Fully instrumented rats rested in the restrainer for 15-30 min before the intravenous administration of either NPY (16, 32 or 64 µg/lkg) or saline. The volume of all injections was 300 µl. Baseline recording was begun 5 min preceding the injection, and measurements were continued for at least 35 min post-injection. All experiments were done at ambient temperature of 24-26°C.

Results and Discussion

The changes in all blood flow parameters can be described as having an immediate dynamic component followed by a prolonged static component. NPY caused an immediate increase in tail BF_T (10-40 ml/100 ml tissue/min) that normally peaked within 1-3- min. Tail volume (i.e., tail blood volume) also showed an immediate increase (>1.5%) and followed the same time course as tail BF_T. Tail T_{sk} also increased markedly (0.8-2.25°C), although it lagged the increase in tail BF_T slightly. In contrast, BF_{sk} in the tail and foot declined precipitously to <50% of baseline and leveling out within 1-3 min. There was no large or immediate change in foot T_{sk}.

After the dynamic increase, tail blood flow values (BF_{sk} and BF_T) began to return toward their respective baseline levels, albeit without reaching baseline during 35 mins of post-injection data collection. Although tail BF_{sk} initially was a mirror image of tail BF_T, it typically declined again, completely independent of tail BF_T. Foot BF_{sk} did not return toward baseline, but continued at its new reduced level. Blood flow values did not return to baseline even in experiments in which data were collected for an additional 60 min.

A dose of 128 µg/kg NPY had an effect on blood flow similar to that of 64 µg/kg. At the lower doses of 16 and 32 µg/kg, both the magnitude and the duration of the initial dynamic response was less. Furthermore, blood flow did return to baseline during the 35 min of post-injection observation. As expected, saline, administered i.v. as a control, had no significant effect on any of the parameters measured.

The most important observations made in this study are (a) NPY causes an increase in tail BF_T and (b) NPY induces a decrease in BF_{sk} in the tail and foot. The increased BF_T in the rat tail was unexpected considering the numerous reports that NPY produces either vasoconstriction or no change in vascular tone in blood vessels and reduced blood flow in organs and tissues heretofore studied [3, 4, 8]. The simultaneous increase in tail blood volume and BF_T suggests that some vessels in the tail are dilating in response to NPY. *In vitro* studies in the rat tail artery indicate that NPY causes a slow depolarization and concentration dependent contraction via direct effects on smooth muscle cells. Our observations show that microvascular blood flow was diminished after administration of NPY. This somewhat limits the possible sites of vasodilation to arterioles, venules, veins, and arteriovenous anastomose (AVAs). AVAs, which are shunts between arterioles and venules, are likely candidates. They occur in large numbers in the rat tail and play a role in thermoregulatory heat loss. AVAs divert blood away from the capillaries as the blood follows the path of least resistance. The AVAs allow for a much higher rate of blood flow through a tissue than do the capillaries. The observed increase in Tail T_{sk}, even while BF_{sk} was diminished, lends strong support to the suggestion that NPY increases tail BF_T by dilating AVAs.

The reduction in resistance that accompanies vessel dilation is arguably responsible for the increase in tail BF_T. The reduction in BF_{sk} may be due to either active vasoconstriction or to the decrease in resistance in larger vessels allowing blood flow to bypass the cutaneous capillaries. The observations that (a) tail BF_{sk} decreases when tail BF_T increases and (b) that tail BF_{sk} increases again just as BF_T begins to decline favor the latter explanation. There is another interesting observation about tail BF_{sk}. After initially being a mirror image of BF_T, it typically declines a second time (completely independent of BF_T) and remains low for the duration of the experiment. This may be due to either the effects of systemic or local changes in blood pressure or to direct constriction in this cutaneous vascular bed.

The foot BF_{sk} was found to be rather distinct from that of the tail. Both the BF_{sk} and T_{sk}

of the foot pad are normally higher than in the tail. While foot BF_{sk} declined immediately and remained quite low, foot T_{sk} remained nearly constant. It neither increased as did tail T_{sk} nor decreased in response to the reduced blood flow and suggests the possible vasodilation observed in tail BF_T does not occur in the leg or foot.

Acknowledgements

The research was supported by US Naval Medical Research and Development Command Research and Technology Work Unit 61153N.MR04120.00B.1058. Experiments were conducted according to the principles set forth in the Guide for the Care and Use of Laboratory Animals, Institute of Laboratory Animal Resources, National Research Council, DHHS Publication (NIH) 86-23, (1985). The opinions and assertions contained herein are those of the authors and are not to be construed as official or reflecting the views of the Navy Department of the Naval Service at large.

References

1. LUNDBERG, J.M., TERENIUS, L., HOKFELT, T., ET AL. (1982). Neuropeptide-Y [NPY]-like immunoreactivity in peripheral noradrenergic neurons and effects of NPY on sympathetic function. *Acta Physiol Scand* **116**, 479-80.

2. LUNDBERG, J.M., TERENIUS, L., HOKFELT, T., GOLDSTEIN, M. (1983). High levels of neuropeptide Y in peripheral noradrenergic neurons in various mammals including man. *Neurosci Lett* **42**, 167-72.

3. EKBLAD, E., EDVINSSON, L., WAHLESTEDT, C., UDDMAN, R., HAKANSSON, R., SUNDLER, F. (1984). Neuropeptide Y coexists and cooperates with noradrenaline in perivascular nerve fibers. *Regul Pept* **8**, 225-35.

4. PERNOW, J., LUNDBERG, J.M., KAIJSER, L. (1987). Vasoconstrictor effects *in vivo* and plasma disappearance rate of neuropeptide Y in man. *Life Sci* **40**, 47-54.

5. LUNDBERG, J.M., FRANCO-CERECEDA, A., LACROIX, J-S., PERNOW, J. (1990). Neuropeptide-Y and sympathetic neurotransmission. In: *Allen, J.M., Koenig, J.I., eds. Central and Peripheral significance of neuropeptide-Y and its related peptides. New York: Annuals of the New York Academy of Sciences* **611**, 166-74.

6. PERSSON, P.B., GIMPL, G., LANG, R.E. (1990). Importance of neuropeptide Y in the regulation of kidney function. In: *Allen, J.M., Koenig, J.I., eds. Central and peripheral significance of neuropeptide-Y and its related peptides. New York: Annuals of New York Academy of Sciences* **611**, 156-65.

7. LUNDBERG, J.M., TATEMOTO, K. (1982). Pancreatic polypeptide family [APP, BPP, NPY, PYY] in relation to sympathetic vasoconstriction resistant to alpha-adrenoceptor blockade. *Acta Physiol Scand* **116**, 393-402.

8.   FRANCO-CERECEDA, A., LUNDBERG, J.M. (1987). Potent effects of neuropeptide Y and calcitonin gene-related peptide on human coronary vascular tone in vitro. *Acta Physiol Scand* **131**, 159-60.

9.   LUNDBERG, J.M., PERNOW, J., LOCROIX, S., FRANCO-CERECEDA. A., RUDEHILL, A., HOKFELT, T. (1989). Neuropeptide tyrosine [NPY] and sympathetic cardiovascular control. In: *Mutt, V., Fuxe, K., Hokfelt, T., Lundberg, J.M. eds. Neuropeptide Y. New York: Raven Press* 191-9.

10. CLARK, J.G., KERWIN, R., LARKIN, S., ET AL. (1987). Coronary artery infusion of neuropeptide Y in patients with angina pectoris. *Lancet* **i**, 1057-9.

11   HOKANSON, D.E., SUMNER, D.S., STRANDNESS, D.E. JR. (1975). An electrically calibrated plethysmograph for direct measurement of limb blood flow. *IEEE Trans Biomed Eng* **22**, 25-9.

12. BONNER, R.F., NOSSAL, R. (1990). Principles of laser-Doppler flowmetry. In: *Shepherd A.P., Oberg, P.A. eds. Laser-Doppler Blood Flowmetry . Boston: Kluwer Academic Publishers* 17-45.

Temperature Regulation
Advances in Pharmacological Sciences
© 1994 Birkhäuser Verlag Basel

EFFECT OF PRETREATMENT WITH DELTA-9-TETRAHYDRO-CANNABINOL ON THE ABILITY OF CERTAIN CANNABIMIMETIC AGENTS TO INDUCE HYPOTHERMIA IN MICE

R.G. Pertwee & L.A. Stevenson, Department of Biomedical Sciences, University of Aberdeen, Aberdeen, Scotland.

Summary
Mice that had been pretreated intraperitoneally with delta-9-tetrahydrocannabinol (THC; 20 mg/kg, once daily for two days) showed tolerance to the hypothermic effect of this drug when it was injected intravenously (1 mg/kg) 24 h after the final intraperitoneal injection. The same pretreatment also induced tolerance to the hypothermic effects of the cannabimimetic agents, WIN 55,212-2 and CP 55,940. However, it did not to induce any detectable tolerance to the hypothermic effect of the putative endogenous cannabinoid, anandamide.

Introduction

An important recent advance in cannabinoid research has been the discovery in the brain of a putative endogenous cannabinoid, anandamide (Devane et al., 1992). It has been found that this compound shares the ability of psychotropic cannabinoids to bind avidly to specific cannabinoid binding sites, to inhibit the electrically-evoked twitch response of the mouse isolated vas deferens and to produce hypokinesia, antinociception, ring immobility and hypothermia in mice (Devane et al., 1992; Fride & Mechoulam, 1993). Further experiments are still required to establish the extent of the overlap between the pharmacology of anandamide and that of 'classical' cannabinoids such as delta-9-tetrahydrocannabinol (THC). Accordingly, the purpose of the present experiments was to determine whether mice rendered tolerant to the hypothermic effect of THC would also show tolerance to anandamide. The ability of THC to induce tolerance to the hypothermic effects of two other cannabimimetic agents was also investigated. These were CP 55,940, which is widely used as a probe for the cannabinoid receptor, and WIN 55,212-2, which binds to cannabinoid receptors, possesses certain other cannabimimetic properties and yet has a completely different chemical structure from that of THC (see Pertwee, 1993). Previous experiments have already shown that in vivo pretreatment with THC can significantly reduce the sensitivity of mouse isolated vasa deferentia both to THC itself and to that of anandamide, CP 55,940 and WIN 55,212-2 (Griffin & Pertwee, 1993; Martin et al., 1993; Pertwee et al., 1992b).

Methods

Male MF1 mice were injected once daily for 2 days with THC (20 mg/kg i.p.). Control animals received injections of Tween 80 (40 mg/kg i.p.). The degree of tolerance was determined 24 h after the second injection by measuring the maximum hypothermic effect of each of the test compounds, injected intravenously at time zero. Body temperature was monitored with a thermistor probe (YSI 402) inserted 3 cm into the rectum of unrestrained mice kept at an ambient temperature of 20 to 22°C. Temperature readings were first taken 30 and 5 min before drug administration, next 5 min after drug administration, then at 5 min intervals until +30 min and finally at +45 and +60 min.

Drugs were mixed with 2 parts of Tween 80 by weight and dispersed in saline as described previously for THC (Pertwee et al., 1992a). The volumes for intraperitoneal and intravenous injection were 0.25 ml/25g and 0.20 ml/25g respectively. THC was obtained from the National Institute on Drug Abuse, U.S.A., CP 55,940 from Dr. L.S. Melvin (Pfizer) and WIN 55,212-2 from Dr. S.J. Ward (Sterling Winthrop). Anandamide was synthesized and supplied by Professor R. Mechoulam (University of Jerusalem).

Values have been expressed as means and limits of error as standard errors (n=6). The effect of THC on body temperature has been calculated by subtracting rectal temperature, measured just before drug administration, from rectal temperature measured at the time of peak hypothermia. Degrees of tolerance to the hypothermic effects of THC, CP 55,940, WIN 55,212-2 and anandamide have been determined using symmetrical (2+2) dose parallel line assays (Colquhoun, 1971). The significance of differences between between means have been estimated by Student's t test (two-tail) for paired or unpaired data (P > or < 0.05).

Results

THC, WIN 55,212-2 and CP 55,940 each produced dose-related decreases in rectal temperature (Figure 1). The ability of these drugs to induce hypothermia was markedly attenuated in mice that had received two intraperitoneal injections of THC (20 mg/kg) as described above. Thus, as shown in Figure 2, THC pretreatment completely abolished the hypothermic responses of mice to submaximal doses of THC, WIN 55,212-2 and CP 55,940 (1.0, 0.2 and 0.05 mg/kg i.v. respectively). The same pretreatment induced significant parallel rightward shifts in the log dose-hypothermic response curves of all three drugs. There was a 6-fold shift in the log dose-response curve of THC (95% confidence limits 3.8 and 10.4), a 4.9-fold shift in the log dose-response curve of WIN 55,212-2 (95% confidence limits 3.8 and 7.0) and a 4.6-fold shift in the log dose-response curve of CP 55,940 (95% confidence limits 2.8 and 9.1). Anandamide also produced dose-related decreases in rectal temperature (Figure 1). However, its ability to induce hypothermia was not attenuated by the intraperitoneal pretreatment with THC described in this paper (Figure 2).

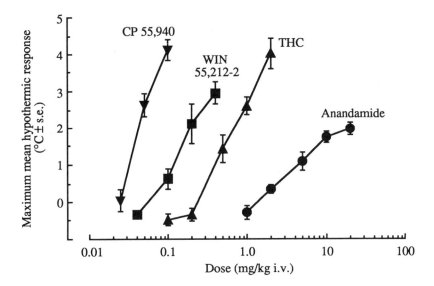

FIGURE 1: Relative hypothermic potencies in mice of CP 55,940, WIN 55,212-2, THC and anandamide.

Discussion

It was found that THC, WIN 55,212-2, CP 55,940 and anandamide all produced dose-related decreases in deep body temperature, confirming previous reports that psychotropic cannabinoids are hypothermic in mice (see Pertwee, 1985). The log dose-hypothermic response curve of anandamide is somewhat shallower than that of THC, CP 55,940 or WIN 55,212-2 (Figure 1) suggesting that anandamide may act as a partial agonist, at least for the production of hypothermia. As anandamide is present in the brain, it will be important to establish its role, if any, in thermoregulation.

The results obtained confirm that tolerance develops rapidly to the hypothermic response of THC in mice (see Pertwee, 1991). In addition, they demonstrate that pretreatment with THC can produce tolerance not only to itself but also to the hypothermic effects of CP 55,940 and WIN 55,212-2. This observation supports the notion that THC, CP 55,940 and WIN 55,212-2 have a similar mode of action, there being good evidence that cannabinoid tolerance is largely pharmacodynamic in nature (see Pertwee, 1991).

The finding that a pretreatment with THC that induces tolerance to its own hypothermic effect and to that of two other cannabimimetic agents does not induce tolerance to anandamide is unexpected, not least because the same pretreatment has been found to induce tolerance to anandamide in murine isolated vasa deferentia (Martin et al., 1993). One factor that may have contributed to the failure to detect anandamide tolerance in the present investigation, could have been the relative shallowness of the log dose-hypothermic response curve of anandamide.

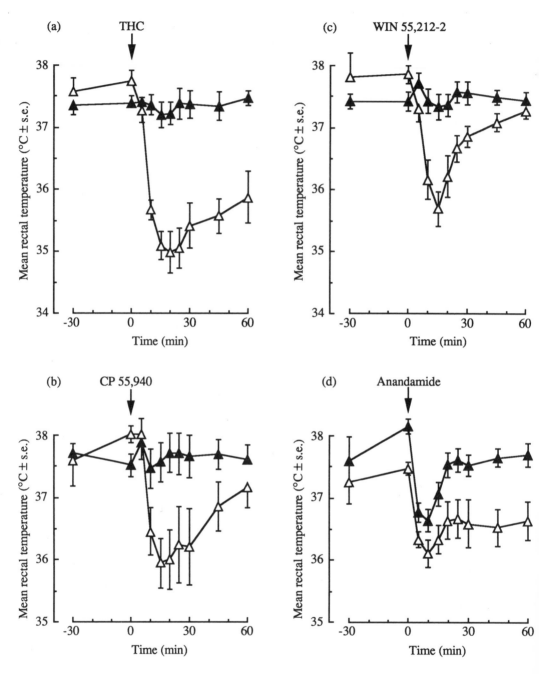

FIGURE 2: The hypothermic effects of (a) THC (1 mg/kg i.v.), (b) CP 55,940 (0.05 mg/kg i.v.), (c) WIN 55,212-2 (0.2 mg/kg i.v.) and (d) anandamide (10 mg/kg i.v.) in mice pretreated once daily for two days with THC (▲; 20 mg/kg i.p.) or Tween 80 (△; 40 mg/kg i.p.). Intravenous injections were made at time zero (n = 6).

Thus the presence of tolerance to any particular drug is indicated by a rightward shift in its dose-response curve, the magnitude of the shift providing a measure of the degree of the tolerance. Because it is more difficult to detect positional changes in log dose-response curves when they are shallow, such curves are less sensitive indicators of tolerance than steep log dose-response curves. In the mouse vas deferens, a tissue in which THC does induce tolerance to anandamide, the log dose-response curve of anandamide is no less steep than that of THC (Devane et al., 1992; Martin et al., 1993; Pertwee et al., 1992a).

Acknowledgements

We thank the Wellcome Trust for financial support, the National Institute on Drug Abuse, U.S.A. for delta-9-THC, Dr. L.S. Melvin (Pfizer) for CP 55,940, Dr. S.J. Ward (Sterling Winthrop) for WIN 55,212-2 and Professor R. Mechoulam (University of Jerusalem) for anandamide.

References

COLQUHOUN, D. (1971). Lectures on Biostatistics. Oxford: Oxford University Press.

DEVANE, W.A., HANUS, L., BREUER, A., PERTWEE, R.G., STEVENSON, L.A., GRIFFIN, G., GIBSON, D., MANDELBAUM, A., ETINGER, A. & MECHOULAM, R. (1992). Isolation and structure of a brain constituent that binds to the cannabinoid receptor. Science 258, 1946-1949.

FRIDE, E. & MECHOULAM, R. (1993). Pharmacological activity of the cannabinoid receptor agonist, anandamide, a brain constituent. Eur. J. Pharmacol. 231, 313-314.

GRIFFIN, G. & PERTWEE, R.G. (1993). Vasa deferentia obtained from mice pretreated with delta-9-tetrahydrocannabinol (THC) show tolerance to delta-9-THC, WIN 55,212-2 and CP 55,940. Br. J. Pharmacol. 109, 105P.

MARTIN, B.R., CHILDERS, S., HOWLETT, A., MECHOULAM, R. & PERTWEE, R.G. (1993) Cannabinoid receptors: pharmacology, second messenger systems and endogenous ligands. In Problems of Drug Dependence, 1993: Proceeding of the 55th Annual Scientific Meeting of the College on Problems of Drug Dependence, Inc.: National Institute on Drug Abuse Research Monograph Series, Research Monograph 133, ed. Harris, L., Rockville: National Institute on Drug Abuse., in press.

PERTWEE R.G. (1985). Effects of cannabinoids on thermoregulation: a brief review. In: Marihuana '84 , ed. Harvey, D.J., pp. 263-277. Oxford: IRL Press.

PERTWEE, R.G. (1991). Tolerance to and dependence on psychotropic cannabinoids. In The Biological Bases of Drug Tolerance and Dependence, ed. Pratt, J.A., pp. 231-263. London: Academic Press.

PERTWEE, R.G. (1993) The evidence for the existence of cannabinoid receptors. General Pharmacol. 24, 811-824.

PERTWEE R.G., STEVENSON L.A., ELRICK D.B., MECHOULAM R. & CORBETT A.
D. (1992a) Inhibitory effects of certain enantiomeric cannabinoids in the mouse vas
deferens and the myenteric plexus preparation of guinea-pig small intestine. *Br. J.
Pharmacol.* **105**, 980-984.

PERTWEE, R.G., STEVENSON, L.A., FERNANDO, S.R. & CORBETT, A.D. (1992b).
The production of cannabinoid tolerance in preparations of mouse vas deferens and of
myenteric plexus-longitudinal muscle of mouse small intestine. *Br. J. Pharmacol.* **107**,
365P.

Temperature Regulation
Advances in Pharmacological Sciences
© 1994 Birkhäuser Verlag Basel

CENTRAL MOTOR COMMAND AFFECTS THE SWEATING ACTIVITY DURING EXERCISE

Norikazu Ohnishi[*], Tokuo Ogawa, Junichi Sugenoya, Keiko Natsume, Yuka Yamashita,
Ritsuko Imamura and Kazuno Imai
Department of Physiology, Aichi Medical University,
[*]Department of Human Sciences, Aichi Mizuho College, JAPAN

Summary
    Contribution of the centrally-generated voluntary motor command signal to the sweating mechanisms during exercise was examined by 2 series of experiments. In Series 1, the increase in sweating rate of the thigh in response to hand-grip was augmented by the curarization and returned gradually along with the recovery process. In Series 2, forearm sweat rate increased in response to exercise and the increase reduced by the intervention of TVR in case of knee extension and was augmented with TVR in case of knee flexion. These observations suggest that the central motor command irradiates to the central sudomotor mechanisms and facilitates thermal sweating during voluntary muscular contraction.

Introduction

    Sweating activity increases immediately after the start of exercise, preceding a rise in body temperature. Such a rapid response has been attributed to some non-thermal factors, "work factors", rather than thermal ones (1,2,3).

    Rapid acceleration has also been observed in the cardiovascular and ventilatory responses and many studies have been designed in order to explain this response (4,5,6,7,8,9,10). Goodwin et al. (1972) applied vibration to the biceps tendon of human subjects performing sustained isometric contractions (6). In case of biceps contraction, central command required to achieve a given tension of the contracting muscle reduced with an element of reflex excitation enhanced by the activation of muscle spindle primary afferents. In case of triceps contraction, on the contrary, more central command was required to achieve a given tension with an element of reflex inhibition of the contracting triceps. They observed that blood pressure, heart rate and pulmonary ventilation altered according to the decreased or increased level of the central command.

McCloskey(1981) employed the partial curarization in order to increase the level of central command required for a given static effort and observed that cardiovascular response to hand-grip exercise with a given strength increased during partial curarization of the limb (8). Those observations suggest that the irradiation of the central motor command contributes to facilitate the respiratory and circulatory activities at an early stage of exercise. We attempted to investigate the influence of the central motor activity on sweating activity as a possible "work factor" and performed 2 series of experiments essentially following the methods of Goodwin et al. (1972) and McCloskey (1981).

Methods

Experiments were carried out on healthy young male subjects in a climatic chamber set at 35-38°C, 40% rh.

In Series 1, partial curarization of the left forearm and hand was induced retrogradely: 3 mg of d-tubocurarine chloride diluted in 20 ml of physiological saline solution was injected into the median cubital vein while a sphygmomanometer cuff around the left upper arm was inflated at a pressure of 160 mmHg. Arterial occlusion was released 10 min after the injection, when the hand was paralyzed nearly completely, and the voluntary movement recovered gradually in 30-40 min. One-min isometric exercise of hand-grip (15-20% of maximal voluntary contraction, MVC) was performed before the occlusion and 10, 20, 30 and 40 min after the release from the occlusion.

In series 2, isometric extension or flexion of the right knee joint of 3-6 min duration was performed repeatedly at intervals of several min (1.8-6.7% of MVC). The vibration with 60-100Hz was applied over the patellar ligament during knee extension which was assumed to augment the contraction of the extensor muscles by inducing the tonic vibration reflex (TVR). The subject was directed to keep watching a strength indicator and to maintain a given strength during isometric exercise with and without TVR. Thus the central motor command required to achieve a given tension was assumedly reduced during the application of vibration. The same maneuver was performed also during knee flexion, which would resist the contraction of flexor muscles and thus require an increase of the central motor command.

Sweating rates were recorded continuously from right and left thigh areas in Series 1, from bilateral forearm areas in Series 2 and also from the right palm in both series by using capacitance hygrometory. Core temperature at the tympanic and/or that in the esophagus, skin temperatures at chest, arm, thigh and leg by using thermistors and heart rate (HR) were recorded continuously.

Results and Discussion

In Series 1, sweating rates of the thighs increased in response to hand-grip exercise. Figure 1 shows typical recordings from a subject. The increase was greater after the partial curarization than before the arterial occlusion and returned gradually during the recovery process from the paralysis. This observation implies that, as the effort to perform exercise decreases, the sweating response reduces. A trend similar to the sweating response was observed in the response of HR in some cases.

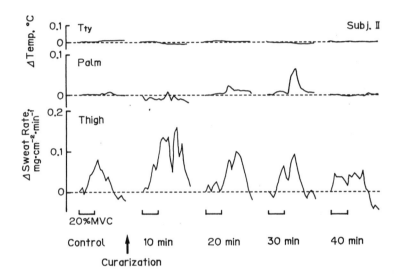

Figure 1: The time course of changes in the sweating rate of the right thigh and palm and tympanic temperature of one subject in Series 1. Each values are showed by the difference from mean value of the record for several minutes before the start of exercise.

In Series 2, forearm sweat rate increased in response to exercise in many cases. Figure 2 shows typical recordings from a subject. The increase in sweating rate reduced by the intervention of TVR in case of knee extension and was augmented with TVR in case of knee flexion . The core temperature failed to increase before the increase in sweating rate. Figure 3 shows the effect of TVR evaluated by the sweating response with TVR minus that without TVR for each of knee extension and flexion. The sweating response with TVR during knee extension was decreased in two of four subjects and that during knee flexion increase in all the subjects. The difference between the sweating response with TVR during the knee extension and that during

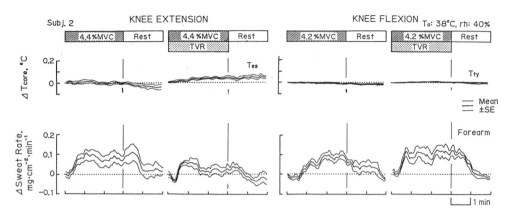

Figure 2: The time course of mean and standard error of changes in the forearm and palmer sweating rates and core temperature (esophageal temperature during knee extension and tympanic temperature during knee extension) in response to the knee extension (left column) and flexion (right column) with and without TVR in Series 2. Each values are revealed as the difference from the value before the start of exercise.

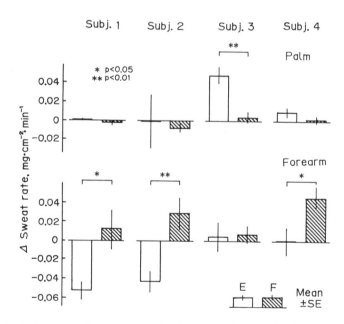

Figure 3: The effect of TVR on the forearm and palmer sweating activity in response to the knee extension (E) and Flexion (F) in Series 2. Each parameters are expressed as the mean and standard error of the subtraction values of sweating rate in response to the exercise without TVR from that with TVR.

the knee flexion was significant (p<0.05). Those increase or decrease in the sweating responses with TVR were considered to occur along with the increase or decrease in the levels of central motor command. Similar tendency was also observed in HR. On the other hand, there was no difference between palmer sweating response with TVR during knee extension and that during knee flexion except for one case.

Mental activity could affect the thermal sweating activity in initial stage of exercise (3). In the present study, palmer sweating increased in some cases but the feature of the increase in palmer sweating was different from that observed in the thigh during the recovery process in Series 1 (Fig. 1) and that observed in the forearm with TVR in Series 2 (Fig. 3). Irradiation of central motor activity might facilitate thermal sweating in initial stage of exercise in cooperation with mental activity. Similar tendency in HR was also observed in the present study. These observations suggest that the sweating activity as well as cardiovascular and respiratory activities, possibly autonomic activity as a whole, increases in association with an increase in the central motor out flow to perform exercise.

It is concluded from the above observations that irradiation of central motor command to the central sudomotor mechanisms takes place and plays a role in facilitation of thermal sweating during voluntary muscular contraction.

References
1 Beaumont W van, Bullard RW. Sweating: it's rapid response to muscular work. Science 1963;141:643-646.
2 Beaumont W van, Bullard RW. Sweating: exercise stimulation during circulatory arrest. Science 1966;152:1521-1523
3 Miyagawa T, Ogawa T, Asayama M, Yamashita, Y. Sweating response to abrupt changes in work load. J Physiol Soc Jpn 1985;47:17-24.
4 Krough A, Lindhard J. The regulation of respiration and circulation during the initial stages of the muscular work. J Physiol (Lond) 1913;47:112-136.
5 Goodwin GM, McCloskey DI, Mitchell JH. Cardiovascular and respiratory responses to change in central motor command during isometric excercise at constant muscular tension. J Physiol (Lond) 1972;226:173-190.
6 Mitchell JH, Payne FC, Saltin B, Schibye B. The role of muscle mass in the cardiovascular response to static contractions. J Physiol (Lond) 1980;309:45-54.
7 McCloskey DI. Centrally-generated commands and cardiovascular control in man. Clin Exp Hypertension 1981;3(3):369-378.
8 McCloskey DI, Mitchell JH. Reflex cardiovascular and respiratory responses originating in exercising muscle. J Physiol (Lond) 1972;224:173-186.
9 Wyss CR, Ardell JL, Scher AM. Rowell LB. Cardiovascular responses to graded reductions in hindlimb perfusion in exercising dog. Am J Physiol 1983;245:H481-H486.
10 Rotto DM, Stebbins CL, Kaufman MP. Reflex cardiovascular and ventilatory responses to increasing H^+ activity in cat hindlimb muscle. J Appl Physiol 1989;67:256-263.

Temperature Regulation
Advances in Pharmacological Sciences
© 1994 Birkhäuser Verlag Basel

SELECTIVE BRAIN COOLING IN THE HORSE DURING EXERCISE

F.F. McConaghy, J.R.S. Hales[1], & D.R. Hodgson.
Department of Animal Health, University of Sydney, Australia.

SUMMARY

To investigate the possibility of selective brain cooling in the horse, a species with a high capacity for exercise, which sweats & lacks a carotid rete, we monitored hypothalamic (T_{hy}), central venous (T_{mv}) & rectal (T_{re}) temperatures in two geldings during exercise, with fanning, on a treadmill. T_{mv} increased 3.5 – 4.1°C (to c.42°C), T_{hy} was a maximum of 1.3 ± 0.4°C lower, and T_{re} was c.1°C below T_{hy}. The existence of selective brain cooling in the horse, an animal that closely resembles man in thermoregulatory function, supports the contention that selective cooling of the brain can occur without a carotid rete.

INTRODUCTION

Substantial heat loads are generated during exercise as a result of the inefficiency of converting chemical to mechanical energy. Exercise hyperthermia potentially limits physical performance (see 1) & if severe enough can lead to the development of heat stroke. To reduce the detrimental effects of hyperthermia a number of animal species can selectively cool the brain via a vascular countercurrent heat exchanger, the carotid rete, which functions maximally during exercise (see 2). Evidence that the human brain may be protected from critical levels of hyperthermia particularly during exercise despite the absence of a carotid rete (3) is controversial, largely because investigations have usually employed tympanic membrane temperature as an index of brain temperature (4). Recent use of brainstem potentials as an index of changes in brain temperature failed to show any significant selective brain cooling with oesophageal temperature raised from 37.1°C to 39.0°C (5). To permit direct measurement of brain temperature a suitable animal model for

---

[1]School of Physiology & Pharmacology, University of NSW.

man has been sought. Hominoidae & Equidae are the only families which depend principally on the evaporation of sweat for heat dissipation, both are capable of exercising at high intensities & neither possesses a carotid rete.

METHODS

Two trained geldings were used. Four weeks prior to the experiments a 14swg stainless steel tube was surgically implanted with its tip in the pre–optic/anterior hypothalamic region (verified radiographically). The horses exercised on a treadmill with 10 percent slope, by walking at 1 m/s for c.3 min, immediately followed by trotting at 4 m/s for c.5 min & then cantering at 7 m/s until T_{mv} reached c.42°C. During exercise two fans of 1m diameter blew air primarily over the head at c.4 m/s. Dry bulb $\approx$ 21.3°C & wet bulb $\approx$ 16°C.

Copper–constantan, 38 swg thermocouples and a potentiometric recorder were used to measure temperatures at 30sec intervals to within 0.1°C. T_{hy} was measured in the guide tube. T_{mv} was measured via a catheter passed down the jugular vein to the level of the right ventricle (verified by blood pressure manometry). T_{re} was recorded using a 40cm probe following manual evacuation of faeces.

RESULTS

Baseline T_{hy} & T_{re} were usually within 0.1°C of each other, viz., T_{hy} = 37.9 ± 0.2°C & T_{re} = 37.8 ± 0.2°C, T_{mv} was slightly lower at 37.5 ± 0.3°C. The representative pattern of responses is shown in Fig. 1. In three experiments on each of the two horses, body temperatures did not change significantly during walking. During trotting T_{mv} increased 0.7 ± 0.2°C to 38.2 ± 0.3°C, T_{hy} increased 0.3 ± 0.1°C to 38.1 ± 0.2°C & T_{re} increased 0.1 ± 0.01°C to 37.9 ± 0.2°C. During cantering T_{mv} increased 3.6 ± 0.7°C (range 3.5 – 4.1) to 41.9 ± 0.7°C, T_{hy} increased 2.9 ± 0.8°C (range 1.7-4.4) to 41.5 ± 0.1°C & T_{re} increased 1.8 ± 0.7°C (range 0.9–3.1) to 40.2 ± 0.6°C.

T_{mv} – T_{hy} reached a mean max.= 1.3 ± 0.4°C, with an individual max.= 2°C on one occasion. T_{re} was c.1.9 ± 0.6°C below T_{hy} during exercise with a maximum mixed venous–rectal difference of 3.2°C.

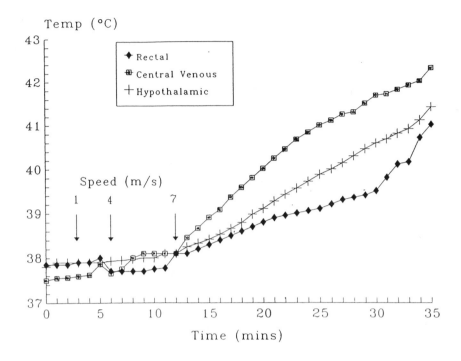

Fig. 1 Effects of exercise, with fanning, on deep body temperatures
of the horse. Representative results from one run.

DISCUSSION

The brain of the horse was maintained at a lower temperature than central blood during exercise, demonstrating that selective cooling occurred without the existence of a carotid rete. In animals which possess a carotid rete, venous blood cooled by evaporation from the walls of the nasal passages, drains through the cavernous sinus where countercurrent heat exchange results in cooling of carotid arterial blood flowing to the brain (2). Similar mechanisms for brain cooling have been documented in a variety of other species lacking a carotid rete, eg. rabbit, rhesus monkey & pigeon; evaporative cooling from the cornea is also important in the latter.

The horse nasal passage is highly vascular (6), & as with many other species (7), could provide a substantial volume of cooled blood during exercise. Blood flow in this region of the horse has not been measured, however hyperaemia indicative of increased blood flow is visible at the nasal mucosa during exercise. Also sweating of the muzzle & around the ears & eyes is profuse & initiated early (8).

Thus, blood cooled by evaporation from the skin of the head, nasal turbinates & possibly the corneas may assist in cooling the brain during exercise in the horse; its internal carotid artery lies within a venous sinus at the base of the brain (9), which may perform a similar function to the carotid rete. Cooled venous blood could be variably directed into this sinus during hyperthermia in the horse, as documented in other species (eg.10). In man, the anatomical arrangement of vessels at the base of the brain could possibly result in heat exchange between carotid arterial & facial venous flows resulting in selective brain cooling (3). The surface of the angularis oculi vein can be cooled by as much as 2°C during forced facial heat loss (by fanning) (11). The level of brain cooling that can occur without a specialised mechanism, such as the carotid rete would logically be lower. The data presented here demonstrates a degree of cooling much less than the c.3–5°C seen in animals with a highly developed carotid rete (eg. 12). Man, lacking a carotid rete, is liable to display a limited degree of selective brain cooling, possibly similar to that for the horse. However, this could not be demonstrated, perhaps because the core temperature of subjects could be raised to only 39.1°C, at this level in our horses T_{mv} – T_{hy} was usually < 0.5°C, which could have been barely detectable & therefore appeared insignificant in Nielsen & Jessen's experiments (5).

The extreme lag in T_{re} below T_{mv} in the horse is in marked contrast to the situation in other species. It is well known that T_{re} lags somewhat behind other core temperatures, such as tympanic or oesophageal, during exercise in man, possibly due to the redistribution of heat from the muscles to the rest of the body, but it is usually adequate (13). T_{re} in the horse is obviously of little value for studies of core temperature during relatively short term moderately intense exercise (14), yet this site is used almost exclusively in clinical practice. It is highly probable that the severity of body temperature elevations following heavy exercise will be underestimated in a variety of clinical situations. A number of reasons for this lag in T_{re} are apparent, including possibly poor contact between the thermocouple probe & the rectal wall, & the reduction in rectal blood flow during exercise (15). In addition, direct local heating from nearby pelvic muscles may contribute to rectal heating in man (16), but in horses there is a greater distance between the rectum & pelvic muscles.

Determination of the source of brain cooling and underlying control mechanisms in the horse requires further investigations.

REFERENCES

1. Olschewski, H. & Brück, K. (1988). Thermoregulatory, cardiovascular and muscular factors related to exercise after precooling. *J. Appl. Physiol.* 64: 803–811.

2. Baker, M.A. (1982). Brain cooling in endotherms in heat and exercise. *Ann. Rev. Physiol.* 44: 85–96.

3. Cabanac, M. (1986). Keeping a cool head. *News in Physiol. Sci.* 1:41–44

4. Brengelmann, G.L., 1987. In: *Man in Stressful Environments.* Eds. K. Shiraki & M.K. Yousef. Springfield: Thomas, p.5–22.

5. Nielsen, B. & Jessen, C. (1992). Evidence against brainstem cooling by face fanning in severely hyperthermic human. *Pflugers Arch.* 422: 168–172.

6. Hare, W.D. (1975). Equine respiratory system. In: *Sisson & Grossman's; The Anatomy of the Domestic Animals. 5th ed.* Ed. P. Getty. Philadelphia: WB Saunders 1975, p 498.

7. Pleschka, K. (1984). Control of tongue blood flow in regulation of heat loss in mammals. *Rev. Physiol. Biochem. Pharmacol.* 100: 75–120.

8. Evans, C.L. & Smith, D.F.G. (1956). Sweating responses in the horse. *Proc. Roy. Soc. B. (London).* 145: 61–83.

9. Popesko, P. (1979). *Atlas of Topographical Anatomy of the Domestic Animals. Vol. 1, 3rd Ed.* Philadelphia: W.B. Saunders., p.129.

10. Magilton, J.H. & Swift, C. (1967). Thermoregulation of the canine brain by alar fold and interarteriovenous heat exchange system. *Physiologist.* 10: 241.

11. Cabanac, M. & Caputa, M. (1979). Natural selective cooling of the human brain: Evidence of its occurrence and magnitude. *J. Physiol. (Lond.)* 286: 255–264.

12. Taylor, C.R. & Lyman, C.P. (1972). Heat storage in running antelopes: Independence of brain and body temperatures. *Am. J. Physiol.* 222: 114–117.

13. Saltin, B. & Hermansen, L. (1966). Esophageal, rectal, and muscle temperature during exercise. *J. Appl. Physiol.* 21: 1757–1762.

14. Hodgson, D.R., McCutcheon L.J., Byrd S.K., Brown, W.S., Bayly, W.M., Brengelmann, G.L. & Gollnick, P.D. (1993). Dissipation of metabolic heat in the horse during exercise. *J. Appl. Physiol.* 74: 1161–1170.

15. Manohar, M. (1987). Furosemide and systemic circulation during severe exercise. In: *Equine Exercise Physiology 2.* Eds. J.R. Gillespie & N.E. Robinson NE. Davis, California: ICEEP Pub., p.132–147.

16. Aulick, L. H., Robinson, S. & Tzankoff, S. P. (1981). Arm and leg intravascular temperatures of men during submaximal exercise. *J. Appl. Physiol.* 51: 1092–1097.

Temperature Regulation
Advances in Pharmacological Sciences
© 1994 Birkhäuser Verlag Basel

SKIN BLOOD FLOW DURING SEVERE HEAT STRESS:
REGIONAL VARIATIONS AND FAILURE TO
MAINTAIN MAXIMAL LEVELS

J.R.S. Hales[1], *B. Nielsen, & M. Yanase*[2],

August Krogh Institute, University of Copenhagen,

Denmark

SUMMARY

Elite athletes exercised to exhaustion in a hot dry environment. Photoplethysmograph, laser-Doppler (LD) & venous occlusion plethysmographic (VOP) indices of skin blood flow (BF) showed:
(a) Skin BF, after attaining a peak, decreases as exhaustion approaches (which may be important in the etiology of heat exhaustion).
(b) Different body regions need to be considered to understand the entire response.
(c) Different BF techniques may lead to different conclusions.

INTRODUCTION

At rest in a hot environment or during exercise, increased skin BF is essential for increased heat loss to control body temperature; skin blood volume is also increased. It is well established (1) that, due to cardiovascular demands competing with thermoregulatory requirements, skin BF in a hot environment is lower during exercise that at rest with comparable heat loads. However, it has not been clear whether, as exhaustion or heat stroke approaches, skin BF actually decreases after reaching a peak. This occurs inexperimental animals & skin BF appeared to be lower in collapsed compared with healthy fun-runners (see 2). It is therefore possible that failure to maintain maximal levels of skin BF

---

[1]School of Physiology & Pharmacology, University of New South Wales, Sydney.
[2]Department of Physiology, Osaka University.

during severe heat stress could be a critical event in the onset of heat stroke. The first aim of the present study was to establish the human skin BF responses to exhaustive heat stress.

Secondly, since studies contributing most to the current understanding of control of skin BF during exercise or temperature stresses have employed forearm VOP, we sought to check (a) the validity of using VOP by comparisons with modern photoplethysmographic & LD techniques, & (b) the validity of using the forearm as an index of whole body skin BF.

METHODS

Full details, excluding skin BF, have been reported (3) & therefore only an outline follows. Four elite male endurance athletes exercised on a Krogh bicycle ergometer in a hot, dry environment (40 - 42°C dry bulb with 10 -15% RH) at 60% of their maximal oxygen consumption until they were exhausted.

Skin BF was estimated by three techniques: (a) Forearm VOP employing a mercury-in-silastic Whitney strain gauge around the mid-forearm, with the arm raised in a sling to shoulder level & inflatable cuffs at the wrist and elbow (see 1); triplicate measurements were taken each 5 min. (b) A laser -Doppler (Perimed PF2B operating at 632.8 nm) probe with 0.5 mm fibre separation mounted on the outer surface of the forearm adjacent to the strain gauge; "flux" measurements were monitored for 1min/5min. (c) Photoplethysmography, using the "Tissue Perfusion Monitor" detailed elsewhere (3) except that a "Tissue Perfusion Index" (TPI) derived as area of the pulsatile signal curve multiplied by heart rate was employed (rather than the signal curve amplitude x heart rate previously described); TPI was monitored for 1min/5min via six probes, on the forearm (one proximal & one distal to the strain gauge), forehead, chest over the right pectoral muscle & on the tips of the index & mid-finger of the raised arm.

RESULTS & DISCUSSION

Details of several cardiovascular & thermoregulatory parameters except skin BF have been published (3). Endurance time was 48± 1.9 min & oesophageal temperature increased from 36.9 ± 0.17 to 39.7± 0.15°C.

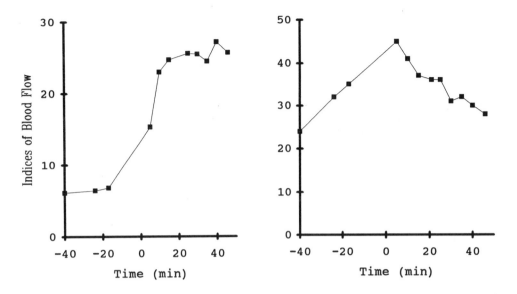

Fig. 1 Representative data from one subject, illustrating venous occlusion plethysmographic measurements of forearm BF (left, ml $100ml^{-1}min^{-1}$) & laser-Doppler blood "flux" in forearm skin (right, mV), during severe heat stress pre- (o) & post- acclimation (•). Entered hot room at -50min & started exercise at 0min.

Patterns of BF Changes

Forearm VOP (Fig. 1) increased within 6 - 16 min to 3 - 9 - times control values & remained at that level, exhibiting no signs of decreasing. The forearm LD (Fig. 1) changes were biphasic, increasing to c. 240% within 5-25 min followed by a decrease to c. 80% of that peak. The forearm TPI (Fig. 2) exhibited a similar biphasic pattern, increasing within 5-20 min to c. 240% of resting levels then decreasing to approximate resting levels. Note in Fig. 2 that resting TPI was similar for the forearm & chest, the forehead was on average 3-times greater, & the finger was highly variable; the finger was usually higher but often covered the entire range of values for other regions. The forehead & chest had patterns of responses similar to the forearm but the magnitudes of changes were smaller, viz., both increasing to c. 190%. Seven finger TPI records revealed no consistent pattern, 2 exhibiting a byphasic pattern but with a 270% peak much later than for other regions (around 35 min vs. 5-20min), 2 increasing throughout to c. 155%, & 3 remaining within the highly variable resting range.

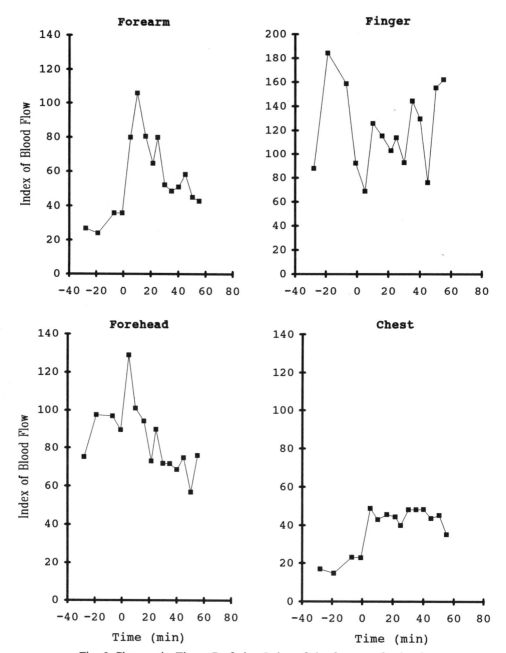

Fig. 2 Changes in Tissue Perfusion Index of the forearm, forehead, chest & a finger during severe heat stress pre - (o) and post-acclimation (•). Representative data from one subject.

These results indicate that in humans, as previously shown in animals, the marked cutaneous vasodilator response to severe heat stress is not maintained; this would reduce heat loss & aggravate the hyperthermia. As discussed in detail elsewhere (2), this 'physiological failure' appears to represent regulatory requirements for body temperature being partly dominated by those for blood pressure; this could be brought about by a heat - induced reduction in central venous pressure (known to occur in animals) acting via low-pressure baroreceptors.

The different behaviour of body regions indicates that interregional differences in the skin BF responses to heat stress should be considered in future studies such as of whole body heat balance, or even detailed characteristics of skin BF regulation need to be checked. Nishiyasu et al. (5) have already reported smaller responses (measured by VOP) in calf than forearm BF.

Differences Between Techniques

The three techniques used do not purport to measure the same parameters of skin BF. VOP is the 'classical' technique (see 1) & while an advantage might be that it is measuring total BF rather than any specific components, a disadvantage is that BF to all forearm tissues is included & not solely skin. There is good evidence that increases in forearm BF during acute, moderate heat stress are restricted to skin, with muscle, fat & bone either constant or decreasing (see 1); similar validation tests have not been made under the numerous conditions in which VOP has been employed (such as the present). With respect to our other BF methods, there is evidence (6) that the Perimed LD normally measures superficial capillary BF but its response can be negated by large changes in perfusion of deeper vasculature, such as in AVA BF. The TPI measures either total local skin BF or a closely related parameter, ie., includes the deeper AVAs (6), however, we were able to process only the pulsatile component of BF with the non-pulsatile component unmeasured (4). While there are few, if any, AVAs in forearm skin, the VOP and TPI data will include arteriolar & venular BF probably not detected by the LD. Thus, our variations in forearm data are suggestive of different compartments of skin or limb BF being involved in responses to severe heat stress, ie., skin capillary vs. arterioles and venules vs. other limb tissues. On this basis, the absence of a decrease in the VOP index of skin BF during severe heat stress (Fig.1) but its presence in both the LD and TPI indices (Fig. 1 & 2) can be interpreted as showing that the decrease occurs in skin capillaries & possibly deeper skin vasculature & is negated in VOP measurements by increased non-cutaneous BF.   Clearly, further investigations are necessary to clarify the meaning of these differences.

REFERENCES

1. Rowell, L.B. (1983). Cardiovascular adjustments to thermal stress. In: *Handbook of Physiology Sec. 3, the Cardiovascular System Vol. III Part 2.* Eds. J.T. Shepherd & F.M. Abboud. Bethesda: Amer. Physiol. Soc., p. 967 - 1023.

2. Hales, J.R.S., R.W. Hubbard, & S.L. Gaffin (1993b). Limitation of heat tolerance. In: *Handbook of Physiology - Adaptation to the Environment.* Eds. M.J. Fregley, C.M. Blatteis & S.L. Senay. New York: Amer. Physiol. Soc./Oxford Univ. Press.

3. Nielsen, B., J.R.S. Hales, S. Strange, B. Saltin & N. J. Christensen (1993). Human circulatory and thermoregulatory adaptations with heat acclimation and exercise in a hot, dry environment. *J. Physiol. (London).* 460: 467 - 485.

4. Hales, J.R.S., F.R.N. Stephens, A.A. Fawcett, K. Daniel, J. Sheahan, R.A. Westerman & S.B. James (1989). Observations on a new non-invasive monitor of skin blood flow. *Clin. Exp. Pharm. Physiol.* 16: 403 - 415.

5. Nishiyasu, T., X. Shi, C.M. Gillen, G.W. Mack & E.R. Nadel (1992). Comparison of the forearm and calf blood flow response to thermal stress during dynamic exercise. *Med. Sci. Sports Exerc.* 24: 213-217.

6. Hales, J.R.S., R.G.D. Roberts, R.A. Westerman, F.R.N. Stephens and A.A. Fawcett (1993). Evidence for skin microvascular compartmentalization by laser-Doppler & photoplethysmographic techniques. *Int. J. Microcirc.: Clin. Exp.* 12: 99 - 104.

HEAT TRANSFER VIA THE BLOOD

Jürgen Werner and Heinrich Brinck

Institut für Physiologie, Ruhr-Universität Bochum, MA 4/59,

D-44780 Bochum, Germany

Summary.
Circulation affects internal heat distribution in three ways by convective heat transfer:
1) it minimizes temperature differences within the body by different perfusion rates,
2) it controls effective body insulation in the skin region by vasomotor activity,
3) countercurrent heat exchange may be a non-negligible factor.
This paper provides numerical information about the efficiency of these processes.

Vessel length to achieve thermal equilibrium

Using a simple model which assumes constant temperature and velocity of the blood throughout the whole cross section of the vessel, it is possible to estimate the heat transfer to the surrounding tissue and the length of the vessel necessary to achieve equilibrium between blood and tissue temperature. The heat balance equation is as follows:

$$\frac{\partial T_B(z,t)}{\partial t} + v_z \frac{\partial T_b(z,t)}{\partial z} = K(T_W(z,t) - T_B(z,t))$$

with $\quad K = 4\alpha/\rho cd$

and $\quad \alpha$ = heat transfer coefficient $\approx 180 \ \mathrm{Wm^{-2}\,^\circ C^{-1}}$

$\quad \rho$ = density $\approx 1060 \ \mathrm{kg \ m^{-3}}$

c = specific heat $\approx$ 3650 Ws kg^{-1} $^{\circ}$C^{-1}

d = diameter

z = direction of flow

t = time

v_z = axial blood velocity

T_B = blood temperature

T_W = wall temperature

The solution of this equation delivers the necessary length z_y to achieve a certain degree of equilibrium (y = 1.0...0.0):

$$z_y = -\frac{v_z}{K} \ln(y)$$

Using the parameters for the different vessel types, it is shown that in the large vessels there is almost no heat transfer to the tissue. However, thermal equilibrium with the surrounding tissue does occur in the terminal arteries. This means that thermally relevant heat transfer should be in these vessel types and not, as assumed in many papers, in the capillaries. This conclusion is compatible with the findings presented in [1], claiming that thermal equilibrium is already reached in vessels of medium order.

Contribution of blood flow to heat endurance

A rough estimation of the contribution of blood flow to homoiothermia and to thermal endurance may be achieved by another relatively simple model [2], approximating the human body to one cylindrical element with equal volume, surface and tissue parameters, taking into account four concentric layers of core, muscle, fat and skin. If such a body had no convective heat transfer via the blood and no evaporative capacity, a basal metabolic heat production of e.g. 70 W and an overall tissue conductivity of $\lambda \approx$ 0.4 W m^{-1} $^{\circ}$C^{-1} could produce a mean core temperature of 37°C, if ambient temperature T_A were about 26°C. Above this limit there would be a linear increase of core temperature according to the increase of environmental temperature. However, even at T_A = 26°C the temperature in the central axis would be significantly above 40°C without any blood flow (see Fig. 1, upper curve). Even basal blood flow dramatically lowers the difference between the temperature in the central axis of the body and the skin surface (Fig. 1). Nevertheless the overall contribution to thermal endurance is but small compared to the efficiency of the evaporative power. A comparable effect could be achieved by initiating evaporative cooling by only 10 W. Increase of cardiac output enables the body to maintain a core temperature of 37°C up to about 32°C air temperature without enhanced evaporation. This would require an increase of blood flow to about 30 ml $(100 \text{ g})^{-1}$ min^{-1} which for a human weighing 70 kg, would mean a cardiac output of about 21 l min^{-1}.

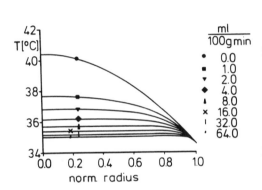

Fig.1: Contribution of blood flow to radial temperature profiles of the body (no activation of evaporative power). Blood flow increasing from upper curve (0 ml g^{-1} min^{-1}) to lower curve (64.0 ml g^{-1} min^{-1}).

Fig. 2: Maximal ambient temperature T_{Amax} maintaining a core temperature of 37°C as function of cardiac output CO, assuming four different cases: increase of skin, core, muscle blood flow and global (homogeneous) increase.

The efficiency of the convective heat transfer depends on the spatial control of blood flow (Fig. 2). Increasing muscle blood flow alone would be of almost no effect, increase of blood flow within the body core would achieve only minor effects, and a homogeneous increase within in the body is significantly less efficient than increase of skin blood flow, a measure which is normally activated by the thermoregulatory system.

So the dominant relevance of convective heat transfer via the blood is to minimize the temperature gradient between core and skin in heat stress, particularly during endurance exercise in the heat, when there is strong competition for circulation between the active muscles to support metabolism and skin to dissipate heat. If the stressors are sufficient, low central venous pressure can reduce skin blood flow, which together with a final lower sweat and evaporative rate, constitute a serious impairment of heat loss mechanisms resulting in a possibly lethal increase of core temperature [3, 4].

Thermal effect of countercurrent networks

During the last decade, particular efforts were made to account for the quantitative contribution of countercurrent heat exchange. Increase of thermal insulation by the shift of venous flow from cutaneous to deep vessels has a very important impact on the conservation of heat. In comparison countercurrent heat transfer is of minor efficiency for heat saving. It depends strongly on the rate of blood flow. For the human upper limb, when the arterial blood flow is minimal, a countercurrent efficiency of maximal 15 % was found [5].

To investigate the thermal processes of a human extremity especially within the muscle tissue near to the skin surface, a three-dimensional model was developed which quantifies the convective heat transport on the basis of a detailed simulation of the vascular architecture using the relevant physical laws [6]. Computation of heat flow via the vessel walls is carried out using the Nusselt-number, the wall temperatures, the blood temperature and the vessel diameters. In contrast to former approaches, no serious assumptions have to be made. The model approximates the three-dimensional temperature profiles within the tissue and the arterial and venous temperatures of thermally significant vessels along the branching countercurrent vessel network. The geometry varies with the distance from the skin surface. The idealized cross-section of an extremity has a three-layer tissue structure (core-muscle-skin) and the adequate structure of peripheral circulation. The core comprises the larger central vessels and the directly adjacent tissue. The central artery and vein are the origin for a countercurrent arterio-venous network in the muscle layer. Neighbouring arteries and veins branch typically eight times. The outer skin layer is provided with separate rising vessels. The three-dimensional heat balance equation with a convective boundary condition at the vessel walls is solved using local central differences with a highly irregular grid. The grid width is small in the vicinity of the vessels, at the maximal distance it is 64-fold. The three

dimensional profiles of tissue temperatures and the one-dimensional profiles of temperatures
of the arterial and venous blood were computed for resting conditions, for maximal exercise
and for a low ambient temperature. In resting conditions, there is a greater temperature
gradient around the arteries than around the veins, and the mean tissue temperature is closer
to the venous temperature, while the mean tissue temperature of the spacing between the
vessels corresponds to the mean of arterial and venous blood temperatures.

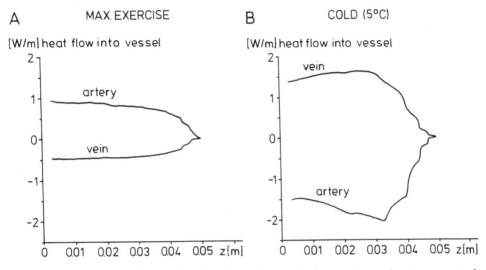

Fig. 3: Heat flow [Wm^{-1}] from the tissue into the vessels for arteries and veins computed
from a three dimensional model of a branching countercurrent network in a human arm
(negative values mean heat flow out of the vessels) for extreme conditions.
A. maximal exercise   **B.** cold (5°C)

In a cold environment, the arterial blood temperature decreases from 37.5°C to about
30°C in the outer section of the muscle tissue. There must be a prearteriole heat transfer
between the blood and the tissue which cannot be neglected especially in the cold
environment and when the perfusion rate is small. At the end of the vascular tree, blood
temperature is equilibrated with the temperature of the surrounding tissue. Fig. 3 shows the
heat flows from the tissue into the arterial and venous vessels, in part A at the comfortable
ambient temperature, but at maximal exercise, in part B at resting conditions in the cold
environment (5°C). In the cold, there is heat loss from the arterial blood and heat gain to
the vein. The heat flow out of the artery is slightly greater than the heat flow into the vein.
At the high metabolic rate, the direction of heat flows is inverse. It is interesting that in spite

of the fact the mean tissue temperature is below both blood temperatures, there is a heat flow out of the vein. This is a consequence of the tissue cooling by the artery in the neighbourhood of the vessel pair.

The mathematical description of convective heat transfer in whole body models is a very intricate problem. The classical "bio-heat approach" [7] assuming that heat transfer is proportional to blood flow and the difference of arterial and tissue temperature, although primarily based on a wrong assumption, has turned out to deliver more realistic results than recent attempts to account for convective heat transfer by an enhanced conductivity index [8], an approach which had already been questioned in [9]. On the basis of the outlined three-dimensional model it was possible to test the assumptions of various approaches and to develop a simple "efficiency factor-model" [10] to be used in whole body thermoregulatory models.

Supported by Deutsche Forschungsgemeinschaft, (We 919/2).

References
[1]   Chen, M.M. and Holmes, K.R.: Microvascular contributions in tissue heat transfer. N.Y. Acad. Sciences 1980; 335: 137-150.
[2]   Buse, M. and Werner, J.: Closed loop control of human body temperature: Results form a one-dimensional model. Biol. Cybernetics 1989; 61: 467-475.
[3]   Hales, J.R.S.: Proposed mechanisms underlying heat stroke. In: J.R.S. Hales and D.A.B. Richards (eds.): Heat stress: Physical exertion and environment. 1987: Excerpta Medica, pp. 85-102.
[4]   Werner, J.: Temperature regulation during exercise. In: C.V. Gisolfi, D. R. Lamb, E.R. Nadel (eds.): Exercise, heat and thermoregulation. 1993: Brown & Benchmark, pp. 49-79.
[5]   Raman, E.R.: Insulation versus countercurrent heat savings. In: W.A. Lotens and G. Havenith (eds.): Environmental Ergonomics. 1992: pp. 131-132.
[6]   Brinck, H. and Werner, J.: Estimation of the thermal effect of blood flow in a branching countercurrent network using a three-dimensional vascular model. J. Biomech. Eng. 1993, in press.
[7]   Pennes, H.H.: Analysis of tissue and arterial blood temperatures in resting forearm. J. Appl. Physiol. 1948; 1: 93-122.
[8]   Weinbaum, S. and Jiji L.M.: A new simplified bioheat equation for the effect of blood flow on local average tissue temperature. J. Biomech. Eng. 1985; 107:131-139.
[9]   Wissler, E.H.: Comments on the new bioheat equation proposed by Weinbaum and Jiji. J. Biomech. Eng. 1987; 109:226-233.
[10]   Brinck, H. and Werner, J.: The thermal effect of blood flow in a branching countercurrent network. In: W.A. Lotens and G. Havenith (eds.): Environmental Ergonomics 1992; 132-133.

FALL IN BODY CORE TEMPERATURE DURING THE PREVIOUS HEAT EXPOSURE TIME IN RATS AFTER SUBJECTION TO HEAT LOADS AT A FIXED TIME DAILY

Osamu Shido, Sohtaro Sakurada and Tetsuo Nagasaka
Department of Physiology, School of Medicine, Kanazawa University, Japan

Summary
The rats were subjected to an ambient temperature of 33°C in the last half of the dark phase daily for 5-16 days. After terminating the schedule, body core temperature, heat production and spontaneous activity of the rats decreased for 3-4 h during the period when the rats had been previously exposed to heat. The results suggest that a time memory for heat stress could be formed in the rats and that thermoregulatory responses to heat may have occurred during the previous heat exposure time without actual heat exposure.

Introduction

Heat exposure has been shown to modify the pattern of nycthemeral variations of spontaneous activity (1) and of endocrine systems (2). However, it is not known how heat exposure affect the pattern of day-night changes in body temperature. In our previous study where rats were subjected to heat exposure for hours at a fixed time daily for 10 consecutive days and then transferred to a constant thermoneutral ambient temperature, body core temperature appeared to fall during the same hours when the rats had been previously exposed to heat (3). The observation suggests that repeated heat exposure at a fixed time daily can modulate the nycthemeral cycle of the thermoregulatory mechanism.

It is well known that schedules of restricted daily meals alter the distribution of an animal's behavioural and metabolic activities. When food access is limited to a few hours at a fixed time once a day, a great increase of locomotor activity is observed during the hours before meal time (4). This anticipatory activity persists for several days after switching to ad lib-feeding (5). The phenomenon is explained by the establishment of a time memory for meals following a single meal schedule (5). Thus, heat exposure given at a fixed time daily may also modify the pattern of day-night changes of the thermoregulatory system through producing a time memory for heat stress.

In the present study, we investigated whether repeated heat exposure at a fixed time of a day causes a specific change in body core temperature around the period when the animals had been exposed to heat, and examined the thermoregulatory mechanisms involved in the occurrence of body core temperature change in heat-exposed rats.

Methods

Male Wistar rats were housed in wire mesh cages and given chow and water *ad libitum* in a 12:12 h light-dark cycle. After a stainless steel guide cannula was implanted into the anterior hypothalamus, they were divided into two groups: the control (CN) rats were constantly kept at an ambient temperature of 24°C. For the heat-exposed (HE) rats, the room temperature was raised at the middle of the dark phase from 24 to 33°C and maintained, and then restored to 24°C just before the start of the light phase. The heat exposure lasted for 5 to 10 consecutive days in experiment 1 and for 16 consecutive days in experiments 2 and 3.

Experiment 1 Nineteen rats in the HE and nine in the CN were used. Five days before the start of the heat exposure schedule, a thermocouple was inserted into the rat's hypothalamus via the guide cannula after passing through a metal spring coil. Then, it was fixed to the rat's head with dental cement. The proximal end of the spring coil was held on the ceiling of the cage. Hypothalamic temperature (T_{hy}) was recorded every min. For the HE, the measurements were started one day before the heat exposure schedule and continued for two days after terminating the schedule. For the CN, the measurements were made for 13 days. The light-dark cycle was maintained during the entire period.

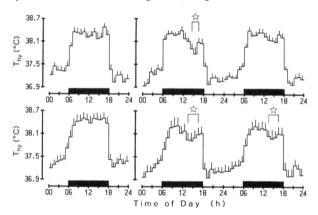

Fig. 1 Changes in T_{hy} in the HE before (left column) and after (right column) the 5-day (upper panel) and 10-day (lower panel) heat exposure schedule. Values are hourly means + SEs. Asterisks, significantly different from corresponding values before the heat exposure schedule; filled bars above abscissa, dark phase of a day.

Experiment 2 Ten rats in the HE and eight in the CN were used. Five days before completing the heat exposure schedule, a thermocouple was set into the hypothalamus as in experiment 1. Three days later, the caged rats were transferred to a gradient type direct calorimeter (7) with a wall temperature of 24°C. The rats were kept inside the calorimeter for 2 days under the same light-dark cycle. For the HE, heat exposure was repeated inside the calorimeter during the same period of the day. After the last heat exposure, the chamber wall temperature was strictly controlled at 24°C and the photo-cycle was kept in constant darkness. T_{hy}, evaporative and nonevaporative heat loss, O_2 uptake, CO_2 production, feeding activity (FA) and body movement (BM) were recorded every min. Metabolic heat production ($\underline{M}$) and total heat loss ($\underline{H}$) were then calculated. The measurements were started 6 h after the end of the final heat exposure and continued for the following 24 h.

Experiment 3 Eight rats in the HE and nine in the CN were used. As in experiment 2, the rats were maintained in the direct calorimeter. After the final heat exposure, the food was removed from the cage. The measurements were made as in experiment 2.

Statistics Statistical evaluations among mean values were assessed by two-way ANOVA followed by a paired or an unpaired Student's t-test.

Results

Experiment 1 Clear day-night variations of T_{hy} were observed in all rats under the 12:12 h photo-cycle. In the CN, the pattern of the nycthemeral changes of T_{hy} were essentially the same throughout the 13 days. After the heat exposure schedule, T_{hy} of the HE significantly fell for 3-4 h in the last half of the dark phase when the rats had been previously exposed to heat (Fig. 1). The fall in T_{hy} persisted for at least two days after completing the 10-day heat exposure schedule.

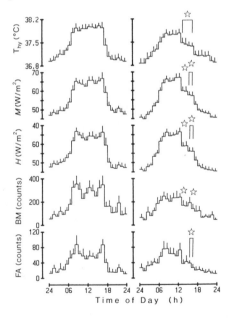

Fig. 2 Changes in Thy, $\underline{M}$, $\underline{H}$, BM and FA for 24 h after terminating the 16-day heat exposure schedule in the CN (left column) and HE (right column). Values are hourly means + SEs. Asterisks, significantly different from corresponding values of the CN.

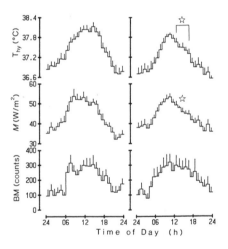

Fig. 3 Changes in Thy, $\underline{M}$ and BM for 24 h after completing the 16-day heat exposure schedule in the CN (left column) and HE (right column) under a starved condition. Values are hourly means + SEs. Asterisks, significantly different from corresponding values of the CN.

Experiment 2   In both groups, clear circadian variations of T_{hy}, $\underline{M}$, $\underline{H}$, BM and FA were shown (Fig. 2).   The patterns of day-night changes of these variables were significantly altered by repetitive heat exposure.   In the HE, the levels of T_{hy}, $\underline{M}$, $\underline{H}$, BM and FA were significantly lower than those of the CN during the previous heat exposure time.

Experiment 3   Clear day-night variations of T_{hy}, $\underline{M}$ and BM were persistent under the starved condition in both groups (Fig. 3).   The patterns of circadian variations in T_{hy} and $\underline{M}$ were different between the CN and HE.   Thy and $\underline{M}$ of the HE were significantly lower than those of the CN for 1-4 h during the previous heat exposure time.   BM of the HE did not differ from that of the CN at any time of the day.

Discussion

The present results clearly confirm that the nycthemeral variations of body core temperature are altered after subjection to repeated heat exposure for hours at a fixed time daily for more than 5 consecutive days.   The T_{hy} of the HE decreased during the period when the rats had been exposed to heat.   The fall in T_{hy} during the specific period persisted for at least two days after terminating the 10 day-heat exposure schedule.   Additionally, the fall in T_{hy} was consistent under both the 12:12 h light-dark cycle and constant darkness.

During heat exposure, the heat loss mechanism is activated and heat production is suppressed to prevent heat accumulation in the body.   These physiological responses to heat for lowering body core temperature occurred daily during the same period per day in rats on the heat exposure schedule.   It is therefore considered that the fall in T_{hy} in the HE during the period of the previous heat exposure time may be attributed to the persistence of thermoregulatory responses to heat even without actual heat exposure.   It appears that a time memory for heat exposure could be formed in the thermoregulatory system of the rats after subjection to repeated heat loaded at a fixed time daily.

Concomitant with the fall in T_{hy}, $\underline{M}$ and $\underline{H}$ decreased during the period of the previous heat exposure time.   When total thermal conductance (C) was calculated, C of the HE was significantly lower than that of the CN only during the specific period.   The decreases in $\underline{H}$ and C do not indicate the activation of the heat loss mechanism but rather suggest the occurrence of a heat conservation response.   Thus, the fall in T_{hy} observed in the period when the rats had been subjected to heat may be predominantly attributed to the suppression of $\underline{M}$.

In the HE, BM and FA decreased during the previous heat exposure time. $\underline{M}$ and body core temperature increase in proportion to an increase in spontaneous activity and with food ingestion.   Thus, the decreases of BM and FA may have, at least partly, contributed to the reductions of $\underline{M}$ and T_{hy}.   Under the starved condition, the decreases in T_{hy} and $\underline{M}$ during the previous heat exposure time were persistent in the HE without any decline of BM.   Furthermore, our previous study showed that, after subjection to heat load in the last half of the dark phase for 10 days, the fall in body temperature during the previous heat exposure time persisted for at least 2 days, whereas the reduction of locomotor activity was observed for only one day (6). It therefore seems that decreases in thermogenic behaviours are not essential for the reductions of $\underline{M}$ or T_{hy} in heat-acclimated rats. In addition to spontaneous activity, basal metabolic rate may have been suppressed in the HE during the period when the rats had been exposed to heat.

In summary, after repetitive heat exposure during the last half of the dark phase for more than 5 days, the body core temperature of the rats decreased during the period of the previous heat exposure time. The characteristic change lasted for at least 2 days after completing a 10 day-heat exposure schedule. It appears that a time memory for the heat exposure was formed in heat-acclimated rats and that thermoregulatory responses to heat may have occurred around the period corresponding to that of the previous heat exposure time even without actual heat exposure. Such changes in heat balance may then have resulted in the fall of body temperature during the specific period.

References

1.   CHOU, B.J., BESCH, E.L. (1974). Feeding biorhythm alterations in heat-stressed rats. *Aerospace Med* **45**, 535-9.

2.   CURE, M. (1989). Plasma corticosterone response in continuous versus discontinuous chronic heat exposure in rat. *Physiol Behav* **45**, 1117-22.

3.   SHIDO, O., YONEDA, Y., NAGASAKA, T. (1989). Changes in body temperature of rats acclimated to heat with different acclimation schedule. *J. Appl. Physiol.* **67**, 2154-7.

4.   NAGASAKA, T., SHIDO, O. (1989). Locomotor activity and heat production of rats on restricted two-hour feeding regimes. *Jpn J Biometeor* **26**, 85.90.

5.   MORI, T., NAGAI, K., NAKAGAWA, H. (1983). Dependence of memory of meal time upon circadian biological clock in rats. *Physiol Behav* **95**, 1195-201.

6.   SHIDO, O., SAKURADA, S., NAGASAKA, T. (1991). Effect of heat acclimation on diurnal changes in body temperature and locomotor activity in rats. *J Physiol (Lond)* **433**, 59-71.

7.   SUGANO, Y. (198). Heat balance of rats acclimated to diurnal 2-hour feeding. *Physiol Behav* **30**, 289-93.

Temperature Regulation
Advances in Pharmacological Sciences
© 1994 Birkhäuser Verlag Basel

TREATMENT OF IMMERSION HYPOTHERMIA BY FORCED-AIR WARMING

Gordon G. Giesbrecht, Marc Schroeder *, and Gerald K. Bristow

Laboratory for Exercise and Environmental Medicine, Faculty of Physical Education and RecreationStudies, and Department of Anesthesia, University of Manitoba, Manitoba, Canada. R3T 2N2; * Department of Anesthesiology, University of California San Francisco, CA 94143-064

SUMMARY Forced-air warming (FAW) was evaluated as treatment for immersion hypothermia. Eight subjects were twice immersed in 8°C water until hypothermic and rewarmed by shivering (SH) or FAW (Bair Hugger® 250/PACU Warming System with 300 Warming Cover). Afterdrop (±SD) during FAW (0.41 °C) was ~30% less than during SH (0.58 °C) ($P<0.05$). Rewarming rate during FAW (2.51 °C·h^{-1}) was not significantly different from SH (2.26 °C·h^{-1}). Skin temperature was 5°C higher during FAW. During SH metabolism increased 100 W and subsequently declined, but immediately decreased with FAW. FAW is a safe, noninvasive treatment and could be used effectively in an emergency medical facility, and possibly in some rescue/emergency vehicles or marine vessels.

INTRODUCTION Many recreation, commercial, and military activities require, or present the risk of, human exposure to cold water, leading to an overall negative heat balance and subsequent decrease in core temperature. There is no consensus concerning the best specific protocol for rewarming victims of immersion hypothermia. It is generally agreed however, that it is desirable to minimize the afterdrop in core temperature (T_{co}) and maximize the subsequent rate of rewarming, provided that the stability of the cardiovascular and respiratory systems are maintained and metabolic imbalances are corrected. Invasive methods such as peritoneal lavage and extracorporeal circulation, are used in severe cases of hypothermia ($T_{co} < $ ~30°C) and meet the above criteria. [1] With mild hypothermia ($T_{co} > $ ~30°C) however, invasive methods are not justified. The most common noninvasive method is the application of some source of external heat. However, warming peripheral thermal receptors reflexly attenuates shivering thermogenesis. [2] Therefore, an external heat source must provide enough exogenous heat to more than replace the consequent loss of shivering heat production, in order to provide more rapid rewarming than shivering itself.

Forced-air warming systems have recently been introduced for prevention or reversal of hypothermia in surgical patients, providing convective heat transfer of 60-70 W to the body.[3]

The present study evaluates forced-air warming as a treatment for hypothermia by quantifying and comparing changes in core body and mean skin temperatures, metabolic heat production, and cutaneous heat flux during two rewarming protocols following immersion hypothermia: 1) shivering thermogenesis; and 2) forced-air warming. Because of the large surface area over which a considerable amount of heat is donated, we hypothesized that forced-air warming would provide a more rapid rewarming rate. Previous studies have demonstrated greater afterdrop with protocols which increase peripheral circulation via warm water immersion [4] or exercise, [2] therefore a greater afterdrop was expected with forced-air warming.

METHODS With approval from our Faculty Human Ethics Committee, eight subjects (6 men, 2 women) were studied after giving written informed consent.

Instrumentation. Esophageal temperature (T_{es}) was measured by an esophageal thermocouple. Cutaneous heat flux ($W \cdot m^{-2}$) and temperature (°C) were measured from 7 sites. Flux values were converted into $W \cdot site^{-1}$ by multiplying by the calculated body surface area [area (m^2) = weight$^{0.425}$ (kg) $\cdot$ height$^{0.725}$(cm) $\cdot$ 0.007184] of each subject and assigning the following regional percentages: head 6%; chest 9.5%; abdomen 9.5%; arm 19%; back 19%; thigh 19.5%; and calf 17.5%. Oxygen consumption (VO_2) was measured with an open-circuit method.

Forced-Air Warming Unit. The forced-air power unit injects warm air (outlet temperature ~43°C) through a connecting hose (flow rate ~31 cfm) into a disposable plastic/paper quilt-like cover The warm air exits slits on the patient side of the cover and provides convective warming to the skin.

Protocol. On two occasions, subjects sat quietly for a period of 10 min, during which baseline values for heart rate, T_{es}, VO_2, skin temperatures, and heat flux were established. They were then immersed to the neck in stirred water at 8°C until a time of 70 min elapsed, or T_{es} reached 33.5°C.

After exiting the cold bath and light towel drying, each subject initially was rewarmed by one of two techniques: 1) shivering in a supine position inside a sleeping bag; or 2) lying supine under a forced-air warming cover with one cotton blanket draped over top (the power unit was started 5 minutes prior to treatment, and set at maximum temperature and flow settings). The order was randomized.

Data Analysis. Data for the two trials were compared using paired t-test. Results are reported as means ± SD; differences were considered significant when P<0.05.

RESULTS The core temperature afterdrop during forced-air warming (0.41±0.23 °C) was about 30% lower than during shivering alone (0.58±0.27 °C, P<0.05) (Fig. 1). The length of afterdrop period was not significantly different between forced-air warming (16.3±5.6 min) and shivering (19.4±4.9 min). The average rewarming rate for forced-air warming (2.51±1.3 °C·hr^{-1}) was not significantly different than during shivering (2.26±0.7 °C·hr^{-1}) (Fig. 2).

Mean skin temperature increased rapidly during the first 10 min of both rewarming protocols and continued to increase gradually during the remainder of rewarming; reaching 31.5 °C by the end of the forced-air warming period. Skin temperature was consistently higher during this protocol than during shivering, by approximately 5°C.

During shivering, metabolic heat production increased 100 W from end-immersion values over the initial 20 min, and subsequently declined gradually during the remainder of this protocol (Fig. 3). In contrast, forced-air warming produced an immediate and continuous decrease in heat production. Cutaneous heat loss decreased substantially upon exiting the cold bath. This transient decrease was followed by a steady increase. During shivering, total heat flux was greater than during forced-air warming by 260 W initially (shivering flux = 30 W; forced-air flux = -230W), and 210 W at the end of the rewarming periods (shivering flux = 50 W; forced-air flux = -160 W).

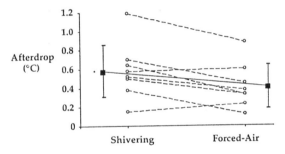

Figure 1. Individual (open) and mean (±SD) (closed) afterdrop values during forced-air warming and shivering. Dotted lines connect data for each subject, for the two treatments.

    * Significantly greater than forced-air (P<0.05).

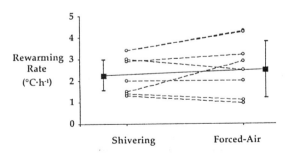

Figure 2. Individual (open circles) and mean (±SD) (closed squares) rates of rewarming during forced-air warming and shivering. Dotted lines connect data for each subject, for the two treatments.

DISCUSSION   The present study is the first to evaluate the efficacy of forced-air warming as a treatment for immersion hypothermia. When compared with the shivering protocol, forced-air warming raised skin temperature with a consequent decrease in shivering heat production. Afterdrop was significantly lower with forced-air warming than shivering. However, average results for rewarming rates were not significantly different between the two protocols.

Numerous studies have evaluated the rewarming efficacy of external heat application. Results vary depending on the amount of heat delivered and the surface area to which the heat is applied. Studies using a single portable STK Heatpac [2] or 4 electric heating pads [5] have not demonstrated any differences in afterdrop or rewarming rates when compared to shivering alone. Similar results have been observed with direct surface-to-surface contact with a human heat donor or a thermal manikin. [6] These heat sources can be considered small to moderate because of limited surface area coverage and/or low heat production. Conversely, rewarming was accelerated when greater sources of heat were applied to larger portions of the body.

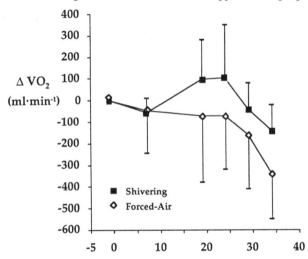

Figure 3. Mean oxygen consumption (±SD) during shivering and forced-air warming.

Compared to shivering alone, afterdrop is greater during immersion in a 26°C water bath with water temperature slowly increased to 42°C. [4] There is no difference in afterdrop when 40-43°C water is pumped through a piped suit [7] and afterdrop is similar [8] or decreased [9] when subjects are immersed in water with initial temperatures of 40-43°C.

We felt that forced-air warming would be a heat source comparable to the initially 26°C water used by Hayward et al. [4] and therefore expected a greater afterdrop compared to shivering. Contrary to our hypothesis, forced-air warming attenuated the afterdrop by ~30%. This result may be explained by some of the factors known to influence this phenomenon: 1) continued loss of core heat into cold peripheral tissue via conduction, [10] 2) core heat lost into colder peripheral tissue by blood convection, [11] and 3) heat produced by metabolism, which counteracts heat lost to peripheral tissues. First, conductive heat loss to cold peripheral tissue may be attenuated as forced-air warming increases peripheral tissue temperature. Second, tissue perfusion would have to change during forced-air warming in order to affect the afterdrop by altering the magnitude of heat extracted from the blood by the colder peripheral tissues (convection). Skin blood flow would likely change little with moderate peripheral warming during forced-air warming in this study (T_{sk} = 31.5°C), because flow largely is determined by central thermoregulatory status at skin temperatures between 20 and 35°C. [12] Therefore, skin flow presumably was minimal as our subjects were hypothermic. Because of the tight coupling between muscle activity (i.e., shivering) and blood flow, peripheral muscular flow would be expected to actually decrease during forced-air warming, consequent to the reflex inhibition of shivering. Regarding the third factor, the decrease in peripheral heat production might actually increase afterdrop under some circumstances, however this does not seem to be the case in the present study. It is therefore likely that forced-air warming attenuates afterdrop by decreasing peripheral muscle blood flow, hence the convective redistribution of heat from the core to the periphery, and possibly through direct warming of peripheral tissue.

As detailed above, greater sources of external heat applied to large portions of the body inhibit shivering heat production, but also provide sufficient heat to speed rewarming. [4, 7-9] We hypothesized that forced-air warming would provide a higher rate of rewarming in a similar manner. As expected, heat production was inhibited with forced-air warming due to increased skin temperature (see Fig. 3), however there was no significant difference between average rewarming rates of the two protocols. Therefore, the amount of heat transferred to the core by forced-air warming, was on average, just enough to compensate for the decrease in endogenous heat production caused by this method.

These results may have important implications for treatment of hypothermic victims both in the field and in medical facilities. The primary concerns with treatment of hypothermic victims are to arrest the decrease in core temperature, stabilize the cardiovascular and respiratory systems, and to correct metabolic imbalances. As long as the core is being consistently warmed, the rate of rewarming is not of particular importance. First, although the absolute difference in afterdrop between forced-air warming and shivering (0.17°C) is small, and would not be of great importance in mild hypothermia, the relative decrease is actually 30%. If similar results were seen in severely hypothermic nonshivering individuals (T_{co} < 30°C), where afterdrop may be up to 4°C, forced-air warming may attenuate afterdrop by up to 1.5°C. This may prevent a victim's core temperature from dropping below the threshold for cold-induced ventricular fibrillation. [1]

Second, maximal shivering may not always be possible or desirable. Shivering thermogenesis is absent or greatly diminished in victims with severe hypothermia, impaired thermoregulatory control (resulting from old age, alcohol ingestion, head or spinal injury, etc.), and diminished metabolic energy substrates. Furthermore, shivering may be undesirable in selected victims having diminished cardiovascular or respiratory reserves. Since forced-air warming decreases metabolic stress of shivering in mild hypothermia, without attenuating the rewarming rate, it would be useful in the above circumstances.

Finally, in severe hypothermia victims are not shivering and endogenous heat production is likely below 50 W at a core temperature of 27°C. [1] In this case some form of exogenous heat application is required. Forced-air warming provides over 200 W directly to the body surface.

In conclusion, forced air warming has the advantage of minimizing the post-cooling drop in core temperature, decreasing shivering and it's associated metabolic stress, and maintaining an adequate rate of rewarming. Forced-air warming is a safe, simple noninvasive method for the treatment of mild hypothermia and would be very efficient during transportation of a severely hypothermic patient. Warming units are readily available, compact, and could be adapted for use in motorized rescue operations. These systems could therefore be used effectively in the emergency medical facility, and even in some rescue/emergency vehicles (ambulances) or marine vessels (military, Coast Guard etc.).

ACKNOWLEDGMENTS. Supported by Man. Health Research Council and Augustine Med. Inc.

REFERENCES

1. Bristow GK: Clinical aspects of accidental hypothermia, Living in the Cold. Physiological and Biochemical Adaptations. Edited by Heller C, Musacchia XJ, Wang L. Stanford University, Elsevier, 1985, pp 513-522.

2. Giesbrecht GG, Bristow GK, Uin A, Ready AE, Jones RA: Effectiveness of three field treatments for induced mild (33.0°C) hypothermia. J Appl Physiol 63:2375-79, 1987.

3. Sessler DI, Moayeri A: Skin-surface warming: heat flux and central temperature. Anesthesiology 73:218-224, 1990.

4. Hayward JS, Eckerson JD, Kemna D: Thermal and cardiovascular changes during three methods of resuscitation from mild hypothermia Res 113:1-13, 1983.

5. Collis ML, Steinman AM, Chaney RD: Accidental hypothermia: An experimental study of practical rewarming methods Aviat Space Environ Med 48:625-632, 1977.

6. Giesbrecht GG, Bristow GK, Sessler DI, Mekjavic IB: Treatment of immersion hypothermia by direct body-to-body contact. FASEB J 7:A441, 1993.

7. Marcus P: Laboratory comparison of techniques for rewarming hypothermic casualties Aviat Space Environ Med 49:692-697, 1978.

8. Daanen HAM, Van De Linde FJG: Comparison of four noninvasive rewarming methods for mild hypothermia Aviat Space Environ Med 63:1070-76, 1992.

9. Romet TT, Hoskin RW: Temperature and metabolic responses to inhalation and bath rewarming protocols Aviat Space Envriron Med 59:630-634, 1988.

10. Webb P: Afterdrop of body temperature during rewarming: an alternative explanation. J Appl Physiol 60:385-390, 1986.

11. Giesbrecht GG, Bristow GK: A second postcooling afterdrop: more evidence for a convective mechanism. J Appl Physiol 73:1253-58, 1992.

12. Rowell L: Thermal Stress, Human Circulation: Regulation During Physical Stress. London, Oxford University Press, 1986, pp 174-212.

FACTORIAL EFFECTS ON CONTACT COOLING

Fang Chen[1] , Håkan Nilsson, Ingvar Holmér

The National Institute of Occupational Health, Department of Environmental Physiology, S-171 84 Solna, Sweden.

Summary

The effects of different physical factors on contact cooling of the finger were studied. A mathematical model was developed to predict the contact skin temperature change under different circumstances. The study proved that the physical and physiological factors, such as the surface material temperature, the mass of the material, the contact pressure, etc., affected the contact skin temperature change.

Introduction

In the daily life, people using hand to perform some tasks in cold situation is quite common. When some meticulous performance is required, people usually prefer to do it with bare hands. From the protective point of view, it is necessary to have a relevant industrial standard to protect the workers from cold injury. Unfortunately, very little research regarding discomfort and possible risks of cold injury under contact cold exposure is available in the literature. Havenith (1) published an investigation related to contact cold exposure. They considered a combined effect of contact cold and cold air exposure. Lotens (2) developed a mathematical model based on the experimental results of Havenith's study to predict the contact skin temperature change. In this model, the palm and back of hand, as well as fingers were considered as a single thermal unit. This model tried to simplify the complicated thermal activities in the human hand and finger. Some important factors, especially the contact pressure, were ignored in the model. Chen (3, 4, 5) reported some preliminary studies about the contact cold by considering only the thermal activities on one finger pad instead of the whole hand. It was shown that en exponential equation with two components was a good description of the

---

[1] F. Chen is a Ph. D student in the Division of Industrial Ergonomics, University of Technology in Linköping, S-581 83, Linköping, Sweden

contact skin temperature change with time. Still, many physical and physiological factors which could influence the change of contact skin temperature are unknown. In present investigation, several factors, such as the mass of the surface material, the finger contact pressure and the surface temperature were studied.

Method

10 male (age 35.4±8.2 years, height 177.0±7.2 cm, weight 80.0±18.7) and 10 female (age 41.4±7.2 years, height 167.0±3.1 cm, weight 63.3+17.0 kg) were participants in the experiments. All the experiments were performed in a small climate chamber. The small chamber was cooled by liquid carbon dioxide evaporation in the chamber. The gas injection was controlled by a valve via a thermistor and a regulation circuit. The air temperature inside the chamber was stabilized at ±1°C, and the air velocity in the operating zone of the hand was less than 0.4 m/s with a relative humidity of 50%. The surface material was put on a balance inside the small chamber for controlling the contact pressure. The temperature of the material was equal to the air temperature inside the small chamber. The subject inserted his (or her) left hand into the small chamber and pressed on the material surface with the index finger.

During the experiment, three physical factors were taken into consideration. These factors were material surface temperature (T_{sm}, -7, 0, +7 °C), contact pressure (P_c, 0.1,0.6 and 1 Kp), the mass of surface material (the big one was the 11 cm cube of aluminum; the small one was only 0.5 cm thickness of surface layer of aluminum). During the experiments, the subjects wore a T-shirt and long trousers. There were 3 minutes of preparation time before the contact cold exposure started. A combination of all the factors was made as a factorial design for the experiments.

Contact skin temperature (T_{sk}) was monitored by taping the copper-constantan thermocouple, which was only 0.5 mm in diameter, on the finger pad of the index finger of the left hand. The surface material temperature (T_{sm}) was measured 0.3 cm under the surface at the point of contact. Contact T_{sk}, T_{sm}, and the air temperature inside the small chamber were monitored simultaneously at 10 records per second during the first 15 sec of contact, followed by records every 2 seconds or every 4 seconds with an IBM compatible personal computer.

The cold exposure time was interrapted whenever the subject's contact T_{sk} reached 0°C, or when the subject felt intolerable cold or pain, or when the contact time was longer than 5 minutes.

Each of the contact Tsk records was analyzed by regression by using the modified Newtonian model (5)

$$T_{sk} = T_{sk_f} + \Delta T \left(A \cdot e^{-\frac{t}{\tau_1}} + B \cdot e^{-\frac{t}{\tau_1}} \right) \qquad (1)$$

Where T_{sk} is the contact skin temperature, T_{skf} is the contact skin temperature when the contact time is infinitly long, DT is the difference between T_{sk} when contact time is zero and T_{skf}, t is the contact time, A and B are two equation parameters, t1 and t2 are two time constant.

Results

The mathematical model accurately described the contact T_{sk} change versus time with a correlation-coefficient up to 0.97 - 0.99. Figure a and b in Figure 1 show the original records of the contact T_{sk} change under different T_{sm} and the regression curve on the same subject.

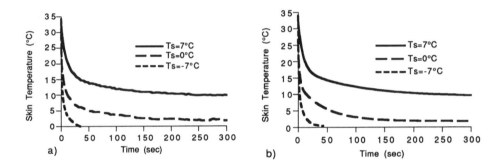

Figure 1. The data of the two panels come from one of the subjects with 1 Kp pressure on the big mass of aluminum surface. a) the experimental records and b) the regression curves.

In equation 1, T_{skf}, parameter A and B, t1 and t2 were constant for each subject under certain conditions of the cold surface contact, but they differed individually. A multiple ANOVA test shows significant effects of different factors in the experiments on these parameters (Table 1).

Table 1. Results of multiple analysis of variance for the parameters of the model

Parameters	Significant factors
T_{skf}	T_{sm}, Mass, P_c, Gender*
A	P_c*
B	T_{sm}, P_c*
t1	Mass*, P_c
t2	P_c, T, Mass

*: $P < 0.05$, others $p < 0.01$

Discussion

The responses to contact cold exposure of the index finger were very repeatable in the same subject. However individual differences were large. The results in Table 1 indicate that nearly all the experimental factors in the design affected the contact cold response of the finger. The analysis of the parameters of the equation also indicates that the physical mechanisms behind the parameters are complex. Some factors have significant effect on nearly all the parameters, such as contact pressure and material surface temperature. The comparison between the experimental records and equation curves (Figure 1 a and b) shows that the mathematical model accurately describes the contact skin temperature changes versus contact time for individual cold exposure records.

It would be desirable to express the values of the parameters of the mathematical model by some linear empirical regression equations based on the experiments. All the condition factors should then be studied during at least three levels in order to find a quantitative effect. Present experiments were not enough to give these empirical equations. Some of the factors, such as surface mass had only two levels in the experimental design. Actually, the environmental factors, as well as the whole body thermal situation should also be taken into consideration (1). The most useful application of an extended model would be to predict the time required for contact skin temperature to reach a certain critical level for any given exposure conditions

Reference

1. Havenith G, van de Linde EJG, Heus R. Pain and thermal sensation and cooling rate of hands while touching cold materials. Europ Journal of Applied Physiology 1992;65:43-51.

2. Lotens WA. Simulation of hand cooling due to touching cold materials. European Journal of Applied Physiology 1992;65:59-65.

3. Chen F. Contact temperature on touchable surfaces. Luleå University, Sweden. 1991,

4. Chen F, Nilsson H, Holmér I. Cooling of finger pad touching different material surface. Proceedings of The Fifith International Conference on Environmental Ergonomics. Maastricht, the Netherlands: TNO-Institute for Perception, 1992:60-61.

5. Chen F, Nilsson H, Holmér I. Cooling responses of finger in contact with an aluminium surface. American Industrial Hygiene Association Journal 1993;(In press)

Temperature Regulation
Advances in Pharmacological Sciences
© 1994 Birkhäuser Verlag Basel

CONVECTIVE AND METABOLIC HEAT IN HUMAN FINGERS

Michel B. Ducharme and Peter Tikuisis
Defence & Civil Institute of Environmental Medicine
CANADA, M3M 3B9

The objective was to investigate the contribution of the convective heat transfer by the blood and the tissue metabolic heat production to the heat loss or gain by the finger during a 3.5 h immersion of the forearm and hand in water at temperatures (Tw) between 20 and 40°C using arterial occlusion of the forearm for the last 30 min. The finger heat loss had decreased during occlusion at $Tw \leq 37$°C but increased at $Tw > 37$°C. At thermal steady state prior to occlusion, the convective heat contributed between 33.3 and 93.8% of the total heat loss from the finger at $20 \leq Tw \leq 30$°C; the remaining being the tissue metabolic heat. At $38 \leq Tw \leq 40$°C, the environment contributed between 88.7 and 96.7% of the total finger heat gain; the remaining being the tissue metabolic heat.

INTRODUCTION

The blood flow in the extremities is sensitive to local skin temperature, to skin temperature elsewhere on the body and to central body temperature. There have been several descriptions of the effects of the general thermal condition of the body and of the temperature of the extremities on the extremities blood flow. Fewer studies have investigated, however, the thermal contribution of the convective heat transfer between the blood and the extremities' tissue to the heat transfer to or from the environment. Previous results have shown that during local (hand) and systemic cold stress, between 70 and 96% of the tissue heat loss from the fingers and hand originated from the convective heat transfer with the blood, the balance originating from the metabolic heat production of the tissue (1, 2, 3, 4). One objective of the present study was to quantify the contribution of the blood in the finger as heat source or heat sink during a wider range of thermal stresses in water, while the rest of the body was thermally comfortable.

Bazett and collaborators (5) were among the first authors to report a precooling of the arterial blood in passage to the extremities under cold environmental conditions. The resulting low cold blood perfusing the extremities is responsible for the near zero convective heat transfer with the tissue causing uncomfortable extremities and possible cold injuries. A second objective of the present study was to investigate if forearm subcutaneous fat could play a protective role for the extremities by improving the convective heat transfer with the blood.

MATERIALS AND METHODS

Subjects. Fourteen healthy male subjects, aged 18-40 years, volunteered to participate in the study. They were divided in two groups: 11 lean (height: 177.3 ± 1.8 cm, mass: 75.5 ± 2.7 kg) and 3 obese subjects (176.5 ± 3.8 cm, 93.1 ± 3.0 kg). Skinfold thickness values were measured to the nearest 0.1 mm by using a Harpender caliper (British Indicator Ltd., England), and represent an average of six measurements taken around the circumference of three segments of the forearm for the 14 subjects. These segments were the distal, the medial and the proximal forearm. The distal segment passes through the radio-carpal joint at the distal end of the forearm, the proximal segment was located 9 cm distal from the olecranon process, and the medial segment was at mid-distance between the distal and the proximal segment of the forearm.

Heat flux and temperature measurements. The heat fluxes from the skin of the finger ($\dot{H}sk$) were continuously monitored during the experiments with 4 waterproofed heat flux transducers [HFTs; model FM-060 (19 mm x 8 mm x 2 mm), Concept Engineering, Old Saybrook, CT] fixed to the skin with surgical tape (Blenderm, 3M, St-Paul, MN). The HFTs were placed equidistant from each other along the circumference of the middle phalanx of the middle finger so that each of the sensors represented approximately an equal volume of that finger. Each HFT was recalibrated before use by a method described elsewhere (6) and measurements were corrected to account for the thermal insulation of the HFT (6). Values reported herein are a simple average of the 4 sites.

The rectal temperature (*Tre*) was measured using a calibrated thermistor probe inserted 15 cm beyond the anus. All heat flux and temperature data were recorded continuously during the experimental sessions by means of a computer-controlled data acquisition system (scanner HP-3497A, computer HP-85, Hewlett-Packard). Mean values over 1-min periods were calculated for all measured variables.

Blood flow measurements. Forearm + hand blood flow ($\dot{Q}$) was measured at steady-state during the water immersion just before the arterial occlusion by the plethysmograph Whitney gauge technique using a mercury-in-Silastic circumference gauge (Parks Electronics Laboratory, Aloha, OR). The gauge was calibrated before each experiment using an accurate custom-made linear calibrator (± 2 μm). For each series of $\dot{Q}$ measurements, a mean value was calculated from five measurements over a 10-min period.

Experimental procedure. Subjects reported to the laboratory on two occasions within a two-week period. Each subject underwent two forearm + hand immersions at two different temperatures: 20 and 38°C. One of the lean subjects underwent 6 additional immersions at 30, 33, 36, 37, 39 and 40°C. Before each immersion, the lightly dressed subjects (T-shirt and casual pants) sat comfortably under thermoneutral conditions (air temperature of 25.0 ± 0.2°C, relative humidity of 40 ± 2%) for 1 hour. During the last 15 min of the pre-immersion period, skin heat loss and rectal temperature were continuously measured. After this period, the subject immersed his left forearm and hand for 3 hours in a well-stirred water bath maintained at a preselected temperature. At the end of the 3-h immersion, an arterial cuff, previously fixed on the left upper arm was inflated between 200 and 250 mmHg for 30 minutes while the forearm and hand were still immersed in water.

Statistical analyses. The student's t-test for the significance of the skinfold thickness among groups was performed using StatViewII Statistical Programme (Abacus Concepts Inc., Berkeley, CA, 1987). The heat flux and rectal temperature data for the two groups of subjects were analysed by a three-factor (2 within and 1 between) analysis of variance for repeated measures using SuperAnova Statistical Programme for General Linear Modelling (Abacus Concepts Inc., Berkeley, CA, 1989). Where applicable, the data are presented as means ± SE. The level of statistical significance was set at $p < 0.05$.

RESULTS

The skinfold thickness of the medial and proximal forearm of the lean subjects were significantly smaller (0.44 ± 0.02 and 0.45 ± 0.02 cm) than for the obese subjects (1.09 ± 0.01 and 1.02 ± 0.13 cm), and no difference was observed for the distal forearm between the two groups.

No shivering or sweating activity was observed during any of the forearm immersions, and *Tre* did not change significantly during the 3.5 h immersions at any temperature. No difference in

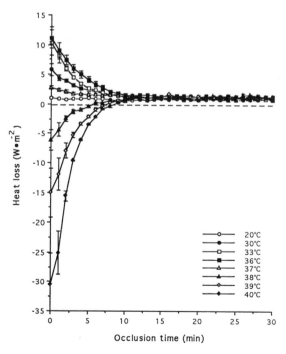

Fig.1. Temporal evolution of the mean (± SE) finger heat loss from a lean subject during the occlusion period following 3 hours of immersion at different water temperatures.

Tre was observed between the lean and the obese groups; the average *Tre* being 37.02 ± 0.05°C. The finger heat losses achieved thermal stability at the end of the 3-h immersion for every water temperature tested.

Figure 1 shows the temporal evolution of the mean finger heat loss from the lean subject tested during the occlusion period for *Tw* ranging between 20 and 40°C. For *Tw* ≤ 37°C, $\dot{H}sk$ decreased during the 30 min occlusion period to an average of 1.05 ± 0.07 W•m^{-2}, and for 37 < *Tw* ≤ 40°C, $\dot{H}sk$ increased to a higher average value of 1.28 ± 0.01 W•m^{-2}. $\dot{H}sk$ reached thermal steady state after ~ 10 min of occlusion for all water temperature tested in all subjects. At thermal steady state during the arterial occlusion, the average $\dot{H}sk$ were assumed to equal the tissue metabolic heat production of the finger, since conductive heat transfer has been found to be negligible in the finger when the forearm is immersed (7).

Figure 2 shows the effect of water temperature on the finger heat transfer at steady state after 3-h of immersion, and the tissue metabolic heat production at steady state during the arterial occlusion for the same lean subject as in Fig. 1. The average metabolic heat production of the finger tissue

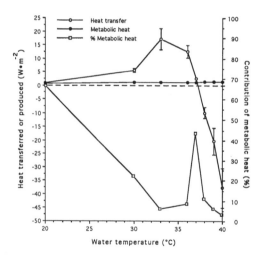

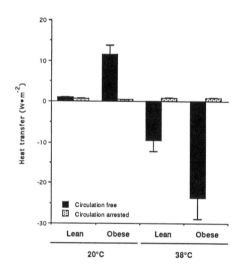

Fig. 2. The effect of water temperature on the mean finger heat transfer, mean finger metabolic heat and mean percentage contribution of the tissue metabolic heat to the total heat transfer from the finger at steady state following 3 hours of immersion for a lean subject. Mean ± SE.

Fig. 3. Finger heat transfer at Tw = 20 and 38 °C for the lean (n = 11) and obese groups (n = 3) during free and arrested circulation. Mean ± SE.

has a tendency to increase with temperature from 0.78 W•m^{-2} at *Tw* = 20°C to 1.3 W•m^{-2} at 40°C due to the Q_{10} effect. The finger heat loss increases with *Tw* between 20 and 33°C from 0.80 ± 0.10 W•m^{-2} to 17.08 ± 2.5 W•m^{-2}, and then decreases with *Tw* between 33 and 37.2°C to 0 W•m^{-2}. For *Tw* above 37.2, the finger gained heat from the environment with a rate that increased linearly with *Tw*.

Figure 2 also shows the effect of the water temperature on the percentage contribution of the tissue metabolic heat to the total heat transfer to or from the finger at steady state. The contribution of the finger metabolic heat to the finger heat loss decreases linearly for $20 \leq Tw \geq 33$ from 66.7 to 6.2% following by an increase to 43.7% at $Tw = 37.0°C$. The remaining portion of the finger heat loss is attributed to the convective heat transfer between blood and tissue. For $37 \leq Tw < 40°C$, the percentage contribution of the finger metabolic heat to the total finger heat gained decreases to 3.4%; the major portion of the heat gained being attributed to the environment.

Figure 3 compares the finger heat transfer at $Tw = 20$ and $38°C$ for the lean and obese groups during free circulation (steady state after 3 hours of immersion) and arrested circulation (steady state during the arterial occlusion period). The finger heat loss during arrested circulation did not differ significantly among temperatures or groups. The finger heat losses during free circulation were larger for the obese group (more heat loss at $Tw = 20°C$ and more heat gain at $Tw = 38°C$) than for the lean group. This can be, at least partially, attributed to a larger forearm + hand blood flow at steady state after 3 hours of immersion for the obese group at both $Tw = 20°C$ (2.16 ± 0.37 ml•min^{-1}•100 g of tissue^{-1}) and $Tw = 38°C$ (6.60 ± 0.24 ml • min^{-1} • 100 g tissue^{-1}) compared to the lean group (1.06 ± 0.08 ml•min^{-1}•100g tissue^{-1} at $Tw = 20°C$; 4.78 ± 0.49 ml • min^{-1} • 100g tissue^{-1} at $Tw = 38°C$).

DISCUSSION

The present study shows that at thermal steady state during immersion of the forearm and hand at Tw ranging between 20 and 40°C, the contribution of the convective heat transfer between the blood and the finger's tissue varies widely from 56% during cold stress to 97% during heat stress; the remaining originates from the tissue metabolic heat production. These results for the high Tw range are supported by Rennie (4) who reported that nearly 96% of the finger heat loss at $Tw = 35°C$ is attributed to convective heat transfer with blood. For the low Tw range, however, the present study does not support the conclusions of Raman and Vanhuyse (3) who observed that 76% of the extremities heat loss originates from the convective heat transfer between blood and tissue at $Tw = 20°C$. This large contribution could be explained by a major conductive heat transfer component between the forearm and the hand, and a larger temperature difference between the arterial blood entering the extremities and the tissue temperature of the extremities since only the hand and not the forearm and hand was immersed as in the present study. Greenfield et al. (1) also reported a large contribution of the convective heat transfer (86 to 92%), but in this case the immersion temperatures ranged between 0 and 6°C, which prompted a cold induced vasodilatation response that lead to the large contribution.

The present study also shows a significant increase in the contribution of the convective heat transfer in the finger to the total finger heat loss or heat gain for the obese group compared to the lean group as shown in Fig. 3. The increase is mainly attributed to an enhanced limb perfusion for the obese group despite the similar Tre in both groups. The larger subcutaneous fat thickness in the forearm of the obese subjects did not lower the forearm heat loss at $Tw = 20°C$ as expected when compared to the lean group. The mean forearm heat loss for the obese group was an average of 82% larger than for the lean group (unpublished results). This is possibly due to the two fold increase of the forearm and hand blood flow in the obese group which enhanced the convective heat transfer between the blood and the forearm' tissue. The larger limb blood flow in the obese group during the partial cold water immersions can be a way for these subjects to thermoregulate their body temperature by using the least resistant pathway for heat loss.

In conclusion, the results of the present study suggest that during immersion of the forearm and hand in water, the blood in the finger has a role of heat source for $Tw \leq 37°C$ and a role of heat sink for $Tw > 37°C$. The contribution of the convective heat transfer to the total finger heat loss or gain is not larger than the tissue metabolic heat at $Tw = 20°C$, but becomes major for $Tw \geq 30°C$.

During immersion of the upper limb in cold water while the rest of the body is at thermal neutrality, the limb blood flow and the subsequent heat losses from the extremities increase with increased forearm subcutaneous fat thickness. This suggests that forearm insulation might be an important factor in the protection of the fingers against cold stress through enhanced blood flow.

REFERENCES

1. Greenfield, A.D.M., J.T. Shepherd, and R.F. Whelan. The loss of heat from the hands and from the fingers immersed in cold water. J. Appl. Physiol. 1951; 112: 459-475.

2. Hong, S.K., C.K. Lee, J.K. Kim, S.H. Song, and D.W. Rennie. Peripheral blood flow and heat flux of korean women divers. Fed. Proc. 1969; 28(3): 1143-8.

3. Raman, E.R., and V.J. Vanhuyse. Temperature dependence of the circulation pattern in the upper extremities. J. Physiol. 1975; 249: 197-210.

4. Rennie, D.W. Body heat loss during immersion in water. In: Human Adaptation and its Methodology (Proc. Internatl. Symp. Environmental Physiol., Kyoto, Japan, Sept. 13-17, 1965), Japan Society for the Promotion of Science 1967: 36-41.

5. Bazett, H.C., L. Love, M. Newton, and C. Eisenberg. Temperature changes in blood flowing in arteries and veins in man. J. Appl. Physiol. 1948; 1: 3-19.

6. Ducharme, M.B., J. Frim, and P. Tikuisis. Errors in heat flux measurements due to the thermal resistance of heat flux disks. J. Appl. Physiol. 1990; 69(2): 776-784.

7. Ducharme, M.B., and P. Tikuisis. The role of blood as heat source or heat sink in human limbs during thermal stress. J. Appl. Physiol. (In press).

FROM FOETUS TO NEONATE - IMPLICATIONS FOR THE ONTOGENY OF THERMOREGULATION

Helen Laburn, Duncan Mitchell and Kathleen Goelst

Department of Physiology, University of the Witwatersrand Medical School, 7 York Road,
Parktown, Johannesburg,
South Africa

Summary

Thermoregulatory defences against cold and heat are well developed in the newborn of many species at birth. Measurements made on foetal body temperature and the foetal thermal environment in sheep suggest that the development of thermoregulatory sensitivity and effector mechanisms is not dependent on thermal cues *in utero*. Thermoregulation may be suppressed in the foetus by a placental factor.

Introduction

At birth, the mammalian foetus makes the transition from the thermally stable, liquid environment of the uterine cavity to a thermally labile, air environment. In a typical birth environment, the neonate, at the moment of birth, experiences a fall in environmental temperature of some 12 - 15°C for most humans, or more than 30°C for some animals born in the field (1). The air environment and the newborn's high surface area to body mass ratio accelerate heat loss. Evaporation occurs from the wet body surfaces and heat is readily conducted away from the body core to skin (2). Per unit body mass, the heat loss of a newborn human infant is four times that of an adult in the relatively innocuous environment of 28°C (3). The challenge to the thermoregulatory system is greater than that which most mammals will ever encounter again. The fact that the challenge seldom is lethal implies that the neonate has adequate thermoregulatory competence at birth. How the foetus acquires the competence to deal with a thermal environment to which it never has been previously exposed is the subject of our review.

Thermoregulatory competence at birth

The body temperature of the neonates of many species falls precipitously after birth. Human babies, may undergo falls of body temperature of 2° to 3°C in the first hour of after birth The most dramatic falls occur in the first few minutes after delivery (4). Some recorded falls may not reflect the real change from the temperature *in utero*; the records start some minutes after delivery (see for instance reference 1) by which time, in cold conditions, the neonate's temperature is likely to have already fallen considerably, so that the actual extent of the post-partum fall would be underestimated. What is required is uninterrupted measurement of temperature through the birth process, only one such measurement has been reported.

Unless small body mass constitutes too large a predisposition to heat loss in very cold environments (1), the fall in body temperature is arrested by thermoregulatory responses available to the full-term neonate. Heat conservation occurs through vasoconstriction (3, 5) and is aided by fur or wool if present, and by appropriate behaviours by both mother and neonate (6). In most species, however, lethal post-birth hypothermia is prevented by a substantial increase in heat production (7, 8), largely from non-shivering thermogenesis (NST) in brown adipose tissue (BAT), with which the newborn of most species are endowed (9).Shivering too may be present in some species. Thyroid hormones (10) and sympathetic nerve activation (11) appear to be necessary for normal NST, so these components of the endocrine and nervous systems must be competent too at birth. In thermoneutral conditions, minimal metabolic rate of the neonate is similar to that of the foetus (about twice that, per kilogram, of the adult) (8), but in response to stimulation of cutaneous or body core thermoreceptors, after birth, metabolic rate rises. If adjusted for surface area or body mass, the metabolic rate of the newborn within hours of birth is two to three times that of an adult of the same species (7, 12). By the end of gestation, therefore, mammals have the machinery to detect a fall in body temperature, and to arrest it by appropriate behaviour, vasoconstriction, and enhanced metabolic heat production.

Though avoiding hypothermia is the typical thermoregulatory competence required of a mammal at birth, in hot environments neonates can avoid hyperthermia too. Figure 1, from the work of Mercer et al. (8), shows rectal temperatures, oxygen consumption and respiration rates of lambs only 3 hours old exposed to low and high ambient temperatures. Above ambient temperatures of about 35°C, there is a steep rise in respiratory frequency such that panting is well-established by four days of age (13). Human neonates only 4 to 12 hours old can more than double evaporative heat loss, due to active sweating (14). Neonates, therefore, also are competent to detect a rise in body temperature, and to implement the corrective thermoregulatory effectors of panting and sweating.

The thermoregulatory effectors employed immediately after birth are not invoked *in utero*. The foetus is isolated from the external environment; it relies on its mother for dissipation of the metabolic heat it generates, most heat exiting via the umbilical/placental circulation, and the rest through the amniotic cavity and uterine wall (15). Equilibrium between foetal heat production and heat loss occurs when foetal body temperature is about 0.5°C above that of the mother. The temperature of the foetus's immediate environment, the amniotic cavity, is approximately mid-way between that of foetus and mother animal (16, 17). Such a thermal relationship, which has been observed in all species studied so far, including humans (4, 18). It is not self-evidently conducive, however, to preparing the foetus to meet its post-natal thermoregulatory demands.

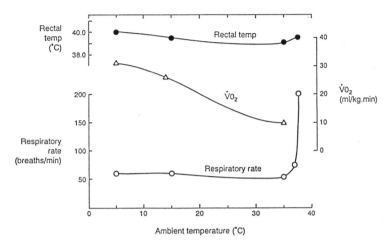

Fig. 1. Mean values for measured at different ambient temperatures (abscissa) in 3-hour old lambs. Graphs redrawn from Mercer et al.1979 (8).

Perinatal plasticity in thermoregulation

Although the changes are not as great as those experienced by the neonate at birth, mammals experience enduring changes in their thermal environment. Such changes may occur through changes in season, habitat, or in exercise pattern. Following such changes, the sensitivity of the thermoregulatory system may alter, and the mammal may develop enhanced capacity for particular thermoregulatory effectors, through the process of adaptation (19). Moreover, the thermoregulatory system can be primed in anticipation of an imminent change in the thermal environment, by the prior application of appropriate thermal stress (19).

How neonatal thermoregulation would be affected by enduring changes in the *in utero* thermal environment is not known, but the question already is being investigated (20). Direct manipulation of the foetal thermal environment has been achieved so far in acute experiments only (15, 21, 22, 23, 24), in which the foetuses were not brought to term, so that the consequences for neonatal thermoregulatory competence are unknown. Raising maternal body temperature, and indirectly that of the foetus, has deleterious effects on foetal skeletal formation and intra-uterine growth (25), both of which could adversely affect the neonate's ability to thermoregulate post-partum, but no-one has investigated thermoregulatory control mechanisms in such newborn. Winter-shearing of pregnant ewes enhances neonatal BAT thermogenic activity *in vitro*, and causes a greater increase in metabolism of newborn lambs at birth, compared to lambs from unshorn ewes (20). These environment-induced changes in metabolic heat production do provide evidence for plasticity in foetal thermoregulation, that is, the adjustment to a new stable state of thermoregulation appropriate for prevailing circumstances.

Soon after birth, young mammals exhibit profound plasticity in thermoregulation. Compared to those reared at 20°C, rabbit pups raised from birth at 33°C show significantly greater drops in body temperature when exposed to 2°C and also reduced thermogenic responses to both pyrogen and noradrenaline (26). In similar experiments, warm-reared rat pups also were unable to maintain body temperature in the cold; in addition, they showed an inverted response to noradrenaline administration (27). The thermoregulatory characteristics of warm-reared rats do not appear to arise from deficiencies in their ability to sense the cold environment (28), so they are likely to arise from changes in thermoregulatory integration or effector activity.

The intra-uterine environment is several degrees hotter than the environments in which the warm-reared rabbit and rat pups were raised. How is it, then, that neonatal mammals are born seemingly competent to cope with cold stress? Despite the thermal clamp between foetus and mother, do variations in the intra-uterine environment occur which could mimic, at least in part, the thermal stimuli with which the newborn organism will have to contend after birth? If the neonate, within hours of birth, displays thermal autonomy, does the foetus also do so, in response to any changes in its thermal environment?

Thermal challenges to the foetus

Using the techniques of radio-telemetry which we have described previously (17) we have measured the first simultaneous measurements of foetal and maternal temperature made for a period of over a month (Fig. 2). the measurements were made on 7 pregnant ewes and their foetuses for the last 34 days prior to lambing. The ewes were kept at 22°C and the temperatures of ewes and foetuses were measured daily at 09:00. Foetal body temperature was approximately 39.5°C throughout the monitoring period and varied little from day-to-day on average. What variations did occur, were due to variations in maternal body temperature, because, as Figure 2 shows, the temperature difference between foetus and mother (F-M) was held constant at approximately 0.6°C except for the immediate post-surgical period, and final days of pregnancy. There is little evidence, therefore, in the normal body temperatures of foetal lambs for foetal thermoregulatory autonomy, nor for the kind of deviations in foetal temperature which might prime the thermoregulatory system for neonatal life.

Figure 3 shows variations in foetal and maternal body temperatures in a 24-hour period, when the foetus was 130 days old approximately. The foetal environmental temperature rose and fell in a circadian pattern, as the mother animal exhibited a typical 24-hour body temperature variation. Foetal body temperature appeared to follow the mother's temperature passively, but with lower amplitude of variation than that of the mother. The temperature of the foetus' environment, the amnion, changed by a mean of 0.76 ± 0.05°C in the 24-hour period. The amplitude of the variation in foetal temperature, about 0.3°C, was scarcely likely to challenge the foetus' thermoregulatory system.

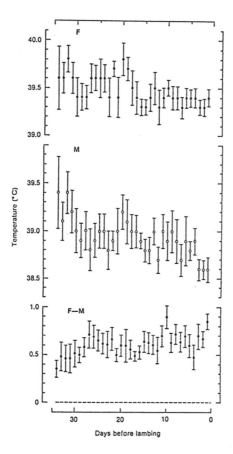

Fig. 2. Foetal (F) and maternal (M) temperatures and foeto-maternal (F-M) temperature difference measured in 7 ewes and their foetuses in late pregnancy. Each point is mean ± SEM. Lambing occurred at day 0. Dotted line at zero °C indicates foetal body temperature = maternal body temperature. From Laburn et al. (17) by copyright permission of the American Physiological Society.

Our results confirmed that the thermal environment of the foetus is largely dependent on maternal body temperature and is remarkably constant, both on a day-to-day basis over the last month of gestation at least, and during a 24-hour period. The priming thermal environment of the developing foetus therefore is uniform and hot, at least when the mother is at rest in a neutral ambient temperature.

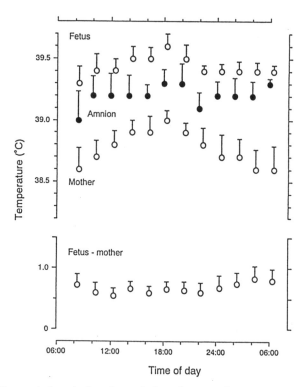

Fig. 3. Circadian variations in foetal, amniotic and maternal temperatures, and difference in foeto-maternal temperatures (F-M) measured over a 24-hour period in late gestation in 7 ewes and their foetuses. Each point is mean ± SEM.

Changes in maternal temperature, could lead to warmer or colder surroundings for the foetus. We were curious to find out how much the foetal environment varied during mild hot and mild cold exposure of the mother animal, of the kind ewes might experience in their natural habitats, and to discover whether such maternal thermal stress presented thermoregulatory challenges for the foetus.

Figures 4 and 5 show the effects of exposing 7 ewes to a heat or cold stress respectively. During 4 hours of exposure to 35°C, 40% relative humidity, ewe body temperature rose, and so did that of the foetus, but foetal body temperature rose less than did maternal temperature, as was evident in the significant decrease in the foeto-maternal temperature difference (F-M) within the first 40 minutes of exposure. The fall in F-M could have arisen because foetal body temperature rose at a slower rate than did maternal body temperature, as a result of foetal thermal inertia. Alternatively, it is possible that, despite the prevailing circumstances, the foetus lost more heat via the umbilical/placental circulation or via the myometrium. When we exposed the ewes to 4°C for 4 hours (Figure 5) we found that, as ewe body temperature fell, foetal body temperature fell to a lesser extent and the foeto-maternal temperature difference tended to widen. It is unlikely that foetal thermal inertia would account for the widening, because of the slow rate of change in temperature; rather we postulated that vasoconstrictor responses in maternal tissues, in response to the cold, affected utero-placental blood flow, decreasing heat loss from the foetus. Whatever the mechanisms, during maternal exposure to both hot and cold environments, circumstances which might have stressed the foetus, the foetus was protected from, rather than exposed to, the thermal challenge.

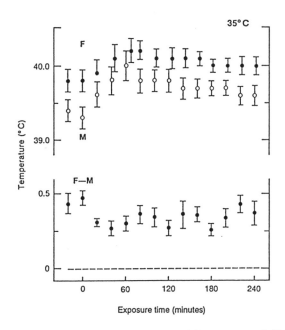

Fig. 4. Foetal (F) and maternal (M) temperature and foeto-maternal (F-M) temperature difference in ewes exposed to 35°C for 4 hours commencing at time 0. Each point is the mean ± SEM for 7 ewes and their foetuses.

From our radio-telemetric of foetal lamb temperature, we conclude that the development of thermoregulatory defence mechanisms in preparation for neonatal life apparently does not require thermal challenge *in utero*. If the foetus is primed by the uterine thermal environment, one would expect the neonate to be pre-adapted to heat stress, although the neonate rarely would be required to utilise its full capacity for active heat loss. There is much to indicate, however, that the foetus actually develops a sensitivity to cold *in utero*; responses to the cold stimuli accompanying birth are initiated within hours of delivery, too soon to be the result of post-partum adaptation processes (29). Indeed, sensitivity to a fall in skin or core temperature is present in the foetus some time prior to parturition because NST can be induced by decreases in foetal temperature, in appropriate circumstances, in foetal brown fat (25).

It would seem, therefore, that the thermoregulatory competence evident in the neonate at birth develops in the absence of thermal challenges to the foetus. Moreover, the conditioning effect of the prevailing thermal environment, which in the neonate, apparently leads to the suppression

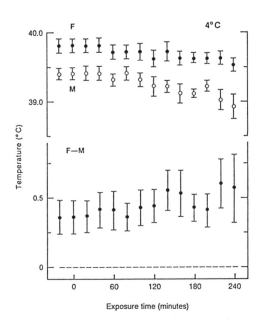

Fig. 5. Foetal (F) and maternal (M) temperatures and the foeto-maternal temperature difference (F-M), in ewes exposed to 4°C for 4 hours. Other details as in Fig. 4.

of thermoregulatory mechanisms which are not employed, is suppressed in the foetus. We need to look further than the thermal environment of the foetus to understand the ontogeny of thermoregulation.

Humoral control of the developing thermoregulatory system

Studies of the ontogeny of the brown fat system reveal that the foetal thermoregulatory system develops not only, or perhaps not at all, under the influence of its thermal context, but under humoral control emanating from the placenta. Brown adipose tissue is present in the last trimester of gestation at least, in sheep (30) but is not activated until after birth, an arrangement which both prepares the foetus for neonatal life and minimises the risk of noxious hyperthermia *in utero*. Gene expression for uncoupling protein, the protein responsible for thermogenesis in BAT (31), also is evident *in utero*, and peaks at or before birth (30).

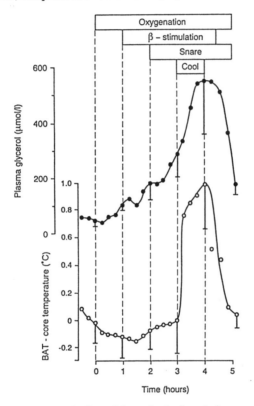

Fig. 6. Activity in foetal BAT as indicated by a rise in foetal plasma glycerol concentrations(°)and differences in BAT temperature versus core body temperature(_) in foetal lambs. Means and SEMs are shown for 5 foetuses. Bars at top indicate the sequence and duration of procedures on the foetuses; ventilation with oxygen, sympathetic stimulation, snaring of the umbilical cord and body cooling. Redrawn from Gunn et al. (25).

In their series of elegant experiments, Gunn et al. (23) have shown the circumstances in which NST can be induced experimentally in foetal lambs. Figure 6 shows the prerequisites for initiation of NST *in utero*; oxygenation of tissues, sympathetic nervous system stimulation, cold stimulation of thermoreceptors, and removal of a placental NST-inhibitor (23), possibly adenosine (24) or prostaglandins (32). Only one of the prerequisites is a thermal one. Critical to the transition from foetal to neonatal status is the presence, *in utero,* of a placental humoral factor controlling thermogenesis in the foetus. Enhanced BAT activity in neonatal lambs after maternal cold exposure may be mediated through maternal or foetal hormonal changes (20).

We suggest that during ontogeny, the foetus develops, in a genetically determined way, the full range of thermoregulatory competencies that the neonate needs, to cope with environmental heat and cold stress. Amongst these competencies would be the ability to increase metabolism, to alter vasomotor tone, to pant or to sweat. None of these effectors is appropriate for foetal thermoregulation, and their implementation *in utero* may be suppressed by a humoral factor or factors, possibly emanating from the placenta. Once the placental link is broken, at birth, and the humoral suppression lifted, the effectors become available.

Conclusions

The normal newborn of most species have thermoregulatory abilities adequate to regulate body temperature at birth, a process which imposes a thermal stress more severe than adults of the species are likely to encounter. Some species (9) activate brown fat thermogenesis in the initial hours and days post-partum, in response to a fall in cutaneous and/or core body temperature. If circumstances so demand, some neonates also have the ability to activate heat losing strategies, such as panting (8) and sweating (14), within hours of birth. The development of sensitive and controlled thermal responses to the relatively cold and dry air environment into which neonates usually are born occurs in the apparent absence of thermal cues *in utero.* We have shown that the thermal environment of the foetal lamb is dictated by maternal body temperature, and varies little from day to day in the last quarter of gestation at least. The foetal temperature has a circadian variation of less than $0.8^{o}C$. When the pregnant animal is exposed to mild warm or cold conditions, the foetus' own thermal inertia, or thermoregulatory strategies used by the mother animal in these circumstances, appear to minimize changes in the foetal thermal environment.

Chronic exposure of the foetus to the high environmental temperature *in utero* does not impair the newborn's ability to respond to cold stress; the plasticity evident in neonates is absent in the foetus. We speculate that a placental factor may suppress activation of all thermoregulatory effectors in the foetus.

Acknowledgements

The Council of the University of the Witwatersrand, together with the Foundation for Research Development funded our work. We thank Debbie Steinberg for assistance with literature searches, and preparation of the text.

References

1. Barlow RM, Gardiner AC, Angus KW, et al. Clinical, biochemical and pathological study of perinatal lambs in a commercial flock. Vet Rec 1987;120:357-362.

2. Adamsons K. The role of thermal factors in fetal and neonatal life. Pediatr Clin North Am, The Newborn I 1966;13:599-619.

3. Brück K. Temperature regulation in the newborn infant. Biol Neonate 1961;3:65-119.

4. Adamsons K, Towell ME. Thermal homeostasis in the fetus and newborn. Anesthesiology 1965;26:531-548.

5. Mount LE. Radiant and convective heat loss from the new-born pig. J Physiol (Lond) 1964;173:96-113.

6. Hull D, Hull J, Vinter J. The preferred environmental temperature of newborn rabbits. Biol Neonate 1986;50:323-330.

7. Dawes GS, Mott JC. The increase in oxygen consumption of the lamb after birth. J Physiol (Lond) 1959;146:295-315.

8. Mercer JB, Andrews JF, Székely M. Thermoregulatory responses in new-born lambs during the first thirty-six hours of life. J Therm Biol 1979;4:239-245.

9. Symonds ME, Lomax MA. Maternal and environmental influences on thermoregulation in the neonate. Proc Nutr Soc 1992;51:165-172.

10. Himms-Hagen J. Brown adipose tissue thermogenesis: role in thermoregulation, energy regulation and obesity. In Schlatent effectors were there Schönbaum E, Lomax P, eds: Thermoregulation; Physiology and Biochemistry. New York: Pergamon Press Inc,1990;327-414.

11. Polk DH. Thyroid hormone effects on neonatal thermogenesis. Semin Perinatol 1988;12:151-156.

12. Gunn TR, Gluckman PD. The endocrine control of the onset of thermogenesis at birth. In Jones CT, ed: Baillière's Clinical Endocrinology and Metabolism. London: Baillière Tindall,1989;869-886.

13. Symonds ME, Andrews DC, Johnson P. The control of thermoregulation in the developing lamb during slow wave sleep. J Dev Physiol 1989;11:289-298.

14. Hey EN, Katz G. Evaporative water loss in the new-born baby. J Physiol (Lond) 1969;200:605-619.

15. Gilbert RD, Schröder H, Kawamura T, Dale PS, Power GG. Heat transfer pathways between fetal lamb and ewe. J Appl Physiol 1985;59:634-638.

16. Gunn TR, Gluckman PD. Development of temperature regulation in the fetal sheep. J Dev Physiol 1983;5:167-179.

17. Laburn HP, Mitchell D, Goelst K. Fetal and maternal body temperatures measured by radiotelemetry in near-term sheep during thermal stress. J Appl Physiol 1992;72:894-900.

18.   Wood C, Beard RW. Temperature of the human foetus. J Obstet Gynaecol Br Commonw 1964;71:768-769.

19.   Brück K, Zeisberger E. Adaptive changes in thermoregulation and their neuropharmacological basis. In Schönbaum E, Lomax P, eds: Thermoregulation; Physiology and Biochemistry. New York:Pergamon Press Inc, 1990;255-307.

20.   Symonds ME, Bryant MJ, Clarke L, Darby CJ, Lomax MA. Effect of maternal cold exposure on brown adipose tissue and thermogenesis in the neonatal lamb. J Physiol (Lond) 1992;455:487-502.

21.   Gluckman PD, Gunn TR, Johnston BM. The effect of cooling on breathing and shivering in unanaesthetized fetal lambs *in utero*. J Physiol (Lond) 1983;343:495-506.

22.   Schröder HJ, Hüneke B, Klug A, Stegner H, Carstensen M, Leichtweiß H-P. Fetal sheep temperatures in utero during cooling and application of triiodothyronine, norepinephrine, propranolol and suxamethonium. Pflügers Arch 1987;410:376-384.

23.   Gunn TR, Ball KT, Power GG, Gluckman PD. Factors influencing the initiation of nonshivering thermogenesis. Am J Obstet Gynecol 1991;164:210-217.

24.   Sawa R, Asakura H, Power GG. Changes in plasma adenosine during simulated birth of fetal sheep. J Appl Physiol 1991;70:1524-1528.

25.   Gericke GS, Hofmeyr GJ, Laburn H, Isaacs H. Does heat damage fetuses? Med Hypotheses 1989;29:275-278.

26.   Cooper KE, Ferguson AV, Veale WL. Modification of thermoregulatory responses in rabbits reared at elevated environmental temperatures. J Physiol (Lond) 1980;303:165-172.

27.   Ferguson AV, Veale WL, Cooper KE. Evidence of environmental influence on the development of thermoregulation in the rat. Can J Physiol Pharmacol 1981;59:91-95.

28.   Dawson NJ, Hellon RF, Herington JG, Young AA. Facial thermal input in the caudal trigeminal nucleus of rats reared at 30°C. J Physiol (Lond) 1982;333:545-554.

29.   Brück K. Thermoregulation: control mechanisms and neural processes. In Sinclair JC, ed: Temperature Regulation and Energy Metabolism in the Newborn. :Grune & Stratton, Inc, 1978;157-185.

30.   Casteilla L, Champigny O, Bouilland F, Robelin J, Ricquier D. Sequential changes in the expression of mitochondrial protein mRNA during the development of brown adipose tissue in bovine and ovine species. Biochem J 1989;257:665-671.

31.   Cannon B, Nedergaard J. The biochemistry of an inefficient tissue: Brown adipose tissue. Essays Biochem 1985;20:110-164.

32.   Gunn TR, Ball KT, Gluckmann PD. Withdrawal of placental prostaglandins permits thermogenic responses in fetal sheep brown adipose tissue. J Appl Physiol 1993:74:998-1004.

EFFECTS OF SEVERAL FACTORS ON THE ENLARGEMENT OF BROWN ADIPOSE TISSUE

Hitoshi Yamashita[1], Tomomi Ookawara[1], Takako Kizaki[1], Mikio Yamamoto[2], Yoshinobu Ohira[3], Toru Wakatsuki[2], Yuzo Sato[4] and Hideki Ohno[1]

Departments of [1]Hygiene and [2]Biochemistry II, National Defense Medical College; [3]Department of Physiology and Biomechanics, National Institute of Fitness and Sports; [4]Research Center of Health, Physical Fitness and Sports, Nagoya University, Japan

Summary

To elucidate the mechanism of brown adipose tissue enlargement, the effects of several factors on the *in vitro* growth of rat brown adipocyte precursor cells (RBAC) and bovine capillary endothelial cells (BCEC) were investigated. In the primary cell culture, norepinephrine (NE) and insulin significantly enhanced the RBAC growth, although they did not stimulate the BCEC growth. In addition, NE and insulin appeared to increase the expression of basic fibroblast growth factor (bFGF) mRNA in RBAC. Considering that bFGF actually stimulated the RBAC growth, NE and insulin may contribute to the RBAC growth through bFGF production by the autocrine mechanism.

Introduction

During chronic exposure to cold, brown adipose tissue (BAT), which is a principal energy source of non-shivering thermogenesis, extremely grows and enhances the thermogenic function[1]. BAT thermogenesis is regulated by the sympathetic nervous system via direct noradrenergic innervation of brown adipocytes. Insulin also plays an important role in the regulation of BAT thermogenesis.

On the other hand, in addition to the growth of brown adipocytes, the BAT enlargement appears to depend on an increase in capillary density, since the supply of oxygen and nutrients to brown adipocytes is limited by the distance from individual capillary units. Actually, Geloen et al. [2,3] have demonstrated that norepinephrine (NE), like cold exposure, markedly enhances the mitotic activity in brown adipocyte precursor cells (interstitial cells and preadipocytes) and endothelial cells forming numerous capillaries. Bronnikov et al. [4] have recently described that brown adipocyte precursor cells respond directly to NE stimulation with an increased DNA synthesis, and that this response is mediated via the classical β_1 receptors. As previously reported in our group [5], the mRNAs for several growth factors (such as basic fibroblast growth factor (bFGF), insulin-like growth factor II (IGF-II) and hepatocyte growth factor) are expressed in

BAT of adult rats. However, the precise mechanism of BAT enlargement still remains obscure. The aim of the present study was to investigate the effects of several factors (including NE and insulin) on the growth of brown adipocyte precursor cells and capillary endothelial cells *in vitro*.

Materials & Methods

Rat brown adipocyte precursor cells (RBAC) were isolated from interscapular BAT of male Wistar rats (2-3 months old) according to the method of Nechad et al. [6] and the primary cells were used in the present study. Bovine capillary endothelial cells (BCEC) were isolated from adult bovine adrenal cortex as described by Folkman et al. [7]. BCEC were subcultured and used at passages 7-11. In the experiments on cell growth, RBAC and BCEC were inoculated with 1×10^4 cells/well into 24-well culture plates containing 0.5 ml of culture medium consisting of Medium 199 (for RBAC) or Dulbecco's Modified Eagle's Medium (for BCEC) supplemented with 10% fetal bovine serum (FBS) and antibiotics (100 U/ml penicillin G, 100 μg/ml streptomycin and 0.25 μg/ml amphotericin B (Gibco Laboratories, Grand Island, NY)). The cells were cultured at 37°C in an atmosphere of 5% CO_2 in air. On day 1-2, after the cells were washed with serum-free medium, the culture medium was replaced with 0.5 ml of various mediums containing *l*-NE (Wako Pure Chemical Industries, Ltd., Tokyo), insulin (Novo Industry, Copenhagen), bovine brain bFGF (R & D Systems, Minneapolis, MN) or recombinant human platelet-derived growth factor-BB (PDGF, Genzyme, Boston, MA). After 4 days, the cells were harvested by tripsinization and enumerated using a particle counter model PC-607 (Erma Inc., Tokyo).

To investigate the effects of NE and insulin on the gene expression of bFGF, after RBAC were cultured until confluence in culture dish, the cells were stimulated by 10^{-7} M NE or insulin for 4 h. Total RNAs of the cultured cells were then isolated by the acid guanidinium thiocyanate-phenol-chloroform method [8]. Polymerase chain reaction (PCR) analysis was performed using oligonucleotide probes (which were designed to include positions 339 to 718 in the sequence of rat bFGF cDNA given by Kurokawa et al. [9]) according to the method described by Arrigo et al. [10]. Amplified products were separated on a 4% polyacrylamide gel. The PCR transcripts with the expected length were confirmed to include the sequence for rat bFGF.

Results & Discussion

As shown in Fig. 1, in the primary cell culture of RBAC, NE significantly enhanced the RBAC growth at the concentration of 10^{-9} - 10^{-5} M in the presence of FBS, whereas NE did not

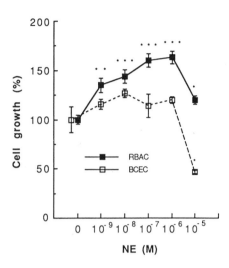

Fig. 1 Effect of NE on the growth of RBAC and BCEC. The cells were cultured in the presence of NE for 4 days as described in Materials & Methods. Points are means ±SEM (RBAC:n=8-12; BCEC:n=4). The value obtained in the control was set to 100%. Significance of difference between control and NE-treated group was assessed by ANOVA followed by Student's t test: *p<0.05, **p<0.01, ***p<0.001.

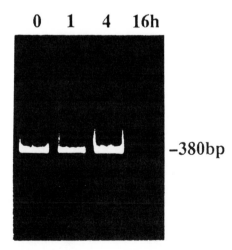

Fig. 2 Time course of bFGF mRNA expression in RBAC stimulated by NE. The confluent cells were exposed to 10^{-7}M NE for the indicated time. PCR analysis was performed using total RNAs of the cells as described in Materials & Methods.

stimulate the BCEC growth. The high concentration (10^{-5} M) of NE appeared to be toxic for these cells. The finding of the present study on the NE effect for RBAC was consistent with that of Bronnikov et al. [4] that NE increases the DNA synthesis in mouse brown adipocyte precursor cells proliferating in primary culture. The same studies performed on insulin gave similar results (data not shown). We next studied the effects of NE and insulin on the expression of mRNAs of several growth factors in RBAC. In consequense, both NE and insulin seemed to increase the expression of bFGF mRNA, although they did not affect that of IGF-I or IGF-II mRNA. Fig. 2 shows the time course of bFGF mRNA expression in RBAC stimulated by 10^{-7} M NE. At 4 h after NE stimulation, the bFGF mRNA expression in RBAC was considerably increased but markedly decreased after 16 h. These results suggested that the bFGF mRNA expression in RBAC is quickly stimulated by NE, with resulting bFGF production. It was also suggested that the decreased expression of bFGF mRNA in RBAC at 16 h after NE stimulation might be caused by the use of bFGF for their own growth (i.e., autocrine mechanism), implying down-regulation. We then evaluated the effect of bFGF on the growth of RBAC (Fig. 3). Actually, bFGF (10 ng/ml) stimulated the RBAC growth up to about 170% of the control, while no effect of PDGF was noted.

On the other hand, Geloen et al. [2,3] have demonstrated that NE markedly enhances the mitotic activity in endothelial cells as well as in brown adipocyte precursor cells. In our study, however, neither NE nor insulin stimulated the BCEC growth or the expression of bFGF mRNA in the cells. Since bFGF is a potent angiogenic factor, bFGF produced by brown adipocytes may affect

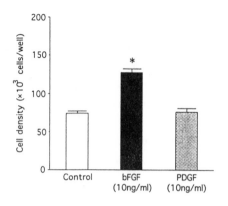

Fig. 3 Effects of growth factors on the growth of RBAC. bFGF or PDGF was added to the RBAC culture at the concentration of 10 μg/ml in the presence of 2.5% FBS. On day 4, the cells were counted as described in Materials & Methods. Mean values are shown with SEM (n=4). Significantly different from control: *p<0.001.

capillaries around them and induce the growth of the endothelial cells. Indeed, the medium conditioned with RBAC for only 4 h in the presence of 10^{-7} M NE nearly doubled the BCEC growth compared with that in the absence of NE. This result suggested that the growth factor (probably bFGF) production in RBAC by NE stimulation may bear some relationship to capillary growth in BAT.

It is thus possible that NE and insulin contribute to the growth of brown adipocytes and endothelial cells through bFGF production in brown adipocytes themselves. Likewise, both NE and insulin appear to play an important role in BAT enlargement as well as in thermogenic function.

Ackowledgments

The authors thank Mr. Masahiko Segawa for his excellent technical assistance.

References

1 Himms-Hagen J. Brown adipose tissue thermogenesis: interdisciplinary studies. FASEB J 1990:4:2890-8.
2 Geloen A, Collet AJ, Guay G, Bukowiecki LJ. ß-Adrenergic stimulation of brown adipocyte proliferation. Am J Physiol 1988:254:C175-82.
3 Geloen A, Collet AJ, Bukowiecki LJ. Role of sympathetic innervation in brown adipocyte proliferation. Am J Physiol 1992:263:R1176-81.
4 Bronnikov G, Houstek J, Nedergaard J. ß-Adrenergic, cAMP-mediated stimulation of proliferation of brown fat cells in primary culture: mediation via ß1 but not via ß3 adrenoceptors. J Biol Chem 1992:267:2006-13.
5 Yamashita H, Sato N, Yamamoto M, Gasa S, Nagasawa J, Sato Y, et al. Is there an intimate interplay between temperature acclimation and exogenous insulin? -with special reference to the participation of brown adipose tissue. In: Sato Y, Poortmans J, Hashimoto I, Oshida Y, editors. Medicine and Sport Science, Vol.37, Integration of Medical and Sports Sciences. Basel: Kargel, 1992:237-42.
6 Nechad M, Kuusela P, Carneheim C, Bjorntorp P, Nedergaard J, Cannon B. Development of brown fat cells in monolayer culture : 1. Morphological and biochemical distinction from white fat cells in culture. Exp Cell Res 1983:149:105-18.
7 Folkman J, Haudenschild CC, Zetter BR. Long-term culture of capillary endothelial cells. Proc Natl Acad Sci USA 1979:76:5217-21.
8 Chomczynski P, Sacchi N. Single-step method of RNA isolation by acid guanidinium thiocyanate-phenol-chloroform extraction. Anal Biochem 1987:162:156-9.

9 Kurokawa T, Seno M, Igarashi K. Nucleotide sequence of rat basic fibroblast growth factor cDNA. Nucleic Acids Res 1988:16:5201.

10 Arrigo SJ, Weitsman S, Rosenblatt JD, Chen IY. Analysis of rev gene function on human immunodeficiency virus type I replication in lymphoid cells by using a quantitative polymerase chain reaction method. J Virol 1989:63: 4875-81.

Temperature Regulation
Advances in Pharmacological Sciences
© 1994 Birkhäuser Verlag Basel

THERMAL PREFERENCE BEHAVIOUR FOLLOWING β3-AGONIST STIMULATION

Harry J. Carlisle, Susan Rothberg and *Michael J. Stock

Department of Psychology, University of California, Santa Barbara, CA 93106 USA
and *Department of Physiology, St. George's Hospital Medical School, Tooting, London
SW17 0RE, UK

Introduction

The β3-adrenergic agonist BRL 35135 (BRL) has been shown reduce the amount of heat obtained by rats in a cold environment using an operant leverpressing task. However, the reduction in heat influx is modest in comparison to the potent thermogenic effects of this drug at a neutral ambient temperature. We therefore assessed the effect of BRL on thermal preference behaviour using a linear thermal gradient. The gradient was constructed from a Plexiglas tube 183 cm in length and 7.5 cm wide; warm (50°C) water was circulated through copper tubing wrapped around the tube and spaced so that a temperature gradient of 7°C at the cold end and 45°C at the warm end was obtained with the entire apparatus in a 5°C environment. Eleven male Sprague-Dawley rats were adapted to the gradient and then given 50-min trials following saline or 2 and 5 µg/kg doses of BRL. The rats preferred an average temperature of 26.7°C during saline trials, which resulted in a post-test colonic temperature (Tc) of 37.5°C. The thermal preference decreased to 19.7°C at 20-30 min after the 2 µg/kg dose if BRL, and to 16.8°C after the 5 µg/kg dose. In spite of the behavioural preference for cooler gradient temperatures following BRL, post-test Tc was 38.6°C after the 2 µg/kg dose and 38.7°C after the 5 µg/kg dose. Thus, rats select cooler temperatures after BRL but not sufficiently so as to mitigate the increase in Tc.

Brown adipose tissue (BAT) contains an atypical class of β-adrenoceptor (termed β3) that can mediate the thermogenic responsiveness of this tissue (1). Pharmaceuticals that stimulate β3 receptors are potently thermogenic when tested at a neutral ambient temperature (Ta), and this effect is the basis for the potential use of these compounds as anti-obesity agents (2, 8, 10). Several of these drugs (BRL 35135, ICI D7114, and Ro 40,2148) have been tested on rats in an operant leverpressing task that permits the animal to obtain infrared heat in a cold environment by pressing a lever in order to activate the lamps (3, 4, 6, 7). As would be expected for thermogenic agonists, these pharmaceuticals tend to reduce the demand for exogenous heat. However, the reduction in behavioural heat influx does not appear to be commensurate with the thermogenic capacity of the drugs as measured by oxygen consumption in a neutral Ta. For example, BRL 35135 (BRL) increases oxygen consumption by 84% at a Ta of 25°C for a dose of 10 μg/kg (5), whereas the reduction in operant responding for heat in the cold is approximately 20% for a comparable dose (7). One explanation for this discrepancy is that BRL may contain some β_2 activity that offsets the thermogenic β_3 effect. Tremor has been noted in clinical trials with BRL (9), and hypokalaemia (10) in animal tests. Both responses have been attributed to effects of β_2-adrenoceptors. In addition, the non-selective β-agonist isoprenaline is potently thermogenic at a neutral Ta, but thermolytic in the cold (3, 4). This paradoxical effect of isoprenaline has been attributed β_2-adrenoceptors because it is blockable with the selective β_2-antagonist ICI 118551 (6). Attempts to uncover a β_2 component in BRL by a similar co-administration of the β_2 antagonist in the operant leverpressing task has failed to identify any such component (6, 7). A second explanation may relate to the operant apparatus. Pressing the lever activates two infrared lamps focused on the rat at the lever. The radiant field covers about two-thirds of the dorsal surface of the animal while the ventral surface remains shaded. Perhaps this arrangement makes it easy to dissipate excess heat by minor postural changes as well as by reducing slightly the demand for external heat. We investigate here the behavioural thermoregulatory responses to BRL using a thermal gradient in order to establish how thermal preference varies under conditions that provide a more uniform thermal environment for the animal, yet permit selection of different temperatures.

Methods

Animals Eleven male Sprague-Dawley rats were obtained from Charles River breeders. The animals were housed singly in wire-mesh cages in a colony room maintained at 22°C and 50% relative humidity. They were fed Purina chow and water *ad libitum*.

Drugs BRL 35135 was a gift from SmithKline Beecham (Epsom, UK). The drug was dissolved in physiological saline (0.9% NaCl), which also served as the control vehicle. BRL doses of 2 and 5 μg/kg were given subcutaneously. All animals received saline and both doses of BRL in a counter-balanced order with 5 days intervening between tests.

Thermal Gradient The gradient was constructed from a Plexiglas tube 183 cm in length and 7.5 cm wide. Copper tubing was coiled tightly around one end of the gradient and then spaced so that the gap between the coils increased with distance from the tight end. The apparatus was placed in a room maintained at 5°C. Warm (50°C) water was pumped to the copper tubing at the tight end and thence through the coils along the length of the gradient. The resulting temperature inside the gradient varied from 7°C at the cold end to 45°C at the warm end.

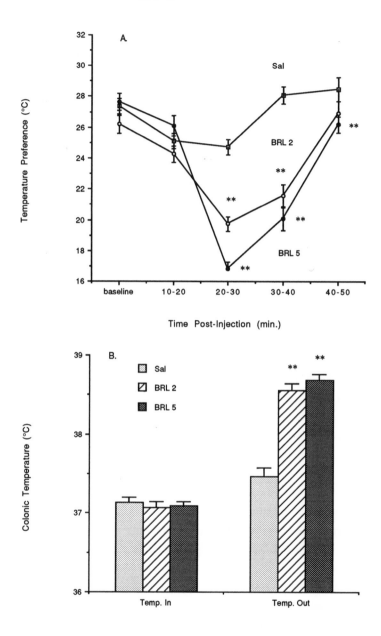

Fig 1. (A): Thermal preference behaviour 20 min prior to (baseline) and 50 min after saline or BRL 2 and 5 µg/kg; (B): Pre- and Post-test colonic temperature (°C.) Values are mean +/- SEM; n= 11. *p< 0.05, **p< 0.01 compared to saline (paired t-test)

The gradient was calibrated by measuring the temperature along the inner length with a Yellow Springs Instrument meter (No. 46) and thermistor probe (No. 402) encased in 25g of clay.The position of the animal could be noted by an observer outside the cold room via an observation window.

Procedure The initial response of rats to the gradient is to move continuously from one end to the other. The animals were first adapted by placing them in the gradient 1hr/day for 5 days. For test trials, the animals were placed in the gradient for 60 min to permit adaptation and to reduce exploratory behaviour. The position of the rat was then measured for 20 min to obtain the baseline thermal preference. The amount of time the animal spent at any location was noted by the observer, and the mean preference was calculated as the temperatures selected weighted for the duration at that location. The rat was then removed, injected with the drug for that test, and replaced for an additional 50 min. Thermal preference was measured between 10 and 50 min after injection, and averaged for 10-min intervals. Colonic temperature (Tc) was measured prior to and at the end of the test with a Physitemp BAT-12 meter and thermocouple probe inserted 7 cm.

Results

Figure 1 shows that BRL shifted the thermal preference to cooler temperatures during the 10-40-min post-injection intervals. The rats preferred a slightly lower Ta after the 5 µg/kg dose than after the 2 µg/kg dose. Thermal preference began to return toward baseline values towards the end of the test. In spite of the preference for cooler temperatures in the gradient, post-test Tc was elevated by 1.1-1.2°C after BRL in comparison to saline. Thus, the animals tolerated the hyperthermic effects of BRL by selecting a lower Ta, but not one that would offset the thermogenic effect of the drug at either the 2 or 5 µg/kg dose.

Discussion

The present results confirm the potent thermogenic properties of the selective β3-agonist BRL by showing that animals will select a lower Ta in a thermal gradient following the drug.However, this preference does not offset the thermogenic properties of BRL, and thus a significantly elevated post-test Tc occurs in spite of the preference for a lower Ta.

In contrast, animals obtain less heat after BRL in an operant leverpressing task, but post-test Tc is not greatly different than after saline (4, 6, 7). The present results indicate that animals will act so as to mitigate excessive thermal stimulation, but not to the extent that a thermogenic stimulus is eliminated by that behaviour. It is of interest in this regard that no mention is made of unpleasant thermal sensations in clinical trials of BRL with human (9).

References

1.   ARCH, J.R.S., AINSWORTH, A.T., CAWTHORNE, M.A., PIERCY, V., SENNIT, M.V., THODY, V., WILSON, C., WILSON, S. (1984). Atypical β-adrenoceptor on brown adipocytes as target for anti-obesity drugs. *Nature* **309**, 163-165.

2.   ARCH, J.R.S., CAWTHORNE, M.A., CONEY, K.A., GUSTERSON, B.A., PIERCY, V., SENNITT, M.V., SMITH, S.A., WALLACE, J., WILSON, S. (1991). β-adrenoceptor-mediated thermogenesis, body composition and glucose homeostasis. In: *Stock M.J. Rothwell, N.J., eds. Obsesity and Cachexia. London: John Wiley.* 241-268.

3.   CARLISLE, H.J., STOCK, M.J. (1991). Effect of conventional (mixed β_1/β_2) and novel (β_3) adrenergic agonists on thermoregulatory behaviour. *Pharmacol. Biochem. Behav.* **40**. 249-254.

4.   CARLISLE, H.J., STOCK, M.J. (1992). Potentiation of thermoregulatory responses to isoproterenol by β-adrenergic antagonists. *Am. J. Physiol.* **263**, R915-R923.

5.   CARLISLE, H.J., STOCK, M.J. (in press). Effect of cold ambient temperatures on metabolic responses to β-adrenergic agonists. *Exp. Physiol.*

6.   CARLISLE, H.J., STOCK, M.J. (in press). Thermoregulatory effects of β adrenoceptors: Effects of selective agonists and the interaction of antagonists with isoproterenol and BRL-35135 in the cold. *J. Pharmacol. Exp. Ther.*

7.   CARLISLE H.J., DUBUC, P.U., STOCK, M.J. (in press). Effects of β-adrenoceptor agonists and antagonists on thermoregulation in the cold in lean and obese Zucker rats. *Pharmacol. Biochem. Behav.*

8.   CAWTHORNE, M.A., SENNITT, M.V., ARCH, J.R.S., SMITH, S.A. (1992) BRL 35135, a potent and selective atypical β-adrenoceptor agonist. *Am. J. Clin. Nutr.* **55**, 252S-257S.

9.   CONNACHER, A.A., BENNET, W.M., JUNG, R.T. (1992). Clinical studies with the β-adrenoceptor agonist BRL 26830A. *Am. J. Clin. Nutr.* **55,** 258S-261S.

10. HOLLOWAY, B.R., STRIBLING, D., FREEMAN, S., JAMIESON, L. (1989). ICI 198157: a novel selective agonist of brown fat and thermogenesis. In: *Bjorntorp, P., Rossner, S., eds. Obesity in Europe '88. London: John Libby* 323-328.

Temperature Regulation
Advances in Pharmacological Sciences
© 1994 Birkhäuser Verlag Basel

SYMPATHETIC TONE AND NORADRENALINE RESPONSIVENESS OF BROWN ADIPOCYTES FROM RATS WITH HIGH LEVELS OF SEXUAL STEROIDS

Marisa Puerta
Department Animal Biology II (Physiology)
Faculty of Biological Sciences
Complutense University
28040 Madrid, Spain

Diet-induced and cold-induced thermogenesis are impaired when plasma levels of progesterone and oestradiol are increased, respectively. To assess their mode of action on brown fat, we measured the respiration rate of brown adipocytes isolated from rats with high plasma levels of either progesterone or oestradiol. In both cases, we found a poor response to noradrenaline when compared with that of adipocytes from control rats. This effect dissociates brown adipose tissue respiration and thermogenesis from the sympathetic drive in the case of oestradiol. We are currently studying if a similar dissociation also takes place in the presence of progesterone. The consequences of this peculiar effect on brown adipose tissue remain to be elucidated.

INTRODUCTION

Heat produced in brown adipose tissue (BAT) of small animals allows them to maintain body temperature when environmental temperature falls bellow the thermoneutral zone. The thermal stimulus gives as a result an increased sympathetic tone which increases the noradrenaline discharge to brown adipocytes. This chemical messenger switches on a chain of steps that ultimately leads to an enhancement in substrate combustion, oxygen consumption and thermogenesis. But under some circumstances, i.e. overfeeding, the intense oxidative power of BAT is used to dissipate the excess of energy intake which otherwise would be deposited as fat with the concomitant metabolic and cardiovascular risks. To differentiate both kinds of stimuli the terms cold-induced and diet-induced thermogenesis have been used.

The life-span of a female rat is marked by cyclic processes i.e. oestrus and pregnancy-lactation cycles. In the latter case, it has been shown that food intake increases from the last part of pregnancy through lactation to weaning, this increase being higher in chronically cold-exposed animals(1). But despite the fact that in pregnant rats at room temperature the two optional stimuli for BAT thermogenesis (namely, enhanced food intake and cold) are present, the BAT-GDP binding, the usual index of BAT thermogenesis, is depressed(2). Since female sexual steroids 17-β oestradiol and progesterone are known to increase in the plasma of pregnant animals, we studied their involvement in the control of BAT function. Rats preacclimated to either thermoneutrality (28°C), to avoid cold-induced thermogenesis, or to cold (6°C), and therefore with an intense cold-induced thermogenesis, were used as experimental models. The experimentally increased plasma levels of progesterone did not increase the BAT GDP-binding of rats at thermoneutrality despite they showed an increased food intake with respect to controls (3). On the contrary, high plasma levels of oestradiol decreased the BAT GDP-binding of cold-acclimated rats despite the low environmental temperature (4). Thus, it became apparent that female sexual steroids can inhibit BAT thermogenesis although in a stimulus-dependent manner, i.e. progesterone inhibits diet-induced thermogenesis whereas 17-β oestradiol inhibits cold-induced thermogenesis.

NA released from the sympathetic terminals that innervate the tissue is the signal that switch on the thermogenic process. Both, substances and physiological situations that modify BAT thermogenesis are known to modify at the same time the sympathetic discharge to BAT. Therefore, there is a general agreement with the concept that changes in BAT activity are mediated through changes in sympathetic discharge to BAT (5). Thus, the influence of sexual steroids on BAT thermogenesis could be carried out by its modulation of the level of sympathetic discharge to brown adipocytes. Nonetheless, the presence of high affinity sites for oestradiol in BAT has been shown (6). Accordingly, oestradiol and perhaps progesterone might act directly upon brown adipocytes interfering with some step of the thermogenic process. Both possibilities have been studied in our laboratory and the results are presented below.

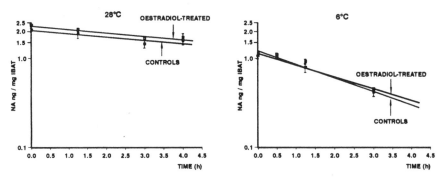

Figure 1. Rate of decrease in noradrenaline content in interscapular brown adipose tissue in control and oestradiol-treated rat acclimated either at 28°C or at 6°C.

MATERIALS & METHODS.

Animal treatments. Female Wistar (210-225 g b.w.) rats were used throughout the experiments. They were preacclimated for two weeks to the corresponding temperature (6°C or 28°C in noradrenaline turnover studies and 21°C in *in vitro* studies). Once acclimated, hormones were administered by means of subcutaneous Silastic capsules filled with the hormone under study (0.5 cm long; inner and outer diameter 1.5 and 3.2 mm, respectively). Control animals received empty implants. All the rats remained at the corresponding temperature until the end of the experiments.

Sympathetic stimulation of BAT. It was assessed measuring the noradrenaline turnover rate in interscapular brown adipose tissue (5). After blocking noradrenaline synthesis with α-methyl-tyrosine (250 mg/kg i.p.), the rats were sacrificed at different time intervals and the noradrenaline content of the interscapular brown fat depot was measured by HPLC using an electrochemical detector.

The noradrenaline content of interscapular BAT at each time point was plotted on a semilogarithmic scale. The slope (*b*) was determined by the method of least-square regression analysis. The total turnover rate was calculated from the fractional turnover rate ($k=b/0.434$) and the initial noradrenaline content of the fat pads.

Brown adipocytes preparation. Isolated adipocytes were obtained from BAT pooled from 3-5 rats by an incubation with collagenase in a shaking bath at 37°C (7,8). After digestion, cells were collected allowing them to float and removing the infranatants. The number of cells in the final suspension was counted in a Neubauer's hemacytometer with trypan blue stain to visualize dead cells.

Determination of cellular respiration rates. Oxygen uptake of aliquots of the cellular suspension was calculated by measuring polarographically the oxygen decay in a closed chamber with an oxygen probe (8). After monitoring basal O_2 consumption, noradrenaline was added to the chamber at increasing concentrations.

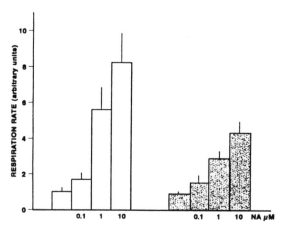

Figure 2. Oxygen consumption of brown adipocytes isolated from oestradiol-treated (shaded bars) and control (open bars) rats in basal conditions and after successive additions of noradrenaline.

RESULTS AND DISCUSSION

Studies on oestradiol effects.

It has been repeatedly observed that changes in the thermogenic activity of BAT run parallel to changes in its sympathetic discharge e.g. cold-acclimation, cold-exposure or overfeeding increase both GDP-binding and noradrenaline turnover (5) whereas lactation or obesity depressed them (9,10). Therefore, the concept that BAT thermogenesis is governed by the sympathetic nervous system has emerged (5). Accordingly, it seemed reasonable to expect a decreased noradrenaline turnover rate in rats with high plasma oestradiol since they show a diminished GDP-binding (4). But contrary to our expectations, the noradrenaline turnover rate of female rats

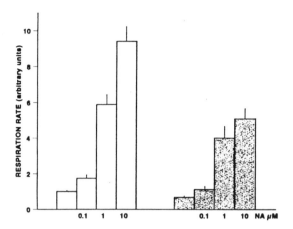

Figure 3. Oxygen consumption of brown adipocytes isolated from progesterone-treated (shaded bars) and controls (Open bars) rats in basal conditions and after successive additions of noradrenaline.

with high plasma levels of oestradiol did not differ from that of control animals. This result was found both in rats with only a basal level of sympathetic discharge (thermoneutral conditions, 28°C) and in rats with an intense sympathetic activity (cold-acclimated rats, 6°C) (Fig.1). Therefore, the inhibition of BAT thermogenesis by oestradiol in cold-acclimated rats seems to be brought about through a mechanism independent of the degree of sympathetic stimulation to brown adipocytes. As far as we know, this is the only case of dissociation between the sympathetic nervous system and BAT thermogenesis reported in rats or mice in which, as above mentioned, a direct relationship between them is the rule. However, it is not an entirely new observation since it was previously detected in the Syrian hamster during pregnancy and lactation (11) and when feeding a high fat diet (12). The fact that the dissociation between BAT thermogenesis and sympathetic activity takes place a) in the hamster during pregnancy, b) in the rat when plasma oestradiol levels are similar to those reached during late pregnancy, together with c) the known reduction in GDP-binding in pregnant rats, mice and hamster, suggests that oestradiol might be responsible for the inhibition of BAT thermogenesis during pregnancy.

The independence of oestradiol effect on BAT from the sympathetic stimulation together with the known presence of oestradiol receptors in brown fat depots prompted us to study the response to noradrenaline of brown adipocytes isolated from rats with high plasma levels of oestradiol as a way to check an alteration in the physiology of the brown adipocytes. Fig. 2 depicts the results obtained. When brown adipocytes from control rats were stimulated with increasing doses of noradrenaline, their oxygen consumption was increased in a dose-dependent manner. But the response was attenuated in adipocytes isolated from oestradiol-treated rats. In other words, the responsiveness to noradrenaline was depressed by the presence of high plasma oestradiol.

As a whole, our studies show that plasma oestradiol modulates the thermogenesis of brown adipose tissue but in an unusual manner i.e., it decreases the noradrenaline responsiveness of brown adipocytes, instead of regulating BAT activity by modulating its sympathetic tone. To the best of our knowledge this is the only reported case of such a mode of action but it could be not a particular one since, as we shall see later, it has been also found for progesterone. On the other hand, oestradiol is generally considered as a hormone that reduces energy deposition because it has anorexic effects (13). But its capacity for lowering the noradrenaline responsiveness of brown adipocytes reveals that it can also modulate energy expenditure. This conservative role might counterbalance the decreases in energy gains that its anorexic effect produces. Furthermore, it might also allow some weight gains, as observed during the last part of pregnancy (1).

Studies on progesterone effects.

The lack of diet-induced thermogenesis when food intake is enhanced by high plasma progesterone can be again explained *a priori* by two theoretical possibilities i.e. either the sympathetic nervous system does not increase its activity despite the increase in food intake or it does increase it, but the brown adipocytes response to the enhanced release of noradrenaline is impaired. In view of the results obtained with oestradiol studies, the second possibility was experimentally checked in the first place. As Fig. 3 shows, when brown adipocytes were isolated from control animals and from rats with high plasma levels of progesterone, the same dose of the neurotransmitter evoked a lower increase in the respiration rate of adipocytes isolated from progesterone-treated rats than in those isolated from untreated controls. Therefore, as previously found in oestradiol studies, progesterone seems to act directly upon brown adipocytes decreasing their responsiveness to noradrenaline. This result together with the unaltered thermogenic activity (GDP-binding) found in the BAT of rats with high plasma levels of progesterone previously reported (3) suggest that the increased food intake produced by high plasma progesterone levels might be followed by an activation of the sympathetic discharge to BAT. However, the concomitant diet-induced thermogenesis would not take place due to the diminished response to noradrenaline of the brown adipocytes. Although we are currently studying the noradrenaline turnover rates in brown adipose tissue of progesterone-treated rats, we lack data enough to sustain a definitive conclusion.

Concluding remarks.

The experimental data concerning the thermogenic activity of brown adipose tissue when plasma levels of the female sexual steroids oestradiol and progesterone are high reveal that both steroids have an inhibitory effect on it. The studies undertaken with brown adipocytes isolated from females with high plasma levels of either progesterone or oestradiol reveal that they show a decreased responsiveness to noradrenaline when compared with the responsiveness of adipocytes isolated from control rats. This effect dissociates brown adipose tissue respiration and thermogenesis from the sympathetic drive in the case of oestradiol. We are currently studying if a similar dissociation also takes place in the presence of progesterone.

Acknowledgements.- I am grateful to my collaborators Dr Abelenda, Dr Nava, A Fernández, C Castro and C Venero. Thanks are also given to the CICYT for Grant SAL-91 0502 and to the Complutense University for the fellowship of A Fernández.

REFERENCES
1   Abelenda M, Puerta ML. Inhibition of diet-induced thermogenesis during pregnancy in the rat. Pflügers Archiv 1987:409:314-7.
2   Trayhurn P, Douglas JB, Mc Guckin MM. Brown adipose tissue thermogenesis is "suppressed" during lactation in mice. Nature 1982:298:59-60.
3   Nava MP, Abelenda M, Puerta ML. Cold-induced and diet-induced thermogenesis in progesterone-treated rats. Pflügers Archiv 1990:415:747-50.
4   Puerta ML, Nava MP, Abelenda M, Fernández A. Inactivation of brown adipose tissue thermogenesis by oestradiol treatment in cold-acclimated rats. Pflügers Archiv 1990:416:659-62.
5   Young JB, Saville E, Rothwell NJ, Stock MJ, Landsberg L. Effect of diet and cold exposure on norepinephrine turnover in brown adipose tissue of the rat. The Journal of Clinical Investigation 1982:69:1061-6.
6   Wade GN, Gray JM. Cytoplasmic 17β-(^{3}H)estradiol binding in rat adipose tissues. Endocrinology 1978:103:1695-1701.
7   Mohell N, Connolly E, Nedergaard J. Distinction between mechanisms underlying α_1- and β adrenergic respiratory stimulation in brown fat cells. American Journal of Physiology 1987:253:C301-8.

8   Bukowiecki LJ, Follea N, Lupien J, Paradis A. Metabolic relationship between lipolysis and respiration in rat brown adipocytes. The role of long chain fatty acids as regulators of mitochondrial respiration and feed-back inhibitors of lipolysis. Journal of Biological Chemistry 1981:256:12840-8.
9   Trayhurn P, Wusteman M.C. Sympathetic activity in brown adipose tissue in lactating mice. American Journal of Physiology 1987:253:E515-20.
10   Goodbody AE, Trayhurn P. Studies on the activity of brown adipose tissue in suckling pre-obese ob/ob mice (BBA-41079). Biochimica Biophysica Acta 1982:680:119-26.
11   Trayhurn P, Wusteman MC. Apparent dissociation between sympathetic activity and brown adipose tissue thermogenesis during pregnancy and lactation in golden hamsters. 1987:65:2396-9.
12   Hamilton JM, Mason PW, McElroy JF, Wade GN. Dissociation of sympathetic and thermogenic activity in brown fat of Syrian hamsters. American Journal of Physiology 1986:250:R389-95.
13   Richard D. Effects of ovarian hormones on energy balance and brown adipose tissue thermogenesis. American Journal of Physiology 1986:250:R245-9.

THE TAIL OF THE RAT IN TEMPERATURE REGULATION: EFFECT OF ANGIOTENSIN II

M.J. Fregly, N.E. Rowland and J.R. Cade
Departments of Physiology and Psychology, Colleges of Medicine and Liberal Arts and Sciences, University of Florida, Gainesville, FL 32610, USA.

Summary

The importance of the tail in thermoregulation of the rat was shown by comparison of the colonic temperatures (T_c) of tailed and tailless rats following administration of graded doses of the beta-adrenoceptor agonist, isoproterenol (ISO). The hyperthermic responses of tailless rats was higher and lasted longer than that of tailed controls. Angiotensin II (AngII), a potent vasoconstrictor agent, also increased tail skin temperature (T_{sk}), but reduced T_c in a dose-related fashion. This apparently aberrant vasodilatory response of tail skin vasculature to a vasoconstrictor agent is mediated by way of AT-1 receptors, since it is blockable by losartan potassium, a specific AT-1 receptor antagonist. The mechanism beyond the AT-1 receptor by which AngII induces a hypothermic response in the rat remains to be elucidated.

Introduction

The tail of the rat plays an important role in thermoregulation. Although its surface comprises only 4-6% of total body surface area, 17% of the total heat produced by the rat can be lost from the tail (1). The objective of the first study was to compare the abilities of tailed and tailless rats to maintain their colonic temperatures (T_c) after administration of ISO. Treatment with ISO rapidly increases metabolic rate, after which T_{sk} increases in order to dissipate the extra heat produced (2).

The objective of the 2nd experiment was to establish a dose-response relationship between the dose of AngII and both the increase in T_{sk} and decrease in T_c. The third experiment assessed the specificity of the response by blocking AngII receptors with the non-peptide AngII receptor antagonist, losartan potassium.

Methods

Male rats of the Sprague Dawley strain were used throughout. They were kept in a room maintained at 26±2°C and illuminated on a 12:12 hour light-dark cycle. Food (Purina Laboratory Chow) and tap water were provided ad libitum.

T_{sk} and T_c were measured at an ambient temperature of 26±2°C in unanesthetized rats that were restrained in lucite tunnel-type cages. The cages allowed the rats to rest comfortably within them, but restricted them from turning from head to tail. T_c was measured by a recording potentiometer via a thermocouple inserted 5 cm into the colon. An additional thermocouple was placed on the dorsal surface at the base of the tail for measurement of T_{sk}. Baseline control measurements were made for 30 min after which the designated agent (s) was administered s.c.

Experiment 1

Eighteen rats whose tails had been amputated within 1 cm of the base 4 weeks earlier were used. Eighteen rats with tails were used as controls. No animal was used in any study more than once during a 7 day period.

The experiments were carried out at $26\pm1°C$. Tailed and tailless rats were injected s.c. with d, l-ISO (Isuprel[R]) in doses of either 12.5, 25, 50, or 100 µg/kg b.w.

Experiment 2

At the end of the control period, 6 rats were administered AngII (75 µg/kg, s.c.). The remaining 6 rats received the nonpeptide AngII receptor antagonist, losartan potassium (DuP 753) (10 mg/kg, s.c.), 15 min prior to administration of the same dose of AngII.

Results and Discussion

Prevention of heat loss from the tail results in an increase in T_c following administration of isoproterenol (4). Apparently the rat is unable to use other areas of its body, such as scrotum, ears, and paws, to dissipate a sufficient amount of heat to prevent an elevation in T_c during acute thermal stress, and must depend instead on both vasodilation of the tail and spreading of saliva for maintenance of T_c.

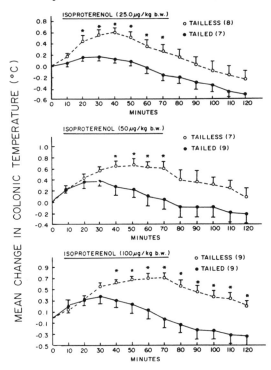

Figure 1 Mean changes in T_c of tailed and tailless rats after administration of either saline or ISO. Numbers in parenthesis indicate sample size. One SE is set off at each mean. * = $P<0.05$. Only responses to the 3 higher doses are shown.

Administration of ISO, ranging in dose from 12.5 to 100 µg/kg, was accompanied by increases in T_C of both tailed and tailless rats (Figure 1 shows the 3 higher doses only). In response to increasing doses of ISO, the T_C of tailed rats showed a small increase above the pretreatment level, but the duration of the hyperthermic effect did not change with increased dosage. On the other hand, the tailless rats showed a large increase in T_C at all doses of ISO and a major increase in the duration of the hyperthermia at the higher doses (50.0 and 100.0 µg/kg). The duration of the responses of T_C was related to the dose of ISO administered (37.5-300 ug/kg) to intact rats (5).

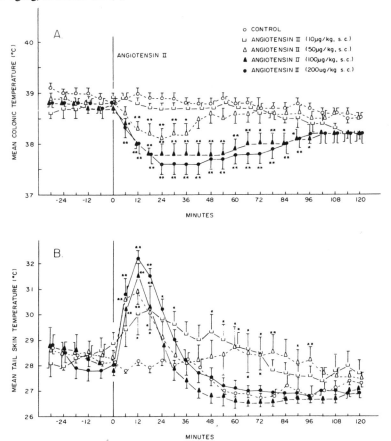

<u>Figure 2</u> Mean T_C (A) and T_{sk} (B) of rats administered either saline or graded doses of AngII at time 0 are shown. One SE is set off at each mean. * = P<0.05; ** P<0.01 compared to control group.

Several investigators have shown that vasodilation of the tail of the rat does not occur until T_C reaches 39.0 to 39.2°C (2, 6, 7). Hence, a possible explanation for the difference in the responses of tailed and tailless rats might be attributed to a difference in heat production rather than heat loss. As reported by others (2, 3, 5), both groups increased their metabolic

264 M. J. Fregly et al.

heat production within 10 min and to the same extent following administration of ISO (50 ug/kg). Hence, the differences in T_c seen in this study were probably related to differences in thermolytic rather than thermogenic mechanisms. These results also suggest that the effect of AngII on T_{sk} was dose-dependent (Figure 2A). The maximal response occurred in 12 min and returned to control level by 24 min.

Subcutaneous administration of AngII also induced a dose-dependent decrease in T_c (Figure 2B).

The AngII, AT-1 receptor antagonist, losartan potassium, administered at a dose that antagonized the dipsogenic response to AngII in an earlier study (8), prevented the increase of T_{sk} and reduction of T_c in response to administration of AngII.

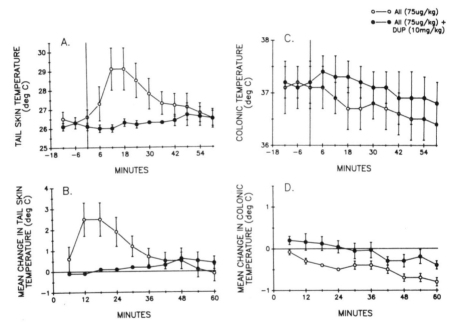

Figure 3 Effect of losartan potassium (DuP 753) on the response of T_{sk} and T_c to AngII. Changes in T_{sk} (A) and T_c (B) are shown. One SE is set off at each mean. AngII was administered at time 0 while losartan potassium was administered 15 min prior to AngII.

Losartan alone had no significant effect on either T_{sk} or T_c of rats during the first 15 min of the experiment (Figure 3A, B). Thus, the results of this study suggest that AngII acts to increase T_{sk} via AT-1 receptors.

The tail of the rat is important in short-term temperature regulation. Administration of either ISO or AngII can increase T_{sk}. The former compound also increases both metabolic rate and T_c while AngII induces the opposite responses. The results suggest that administration of AngII to the rats is interpreted as an elevation of body temperature. The rat then makes all of the appropriate responses to reduce body temperature (decrease metabolic rate; increase T_{sk}). This suggests that acute administration of AngII may reset the thermostat of the rat downward. Additional studies will be required to verify this possibility.

Acknowledgements

The authors acknowledge the contributions of Drs. C.C. Barney, D.E. Spiers and K.M. Wilson. Supported by grant HL39154-07 from NHLBI, NIH.

References

1.   RAND, R.P., BURTON, A.C., ING, T. (1965). The tail of the rat in temperature regulation and acclimatization. *Can. J. Physiol. Pharmacol.* **43**, 257-267.

2.   FREGLY, M.J., BARNEY, C.C., KELLEHER, D.C., KATOVICH, M.J., TYLER, P.E. (1980). Temporal relationship between the increase in metabolic rate and tail skin temperature following administration of isoproterenol to rats. In: *Cox, B., Lomax, P., Milton, A.S., Schonbaum, E. Eds. Thermoregulatory Mechanisms And Their Therapeutic Implications, Karger:Basel* pp.12-18.

3.   LEBLANC, J., VALLIERS, J., VACHON, C. (1972). Beta receptor sensitization by repeated injections of isoproterenol and by cold adaptation. *Am. J. Physiol.* **22**, 1043-1046.

4.   SPIERS, D.E., BARNEY, C.C., FREGLY, M.J. (1981). Thermoregulatory responses of tailed and tailless rats to isoproterenol. *Can. J. Physiol. Pharmacol.* **59**, 847-852.

5.   LITTLE, R.A., STONER, H.B. (1968). The measurement of heat loss from the rat's tail. *Quart. J. Exp. Physiol.* **53**, 76-83.

6.   MYHRE, K., HELLSTROM, B. (1965). Thermoregulation in exercising white rats. *Can. J. Physiol. Pharmacol.* **43**, 279-287.

7.   HELLSTROM, B. (1975). Heat vasodilation of the rat tail. *Can. J. Physiol. Pharmacol.* **53**, 202-206.

8.   FREGLY, M.J., ROWLAND, N.E. (1991). Effect of a nonpeptide angiotensin II receptor antagonist, DuP 753, on angiotensin-related water intake in rats. *Brain Res. Bull.* **27**, 97-100.

Temperature Regulation
Advances in Pharmacological Sciences
© 1994 Birkhäuser Verlag Basel

SELECTIVE VULNERABILITY OF RAT HIPPOCAMPUS IN HEAT STRESS

H. S. Sharma[1-4], J. Westman[2], F. Nyberg[3], C. Zimmer[1], J. Cervós-Navarro[1]
and P. K. Dey[4]

Department of Neuropathology[1], Free University Berlin, Germany;
Department of Human Anatomy[2], and Pharmacology[3], University of Uppsala,
Biomedical Centre, Uppsala, Sweden; Department of Physiology[4],
Institute of Medical Sciences, Banaras Hindu University, Varanasi-221 005, India

SUMMARY

Subjection of *conscious* young animals to 4 h heat stress (HS) at 38°C in a biological oxygen demand incubator (BOD) resulted in a selective cellular damage of hippocampus in the CA4 sector. Disruption of blood-brain barrier (BBB), collapsed capillaries, perivascular edema, synaptic damage and vesiculation of myelin sheaths are frequent findings in this sector as compared to other regions in the hippocampus. These vascular and cell changes were not observed in heat stressed animals pretreated with p-chlorophenylalanine (an inhibitor of serotonin synthesis) or anaesthetised with urethane. Our observations show that the CA4 sector of the hippocampus is selectively vulnerable to heat and serotonin appears to contribute to these cell changes.

INTRODUCTION

Heat stress (HS) and associated hyperthermia is a serious clinical problem in many parts of the tropical world. The victims of HS are mainly children below the age of 15 years. Sporadic case reports show neuronal damage, edema and haemorrhages in the cerebral cortex, cerebellum, thalamus and brain stem [1, 2]. However, the cell changes in hippocampus following HS is still unknown. Hippocampus is intimately connected with the limbic system and participates in the information processing of the CNS [3]. It seems quite likely that the neural mechanisms of hippocampus may selectively be activated by HS. The present study was thus undertaken to examine pathophysiological alterations in the hippocampus in experimental HS. We earlier, observed a profound increase in plasma and brain serotonin accompanied with a considerable decrease in cerebral blood flow (CBF) and an increase in blood-brain barrier (BBB) permeability in conscious animals subjected to a 4 h HS at 38° C [4, 5]. Since hippocampus receives a dense serotonergic innervation [6], a possibility exists that breakdown of the BBB permeability and vasogenic edema may result in cell changes following HS. In present investigation we examined the BBB permeability, edema, and cell changes in the hippocampus of heat stressed rats. Additional group of rats were pretreated with p-chlorophenylalanine (p-CPA) or anaesthetised with urethane before heat exposure in order to find out a role of serotonin and/or hyperthermia in hippocampal damage.

MATERIALS AND METHODS

Animals. Experiments were carried out on 28 Wistar male rats (body weight 90-100 g, age 8-9 weeks) housed at controlled ambient temperature (21±1 °C) with a 12 h light and a 12 h dark schedule. The rat feed and tap water were supplied ad libitum before the experiments.

Heat exposure. Animals were exposed to heat stress (HS) in at 38 °C for 4 h a biological oxygen demand (BOD) incubator (relative humidity 45-50 %, wind velocity 20-25 cm/sec) [2, 7].

Parameters measured

Stress symptoms and physiological variables. Rectal temperature, behavioural salivation and prostration were examined in each animal. The mean arterial blood pressure (MABP), arterial pH and blood gases were monitored at the end of heat exposure (Table 1) [2, 4].

Blood-brain barrier permeability and edema. The BBB permeability (n=8) in hippocampus was examined using Evans blue (0.3 ml of a 2% solution) and ^{131}I-sodium (10µCi/100g) as described earlier [2, 4]. The water content of the hippocampus and the volume swelling were calculated from the differences between dry and weight wet of the sample in control and experimental groups [5].

Morphology. The cell changes in the hippocampus were examined using light and electron microscopy. The brain was fixed *in situ* by perfusion of a fixative containing 4 % paraformaldehyde, 2 % glutaraldehyde in 0.1 M phosphate buffer (pH 7.4) preceded with a brief saline rinse. After fixation, the brain was removed and coronal sections containing hippocampus (bregma -3.3 mm and -3.8 mm) were cut and embedded in paraffin. Three µm thick paraffin sections were stained for glial fibrillary acidic protein immunoreactivity (a specific marker for astrocytes), Nissl or Haemotoxylene and Eosin for light microscopy using commercial protocol [2, 5, 7].

For electron microscopy, small tissue pieces of hippocampus were post-fixed in osmium and embedded in epon. One µm thick epon sections were cut, stained with toludine blue and examined under light microscope. The ultrathin sections from CA 4 regions were cut, stained with uranyl acetate and lead citrate and examined under a Phillips transmission electron microscope [2, 5].

Drug treatment. In one groups of rats, p-CPA (a serotonin synthesis inhibitor, 100 mg/kg/day for 3 days) (n=10) was injected intraperitoneally [8]. On the fourth day these animals were subjected to HS. The other group was anaesthetised with urethane (1.5 g/kg, i.p.) (n=10) before heat exposure. All the above parameters were examined in these animals.

TABLE 1. Stress symptoms, physiological variables, blood-brain barrier permeability, edema and cell changes in heat stressed animals and their modification with p-CPA or urethane. Animals were subjected to heat stress at 38 °C for 4 h in a BOD incubator (for details, see text).

Type of the Experiment	Intact controls	4 h heat stress at 38 ° C		
		Urethane anaesthetised	Untreated	p-CPA treated
A. Stress symptoms	n=5	n=5	n=8	n=5
1. Δ ° C rectal temperature	+0.23±0.05	-0.46±0.08	+3.88±0.23***	+2.89±0.28***
2. Salivation	nil	nil	++++	+++
3. Behavioural prostration	nil	nil	++++	+++
B. Physiological variables	n=5a	n=5a	n=5a	n=5a
1. MABP torr	100±6	110±4	78±6***	82±4***
2. Arterial pH	7.42±0.04	7.38±0.02	7.36±0.06	7.38±0.07
3. PaO2 torr	80.56±0.34	80.78±0.28	81.88±0.21	80.54±1.24
4. PaCO2 torr	34.37±0.32	33.66±0.55	32.33±0.84	33.38±0.76
C. BBB permeability (Hippocampus)	n=5a	n=5a	n=8a	n=5a
1. Evans blue albumin mg %	0.24±0.04	0.22±0.06	1.89±0.32***	0.65±0.25**
2. 131I-sodium %	0.34±0.08	0.28±0.06	2.38±0.44***	0.78±0.24**
D. Edema (Hippocampus)	n=5a	n=5a	n=5a	n=5a
1. Water content %	79.67±0.34	78.97±0.23	82.42±0.23***	80.24±0.46
2. Volume swelling %	nil	-3	+15	0

Values are expressed as Mean±SD, *** = P <0.001, ** P<0.01 Student's unpaired t-test, MABP =mean arterial blood pressure, ++++ = severe, +++ = moderate, a = same group of animals.

Control group. Normal animals (n=10) maintained at room temperature (21±1°C) were served as intact controls.

Statistical analysis. The quantitative data obtained from control and experimental groups were analysed using unpaired student's t-test. p-value less than 0.05 was considered to be significant.

RESULTS

Stress symptoms and physiological variables. Subjection of *conscious* young animals to a 4 h HS resulted in marked hyperthermia and behavioural symptoms (Table 1). These animals showed a mild hypotension after HS. The arterial PaO_2 was significantly increased, whereas the arterial pH and $PaCO_2$ did not differ from the control group. Pretreatment with p-CPA attenuated the hyperthermic response, but did not affect the symptoms or physiological variables. No change in the symptoms or physiological variables were seen in urethane anaesthetised animals (Table 1).

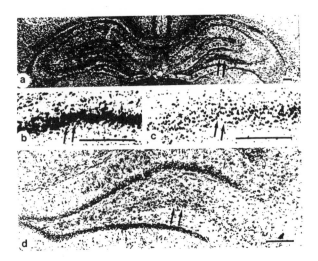

Figure 1. Light micrograph of hippocampus from one normal and one untreated heat stressed rat. The cell changes in the heat stressed animals are more marked in the CA 4 sector (arrows) (a). Neuronal population in CA 4 sector appears quite dense in a normal animal (b) whereas, in a heat stressed animal, loss of neurons, damaged cells and sponginess is quite frequent in this sector (c). However, GFAP positive astrocytes are present in all the regions of hippocampus in a heat stressed animal (d) (bar = 100 µm).

Blood-brain barrier permeability and edema. A marked increase in the extravasation of Evans blue and ^{131}I-sodium was noted in the hippocampus of conscious animals subjected to HS. The water content of the hippocampus was increased by 3 % in this group (Table 1). Pretreatment with p-CPA markedly reduced the increased BBB permeability and water content of the hippocampus. No increase in the BBB permeability or water content was seen in urethane anaesthetised animals after HS (Table 1).

Cell changes. Morphological examination of hippocampus in conscious animals subjected to HS showed marked neuronal loss, damaged neurones and edematous changes in the CA 4 subfield (Figure 1 b, c). The other subfields like CA 3, CA 2, CA 1, hilus and dentate gyrus were mainly intact. Immunostaining for GFAP showed the presence of reactive astrocytes in all regions of the hippocampus (Figure 1 a, d). Electron microscopy of hippocampus CA 4 subfield showed collapsed capillaries, perivascular edema, synaptic damage and disruption of myelin sheaths in the CA 4 sector (Figure 2). These cell changes were absent in animals treated with p-CPA or urethane before HS.

DISCUSSION

The salient new findings of the present study show that HS induces specific cell changes in hippocampus of *conscious* animals but not in *anaesthetised* rats. This indicates that stress associated with hyperthermia may play an important role in hippocampal damage following heat exposure. These cell changes were mainly confined to the CA 4 subfield of the hippocampus indicating that this sector is particularly vulnerable in HS.

Figure 2. High power electron micrograph from the CA 4 sector of hippocampus from one untreated heat stressed rat. Complete collapse of vessel with lanthanum (dark black particles) present into the lumen (arrows) is very common in this region. Synaptic damage (S) and perivascular edema (*) is clearly visible.

bar = 1 μm

The probable mechanisms of a selective vulnerability of the CA 4 subfield of the hippocampus in HS is unclear. It seems likely that a selective release of serotonin in this particular subfield plays an important role. The CA 4 subfield of hippocampus receives dense serotonergic input from the dorsal raphe complex located in the mibrain [3, 6]. Release of serotonin following HS may induce a breakdown of the BBB, vasogenic edema and cell changes [9, 10]. An absence of BBB breakdown, edema and structural changes in the p-CPA treated animals supports this hypothesis. A protective effect of urethane on cell changes may be due to inhibition of stress component of heat exposure. Absence of symptoms and hyperthermic response in urethane anaesthetised animals following HS are in line of this idea.

Our results further show that activation of astrocytes as evident with increased GFAP immunoreactivity could be non-specific to cell damage in the hippocampus. However, further studies are needed to clarify this point [7].

In summary, our results show that (i) HS and associated hyperthermia induces cell damage in the hippocampus in *conscious* animals, (ii) the CA 4 sector of the hippocampus is particularly vulnerable to HS, and (iii) serotonin plays an important role in the pathophysiological mechanisms of hippocampal damage in HS, not reported earlier.

Acknowledgements. This investigation is supported by grants from Alexander Humboldt Foundation, Bonn, Germany; Swedish Medical Research Council Project nos. 02710, 04X-9459, the Torsten and Ragnar Söderberg Foundation; and the University Grants Commission, New Delhi, India. The expert technical assistance of Flink Kärstin, Ingmarie Olsson, Elisabeth Scherer, Katja Deparade, Häna Plückhan, Franziska Drum and secretarial assistance of Aruna Misra, Katherin Kern and Angela Ludwig are appreciated with thanks.

REFERENCES

1. Sterner S (1990) Summer heat illnesses. Conditions that range from mild to fatal. Postgrad Med 87: 67-73.

2. Sharma H S, Cervós-Navarro J (1990) Brain oedema and cellular changes induced by acute heat stress. Acta Neurochir (Wien) Suppl. 51: 383-386.

3. Brodal A (1981) Neurological Anatomy. In relation to clinical medicine. 3rd Edition, Oxford University Press, Oxford, pp 683-690.

4. Sharma H S, Dey P K (1987) Influence of long-term heat exposure on regional blood-brain barrier permeability, cerebral blood flow and 5-HT level in conscious normotensive young rats. Brain Research 424: 153-162.

5. Sharma H S, Kretzschmar R, Cervós-Navarro J, Ermisch A, Rühle H -J, Dey P K (1992) Age related pathophysiology of the blood-brain barrier in heat stress. Prog Brain Res 91: 189-196.

6. Köhler C (1982) On the serotonergic innervation of the hippocampal region: an analysis employing immunohistochemistry and retrograde fluorescent tracing in the rat brain. In: Cytochemical Methods in Neuroanatomy (Eds V Chan-Palay, S L Palay), Allan R Liss, Inc, New York, pp 387-405.

7. Sharma H S, Zimmer C, Westman J, Cervós-Navarro J (1992) Acute systemic heat stress increases glial fibrillary acidic protein immunoreactivity in brain: Experimental observations in conscious normotensive young rats. Neuroscience 48: 889-901.

8. Koe B K, Weissman A (1966) p-chlorophenylalanine: a specific depletor of brain serotonin. J Pharmacol Exp Ther 154: 499-516.

9. Sharma H S, Olsson Y, Dey P K (1990) Changes in blood-brain barrier and cerebral blood flow following elevation of circulating serotonin level in anaesthetised rats. Brain Res 517: 215-223.

10. Sharma H S, Olsson Y (1990) Edema formation and cellular alterations following spinal cord injury in the rat and their modification with p-chlorophenylalanine. Acta Neuropathol (Berlin) 79: 604-610.

*****

Temperature Regulation
Advances in Pharmacological Sciences
© 1994 Birkhäuser Verlag Basel

DEVELOPMENT OF TEMPERATURE REGULATION IN PRECOCIAL CHICKS: PATTERNS IN SHOREBIRDS AND DUCKS

G. Henk Visser[1,2,3] and Robert E. Ricklefs[4]
[1]University of Groningen, Zoological Laboratory, P.O.Box 14, 9750 AA Haren, The Netherlands;
[2]Centre for Isotope research, Nijenborgh 4, 9747 AG Groningen, The Netherlands;
[3]University of Utrecht, Department of Veterinary Basic Sciences, Division Physiology, P.O.Box 80.176, 3508 TD Utrecht, The Netherlands;
[4]Department of Biology, University of Pennsylvania, Philadelphia, PA 19104-6018, USA

Summary

We compared the degree of homeothermy, the level of peak metabolic rate, and minimal thermal conductance in shorebird neonates of 15 species (neonatal body mass 4-55 g) with literature data on duck neonates of 11 species (neonatal body mass 17-61 g). In addition, we have studied the development of homeothermy in shorebird chicks during postnatal growth. In shorebird neonates a positive relationship existed between neonatal body mass and degree of homeothermy. In contrast, duck neonates of all sizes appeared to be homeothermic. The observed differences in degree of homeothermy between shorebird and duck chicks could almost entirely be explained by differences in peak metabolic rate.

Introduction

Time budgets of precocial chicks consist of alternating bouts of freely foraging and being brooded by a parent (1). Shorebird chicks grow up in grassland or tundra habitats mainly feeding upon insects. The time that chicks are brooded is unavailable for foraging. For young shorebird chicks, during adverse weather time available for foraging is too limited to enable normal growth (2). Ducklings grow up in aquatic habitats which are thought to impose a significant thermal constraint on small animals. Therefore we expect that special physiological adaptations have evolved in ducklings enabling them to cope with their thermal demands in aquatic environments (3). We investigate to what extent the thermal abilities differ between chicks of shorebirds and ducks in relation to neonatal body mass.

Materials and Methods

In the laboratory, rates of body cooling were determined in chicks of the following European and North American shorebird species (temperate and arctic Charadriidae and Scolopacidae): Least Sandpiper (*Calidris minutilla*; in the figures indicated by 1), Red-necked Phalarope (*Phalaropus lobatus*; 2), Semipalmated Plover (*Charadrius semipalmatus*; 3), Dunlin (*Calidris alpina*; 4), Stilt Sandpiper (*Micropalama himantopus*; 5), Lesser Yellowlegs (*Tringa flavipes*; 7), Short-billed Dowitcher (*Limnodromus griseus*; 8), Common Redshank (*Tringa totanus*; 9), Lesser Golden-Plover (*Pluvialis dominica*; 10), Northern Lapwing (*Vanellus vanellus*; 11), Hudsonian Godwit (*Limosa haemastica*; 12), Black-tailed Godwit (*Limosa limosa*; 13), Whimbrel (*Numenius phaeopus*; 14), and Eurasian Curlew (*Numenius arquata*; 15) ranging in neonatal body mass between 4 and 55 g. Initial (T_i) and final (T_f) body temperatures were rectally measured of individuals when exposed to 18°C (T_a) for 30 min. For each individual an index of homeothermy (H) was calculated using:

$$H = (T_f - T_a)/(T_i - T_a). \tag{eq. 1}$$

The index of homeothermy equals 1 at achievement of homeothermy (4).

In fed chicks of 5 European shorebird species (Ruff [*Philomachus pugnax*; in the figures indicated by 6], Common Redshank, Northern Lapwing, Black-tailed Godwit, and Eurasian Curlew) the peak metabolic rates, and minimal thermal conductances were determined at different ages by indirect calorimetry (5, 6).

Cooling rates of shorebird and duck neonates were simulated in the absence of metabolic heat production using an equation based on the Newtonian cooling model (5):

$$T_f = T_a + (T_i - T_a)\exp\text{-}(tK_{na}/C) \tag{eq. 2}$$

where t stands for time (s), K_{na} for the species-specific whole body dry thermal conductance (W °C^{-1}; see [5]), and C for the heat capacitance of the animal (3.45 J g^{-1}°C^{-1}). For the simulation T_a was taken as 18°C, t as 1800 s (i.e. 30 min), and T_i was assumed to be 40°C. Lastly, the simulated T_f values for each species were used to calculate the index of homeothermy using eq. 1.

We compare our data on shorebird chicks with existing data (3) on neonates of 11 temperate and arctic duck species (Anatidae) with neonatal body mass between 17 and 61 g.

Results and Discussion

In shorebird neonates the relationship between the index of homeothermy (*H*) and neonatal body mass (*M*; g) could be described by:

$$H = 0.073 + 0.464(SE=0.062)\log M, \ r^2=0.84, \ P<0.0001.$$ (eq. 3)

In contrast, duck neonates of all species were homeothermic when exposed to 18°C (Fig. 1). Neonates of some duck species are even reported to be homeothermic at 0°C (3). Thus, larger shorebird chicks have a higher degree of homeothermy than smaller chicks, and duck neonates are more homeothermic than shorebird neonates at a given body mass. In contrast to ducks, chicks of small shorebird species tripled their body mass before being homeothermic at 18°C, whereas in chicks of large shorebird species the body mass increased by only 10% before homeothermy was achieved.

In order to obtain insight in the importance of the level of peak metabolic rate upon the degree of homeothermy, we have simulated the cooling rates of shorebird and duck

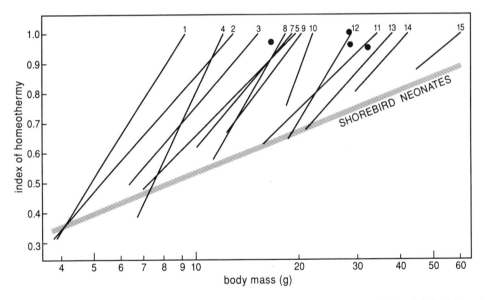

Fig. 1. Index of homeothermy in relation to body mass in neonates of shorebirds (dotted diagonal bar; eq. 3) and ducks (solid circles), and during postnatal development in shorebirds (diagonal lines; the numbers at the top of each line refer to the species mentioned in the Materials and Methods section).

neonates in the absence of metabolic heat production (eq. 2). As can be seen in Fig. 2, in chicks of both taxonomic groups the simulated indices of homeothermy in the absence of metabolic heat production were much lower than the actually measured indices in normal chicks (with metabolic heat production). Additionally, under the simulated conditions, larger chicks do have a higher degree of homeothermy than smaller chicks. Lastly, under the simulated conditions the large difference between shorebird and duck neonates, as observed in Fig. 1, has almost disappeared. Thus, the level of metabolic heat production as well as body mass seems to be decisive for the degree of homeothermy.

In shorebird neonates the relationship between the peak metabolic rate (PMR; W) and body mass (M; g) could be described by:

$$\log PMR = -1.688(SE = 0.1295) + 0.892(SE = 0.1021)\log M, \; r^2 = 0.89, \; P < 0.001 \qquad \text{(eq. 4)}$$

and in duck neonates by:

$$\log PMR = -1.108(SE = 0.2002) + 0.782(SE = 0.1331)\log M, \; r^2 = 0.97, \; P < 0.11 \qquad \text{(eq. 5)}$$

(Fig. 3a; data listed in [5]). Thus, larger chicks have a higher peak metabolic rate than smaller chicks, and within the body mass range of 17-60 g, peak metabolic rates of ducklings were 2.4-2.8 times higher than observed in shorebird neonates. During the early postnatal

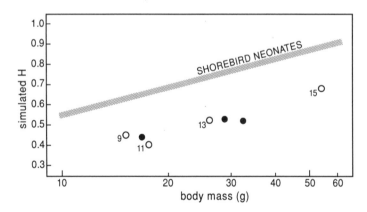

Fig. 2. Simulated indices of homeothermy of shorebird (open circles) and duck neonates (solid circles) in the absence of metabolic heat production when exposed during 30 min at 18 °C. The numbers next to the symbols for shorebirds refer to the species mentioned in the Materials and Methods. The diagonal dotted bar represents the observed relationship between the index of homeothermy and body mass in shorebird neonates (eq. 3).

growth phase, levels of peak metabolic rate of shorebird chicks increased rapidly with body mass (exponents of allometric relationships fell between 1.9-2.3; Fig. 3a).

The relationship between the whole body minimal thermal conductance (K; mW $°C^{-1}$) and body mass (M; g) in shorebird neonates could be described by:

$$\log K = 1.096(SE=0.1394) + 0.373(SE=0.1078)\log M, r^2=0.60, P<0.01 \qquad \text{(eq. 6)}$$

and in duck neonates by:

$$\log K = 1.067(SE=0.1213) + 0.314(SE=0.0774)\log M, r^2=0.67, P<0.01 \qquad \text{(eq. 7)}$$

(Fig. 3b; data published in [5]). Thus, larger neonates have only a slightly higher thermal conductance than smaller neonates (on the basis of the surface to volume relationship of a sphere we would have expected an exponent of 0.67). Within the body mass range of 17-60 g minimal thermal conductances in shorebird neonates were 27-36% higher than in duck neonates. In shorebird chicks during the early phase of postnatal growth the values for the whole body minimal thermal conductances increased only moderately (exponents of the

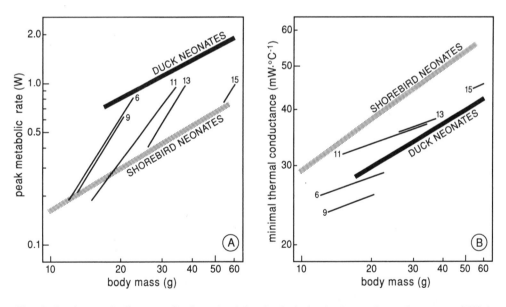

Fig. 3. Peak metabolic rates (3A) and minimal whole body thermal conductances (3B) in neonates of shorebirds (diagonal dotted bar; eqs. 4 and 6) and ducks (solid diagonal bar; eqs. 5 and 7), and during postnatal development in shorebird chicks (diagonal lines; the numbers at the top of each line refer to the species listed in the Materials and Methods).

allometric relationships were 0.19; Fig. 3b).

In conclusion, the high degree of homeothermy in ducklings appears to be the result of high rates of peak metabolic heat production and, to a lesser extent, to a lower thermal conductance. Apparently ducklings follow an energetically "expensive" strategy to achieve homeothermy. Their relative thermal independence enable ducklings to forage for a longer time period, potentially resulting in a higher energy intake. On the other hand, shorebird chicks seem to follow an energetically "cheaper" strategy thereby relying more on parental brooding subsidy, which might improve the over-all efficiency of the parent-offspring unit. The parent and young of precocial species are intimately tied together through the dependence of the chick upon the parent for the maintenance of the body temperature. This relationship varies both during the course of development, between different taxa, and in relation to weather conditions and availability of food.

Acknowledgements

The investigations were supported by the Foundation for Biological Research (BION), which is subsidized by the Netherlands Organisation for Scientific Research (NWO), under Grant #427.181, and the NSF grants DPP 7908571 and BSR 90-07000. K. Baker, L. Clark, D. Goldstein, M. van Kampen, and E. Zeinstra helped in many different ways.

References
(1) Visser GH, Beintema AJ, van Kampen M. Development of thermoregulation in chicks of Charadriiform birds. In: Mercer JB (ed.) Thermal Physiology 1989:725-730. Elsevier, Amsterdam.
(2) Beintema AJ, Visser GH. The effect of weather on time budgets and development of chicks of meadow birds. Ardea 1989;77:181-192.
(3) Koskimies J, Lahti L. Cold-hardiness of the newly hatched young in relation to ecology and distribution in ten species of European ducks. Auk 1964;81:281-307.
(4) Ricklefs RE. Characterizing the development of homeothermy by rate of body cooling. Functional Ecology 1987;1:151-157.
(5) Visser GH, Ricklefs RE. Temperature regulation in neonates of shorebirds. Auk 1993;110 in press.
(6) Visser GH, Ricklefs RE. Development of temperature regulation in shorebirds. Physiological Zoology 1993;66 in press.

Temperature Regulation
Advances in Pharmacological Sciences
© 1994 Birkhäuser Verlag Basel

CENTRAL VENOUS PRESSURE AND CARDIOVASCULAR RESPONSES TO HYPERTHERMIA

Taketoshi Morimoto, Akira Takamata and Hiroshi Nose
Department of Physiology, Kyoto Prefectural University of Medicine
Kyoto 602, Japan.

We studied the relationship between central venous pressure (CVP) and total vascular conductance (TVC) in anesthetized rats at three levels of blood volume: normovolemia (NBV), hypervolemia (HBV), and hypovolemia (LBV), during body heating. TVC showed significant correlation with CVP, and the slope of CVP vs. TVC in LBV was 3- to 4-folds steeper than in NBV or HBV. When rats were vagotomized, the slope of CVP vs. TVC was reduced by about 40% than in the control rats with intact vagi. The results suggest that under hyperthermia, cardiopulmonary baroreflex is involved in the control of TVC, which prevents blood pooling in skin vessel and serves to maintain circulatory function under heat stress.

Thermal stress induces sweating and cutaneous vasodilation for heat dissipation, which causes dehydration and hypovolemia relative to the expanded vascular space. Both of these responses decrease central venous pressure and become a threat for the maintenance of cardiac output and arterial blood pressure under heat stress (1). As a regulatory response, splanchnic vasoconstriction has been known to take place. More recently, Mack et al. (2) applied negative pressure to the lower body of exercising subjects and found reductions of forearm blood flow in proportion to the pressure applied to the lower body, and the reduction induced the increase in body temperature. The result indicates that, in addition to the splanchnic vasoconstriction, skin blood flow is also reduced when the demand for circulating blood to maintain cardiac output is increased.

To study the regulatory mechanism for maintaining circulatory function under heat stress, we studied cardiovascular responses to the elevation of core temperature in rats with different levels of circulating blood volume. The results were analyzed with special reference to the relationship between central venous pressure (CVP) and total vascular conductance (TVC) (Experiment 1). Further, to assess whether cardiopulmonary baroreflex is involved in the regulation of TVC or not, CVP-TVC relationship was determined on vagotomized rats and the relationship was compared with that obtained on the control rats without vagotomy (Experiment 2).

Methods

Experiment 1: α-chloralose-anesthetized rats were divided into three groups: normovolemia (NBV), hypervolemia (HBV; +32% plasma volume by isotonic albumin solution infusion), and hypovolemia (LBV; -16% plasma volume by furosemide administration). Using body surface heating with an infrared lamp, arterial blood temperature (Tb) was raised at a rate of $0.1^{\circ}C/min$ monitoring with a thermocouple probe placed in the mid thoracic aorta. Cardiac output (CO) was determined by a thermodilution method using the thermocouple probe. Arterial blood pressure (AP) and CVP were also measured continuously, and heart rate (HR) was obtained from the pulsations of the AP recording. Stroke volume (SV) was calculated from CO and heart rate, and total vascular conductance (TVC) was calculated from AP, CVP and CO.

Experiment 2: To assess the degree of the involvement of CVP, or cardiopulmonary baroreceptor in the control of TVC, anesthetized rats were divided into four groups: without vagotomy under normovolemia and hypovolemia, and with bilateral cervical vagotomy under normovolemia and hypovolemia, and the same variables as in Experiment 1 were measured during body heating.

Results

Experiment 1: Heart rate responses to the raised Tb were similar among the three groups with varied blood volume. Mean arterial pressure was not affected by blood volume modification or CVP, and was maintained at preheating level until the Tb level of $40^{\circ}C$. Stroke volume was closely correlated with CVP (SV = 0.36 x CVP + 1.17, r = 0.899), and this relationship was not affected by Tb. TVC decreased linearly as Tb was increased. In addition, as shown in Fig. 1, TVC decreased with the fall of CVP, and the slope of TVC vs. CVP in LBV was steeper than in NBV by about 60%, and by about 40% than in HBV (3).

Experiment 2: Circulatory responses in the group without vagotomy were similar as in Experiment 1, whereas the responses to the increased Tb and hypovolemia were reduced in the vagotomized group. In Fig. 2, open circles show the relationship between CVP and TVC of the control groups with intact vagi and closed circles show the results from the vagotomy groups. The slope of the CVP and TVC relationship from the vagotomy groups was reduced by 39% compared to that in the intact vagi groups. The result indicates that the cardiopulmonary baroreflex via vagi contributes about 40% in the control of total vascular conductance during hyperthermia (4).

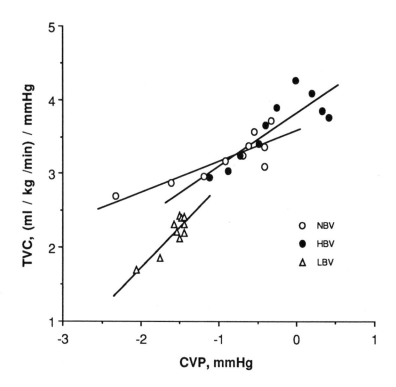

Figure 1. The effect of blood volume modification on the relationship between central venous pressure (CVP) and total vascular conductance (TVC) during body heating in rats. Each point represents mean values of six rats at each body temperature for normovolemia (NBV), hypervolemia (HBV), and hypovolemia (LBV). The regression equations are TVC = 3.6 + 0.4 x CVP, r = 0.847 for NBV, TVC = 3.8 + 0.7 x CVP, r = 0.846 for HBV, and TVC = 3.9 + 1.1 x CVP, r = 0.879 for LBV.

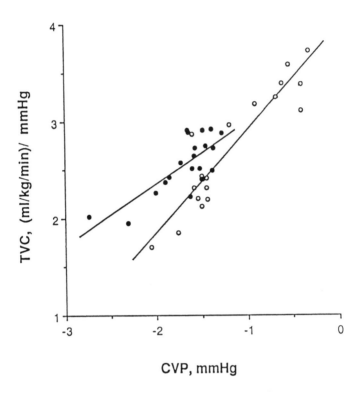

Figure 2. The effect of vagotomy on the relationship between central venous pressure (CVP) and total vascular conductance (TVC) during body heating in rats. Each point represents mean values of six rats at each body temperature for the control group without vagotomy (open circles), and the vagotomized group (closed circles). The regression equations are TVC = 3.6 + 0.4 x CVP, r = 0.925 for the control group, and TVC = 3.8 + 0.7 x CVP, r = 0.757 for the vagotomized group.

Discussion

Heat stress causes redistribution of cardiac output toward cutaneous circulation and central blood volume and cardiac filling pressure are decreased. These heat-induced changes in circulation threaten the maintenance of cardiac output and arterial pressure. As the regulatory responses under this condition, in addition to the splanchnic vasoconstriction, we have shown that the decrease in vascular compliance is very important to maintain CVP, and also that the vascular compliance is regulated with sympathetic nervous system (5).

In addition, we found in the present experiment that a fall in CVP during hyperthermia is associated with a decrease in total vascular conductance. We also found that the slope of linear relationship between CVP and TVC increases with the increase in body temperature. These findings suggest that, under heat stress, cutaneous vasodilation for heat dissipation causes expansion of vascular space and induces hypovolemia relative to the expanded vascular space and reduction of CVP. As the regulatory responses, together with splanchnic vasoconstriction and decrease in vascular compliance, decrease in TVC is observed. The decrease in vascular conductance including cutaneous circulation suppresses heat dissipation and elevates body temperature. In other words, the maintenance of cardiac filling pressure with the constriction of peripheral circulation predominates over thermoregulatory responses under hyperthermia and plasma volume expansion suppresses the reduction in vascular conductance and improves heat tolerance.

The results from the vagotomized rats suggest that under hyperthermia, both cardiopulmonary baroreflex and increased body temperature are involved in the control of vascular conductance.

References
1. Morimoto T. Thermoregulation and body fluids: role of blood volume and central venous pressure. Jpn J Physiol 1990; 40: 165-179.
2. Mack GW, Nose H, Nadel ER. Role of cardiopulmonary baroreflexes during dynamic exercise. J Appl Physiol 1988; 65: 1827-1832.
3. Takamata A, Nose H, Mack GW, Morimoto T. Control of total peripheral resistance during hyperthermia in rats. J Appl Physiol 1990; 69: 1087-1092.
4. Takamata A. Effect of vagotomy on cardiovascular adjustment to hyperthermia in rats. Jpn J Physiol 1992; 42: 641-652.
5. Shigemi K, Morimoto T, Itoh T, Natsuyama T, Hashimoto S, Tanaka Y. Regulation of vascular compliance and stress relaxation by the sympathetic nervous system. Jpn J Physiol 1991; 41: 577-588.

EFFECTS OF BRIGHT AND DIM LIGHT INTENSITIES DURING DAYTIME UPON CIRCADIAN RHYTHM OF CORE TEMPERATURE IN MAN

H. Tokura[1], M. Yutani[1], T. Morita[2] and M. Murakami[2]

[1]Department of Environmental Health, Nara Women's University, Nara, 630 Japan and [2]Human Science Division, Comprehensive Housing R & D Institute, Sekisui House, LTD., Kyoto, 619-02 Japan

Summary
    Circadian rhythms of rectal temperature were individually compared in a bioclimatic chamber with 26°C for several days between two kinds of different light conditions: (1) bright light with 5,000 lux from 6:00 a.m. to 6:00 p.m. (2) dim light with 60 lux from 6:00 a.m. to 6:00 p.m.. Light conditions except the daytime were quite same, i.e., 60 lux from 6:00 p.m. to 10:00 p.m. and almost dark from 10:00 p.m. to 6:00 a.m. in the following day. The maximum and minimum values of circadian rhythm were higher under dim light conditions than under bright ones. Thus,the circadian curves of rectal temperature seemed to shift upwards day and night under dim light conditions.

Introduction

    Light and temperature are most important entraining Zeitgebers for freerunning circadian rhythms in man and animals (1). To our knowledge, it remains to be studied how the entrained circadian rhythm in man could be changed under the influences of different environmental conditions during the daytime. With these in mind, the present paper aimed to know the effects of bright and dim light intensities during the daytime on the circadian rhythm of rectal temperature in man.

Materials & Methods

    One middled-aged and two youg adult females volunteered as subjects. Each subject stayed individually in a bioclimatic chamber with an ambient temperature of 26°C and a relative humidity of 50 % for 3 days under bright light conditions and for another 3 days under dim light conditions. Bright light with ca. 5,000 lux by many fluorescent lamps at 1 m distance from them was turned on at 6:00 a.m. and off at 6:00 p.m.. The bioclimatic chamber had floor area of ca. 30 m^2 with bed, small kitchen, bath, toilet, wash stand, study desk, sofa and a floor lamp. Floor was covered by carpet. The incadescent lamps consisting of 3 desk lamps and 1 floor lamp with light intensity of ca. 60 lux in the middle of the room were turned on at 6:00 a.m. and off

at 10:00 p.m.   both under bright and dim light conditions. The subject had to get up   at 6:00
a.m. and to go to bed at 10:00 p.m.. Rectal and skin temperatures at 7 sites (forehead, chest,
forearm, hand, thigh, leg, foot) were continuously measured throughout day and night.

Two out of three subjects lived under bright light conditions and then   under dim light
conditions a few days later for 3 days, respectively. The other female subject   lived   in opposite
direction: under dim light conditions for first 3 days and   under bright light conditions for
another 3 days.

We applied the best fitting curves (2) for raw measurements of rectal temperatures to get their
maximum and minimum values.

Results

Fig. 1 represents a typical example of circadian rhythm in rectal temperature recorded
continuously for 3 days under bright and dim light conditions in a female subject, respectively.
As seen in the figure, clear circadian rhythm of rectal temperature existed both in bright (right)
and dim (left) light conditions: higher during the daytime and lower at night. However, close
observation disclosed that the level of rectal temperatures was distinctly higher during the
daytime under dim light conditions than under bright light conditions. The night values were also
higher in two out of three nights under dim light conditions. Similar findings were also observed
in other 2 subjects.

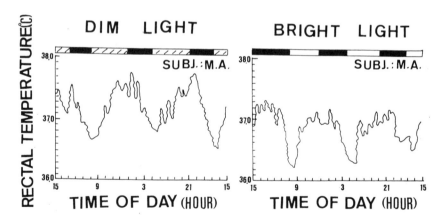

Fig. 1. Circadian rhythm of rectal temperature under the influences of bright (right) and dim (left)
light conditions in a female subject.

Circadian rhythm of rectal temperature was individually compared between two conditions of bright and dim light in Fig. 2. The best fitting curves were drawn for raw measurements of rectal temperature obtained from second 24 hr day. As seen in the figure, it should be noticed here that the circadian curves of rectal temperature shifted upwards slightly day and night in all 3 subjects under dim light conditions than under bright light conditions.

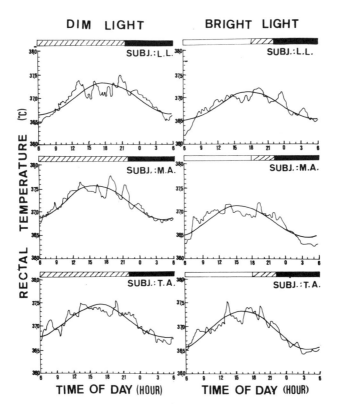

Fig. 2. A comparison of circadian rhythm in rectal temperature between two conditions of bright and dim light in three subjects. Best fitting curves were applied for raw measurements of rectal temperature. Left: dim light. Right: bright light. White: 5,000 lux. Shade: 60 lux. Black: dark.

Judging from the best fitting curves applied for raw temperature data in Fig. 2, the maximum and minimum values were 37.4°C and 36.8°C under dim light conditions, respectively, 37.3°C and 36.5°C under bright light conditions in a male subject (T.A.), 37.3°C and 36.6° under dim light conditions, 37.1°C and 36.5°C under bright light conditions in a femal subject (L.L.),

37.6°C and   36.8°C under dim light conditions, 37.1°C and 36.4°C under   bright light conditions in a female subject (M.M.). Thus, the maximum and minimum values of circadian rhythm in rectal temperature were distinctly   higher in all 3 subjects under dim light conditions. Furthermore, the times that the maximum and minimum values occurred were faster by ca. 20 min under bright light conditions, suggesting that the circadian phase of rectal temperature was slightly in advance under bright light conditions.

Table 1. The maximum and minimum values of circadian rhythm in rectal
temperature and their occurrence times in 3 subjects.

Subject	Light	Maximum Rectal temp.	Time	Minimum Rectal temp.	Time
T.A.	Bright	37.3°C	16 h10 min	36.5°C	4 h 10 min
	Dim	37.4°C	16 h 30 min	36.8°C	4 h 30 min
L.L.	Bright	37.1°C	17 h 50 min	36.5°C	5 h 50 min
	Dim	37.3°C	18 h 10 min	36.6°C	6 h 10 min
M.M.	Bright	37.1°C	15 h 50 min	36.4°C	3 h 50 min
	Dim	37.6°C	16 h 10 min	36.8°C	4 h 10 min

Discussion

What physiological mechanisms are responsible for the most interesting findings that the level of rectal temperature seemed to shift upwards day and night under dim light conditions than under bright light conditions ?

Recently, Ishihara et al. (3) reported that plasma melatonin   during the daytime was higher under bright light conditions of 2,500 lux than under ordinary and dim light conditions of 500 lux and 250 lux. If this were the case in our present experiment, plasma melatonin level during the daytime might be higher   under bright light conditions of 5,000 lux during the daytime. According to Cagnacci et al. (4), circadian rhythms of plasma melatonin and core temperature are inversely linked. Therefore, higher level of plasma melatonin during the daytime under bright light conditions might have induced the rectal temperature lower in our present experiment.

It is well known that a decrease of core temperature is observed when circulating levels of melatonin are high at night (5) and bright light exposure at night reduces both the nocturnal core temperature fall and melatonin secretion (6, 7). However, although environmental conditions at night were quite equal between bright light and dim light conditions during the daytime in our present experiment, the level of rectal temperature at night was lower under bright light conditions during the daytime, suggesting the existence of "after effect" of bright light during the daytime. It seems probable that plasma melatonin level at night might be different under the influences of bright and dim light conditions during the daytime. Simultaneous 24 hr measurements of core temperature and plasma melatonin under the influences of bright and dim light conditions during the daytime is absolutely necessary to make clear the involvement of melatonin for different levels of core temperature at night, which might be affected by two light intensities during the daytime.

Old people always lying in bed throughout the day, taking a nap or closing their eyes often and being almost surrounded by dim light even during the daytime showed higher level of rectal temperature both during the daytime and at night than healthy old people having bright sun light , walking around actively in the facilities and not taking a nap (Tokura et al., unpublished data). These might be interpreted in terms of similar physiological mechanisms hypothesized in our present paper.

References

1. Aschoff J (1979) Z Tierpsychol 49:225-249
2. Takahashi N, Kamiyama A , Ootomo, N (1992) The Structure of Biological
      Rhythms, Fuji Shoin, Sapporo, pp.81-150.
3. Ishihara Y, Tsuboi Y , Fujita S (1993) Jpn J Hyg, 48:332.
4. Cagnacci A, Elliott, J A Yen S S C (1992) J Cli Endocrinol 75: 447-452
5. Wever RA (1989) J Biol Rhythms 4:161-185
6. Lewy A J, Wehr TA, Goodwin FK, Newsome DA, Markey SP (1980)
      Science 210: 1267-1269.
7. Badia P, Cupepper J, Boecker M, Myers B, Harsh J (1990) Sleep Res 19:
      386

Temperature Regulation
Advances in Pharmacological Sciences
© 1994 Birkhäuser Verlag Basel

LONG TERM HEAT ACCLIMATION:ACQUIRED PERIPHERAL CARDIOVASCULAR ADAPTATIONS AND THEIR STABILITY UNDER MULTIFACTORIAL STRESSORS

Michal Horowitz, Walid Haddad *Mara Shochina and Uri Meiri
Division of Physiology Hadassah Schs of Medicine and Dental Medicine, The Hebrew University and *Department of Rehabilitation, Hadassah University Hospital, Mount Scopus, Jerusalem, Israel

SUMMARY

With heat acclimation, effectors efficiency increases. In blood vessels this is exemplified by increased [force generation]/ [sympathomimetic signalling] ratio. Cellular adaptive responses may contribute to this effect; however, their persistence when additional stress is applied is yet unknown. The present investigation shows, in aged rats, that hypohydration, superimposed on heat acclimation blunt acclimation–induced increased force generation in aortic rings and α adrenergic sensitivity. Upon hypohydration, heat stress induces faster and more pronounced splanchnic vasoconstriction. This may suggest that adaptations at the signalling pathway can be subjected to rapid changes under multifactorial stress.

INTRODUCTION

Previous studies from our laboratory have shown that heat acclimation is a continuum of temporally varying processes, shared by both the autonomic nervous system and the peripheral effectors (1). At the early periods of the acclimation regimen accelerated excitability of the autonomic nervous system contributes to the recruitment of transient mechanisms working in concert to alleviate the initial strain, even at the expense of body homeostasis. With prolongation of the acclimation regimen, long–lasting features emerging at the effector level, replacing the need for the evoked transient mechanisms. During this phase, the sum of the varying homeostatic responses is optimized. Under these conditions the [Effector output]/[Autonomic signal] ratio increases, compared to the pre–acclimation level, suggesting increased efficiency of the effector. This is when acclimation has occurred (2,3).

In the cardiovascular system, at the integrative level, adaptations are mostly exemplified by an expanded range of activity of the various mechanisms. For example, lowering of normothermic heart rate allows, during heat stress, greater elevation in heart rate within safety margin. Likewise, increased peripheral tone, observed in humans) allows the supply of greater circulatory volume for heat dissipation (4). In rodents, following heat acclimation, the share of cardiac output distributed to the splanchnic area increases, even in thermoneutral environments, increases. During heat stress vasoconstriction is more pronounced; nevertheless, splanchnic perfusion does not drop much below the preacclimation level. As a result, heat transfer from this metabolically active vascular area to the periphery is not impaired during a longer period of heat stress, thus allowing improved heat endurance (5).

We have recently found that membranal or intracellular biochemical changes in the blood vesseles contribute their share to the latter improved acclimatory response. In isolated blood vessels a greater developed force is accompanied by a left shift of the dose response curve to noradrenaline, suggesting increased responsiveness of the acclimated vasculature to the neurotransmitter (Haddad and Horowitz, in preparation and 3). These findings may imply that alterations in signal transduction may occur in the vasculature, although biochemical changes in contractile proteins can not be refuted. So far this has been studied only in the cardiac muscle. Heat acclimated hearts show a shift of cardiac isoenzymes from the fast V_1 isoform, with high ATPase activity, to the slow V_3 isoform with low ATPase activity. This shift may account for the observed low oxygen consumption in these hearts despite the greater pressure produced (6). These metabolic changes have delayed appearance and delayed disappearance. In contrast, the time course of changes in signal transduction may be rapid and their persistence is unknown.

In a previous study on the effect of hypohydration superimposed on heat acclimation we showed that heat endurance in aged acclimated hypohydrated rats exceeds that of young acclimated hypohydrated rats. Prior to hypohydration, the young acclimated rats resisted heat stress better than the aged. We also showed in the hypohydrated aged rats that thermally evoked vasomotor responses (in terms of temperature thresholds and sensitivity) were improved in the aged compared to the young rats. These aged rats showed better ability than the young ones to regulate blood volume during heat stress (7). Since most of these changes are initiated via transmitter–receptor coupling, we hypothesized that cardiovascular system of aged acclimated–hypohydrated rats can provide a suitable experimental model for studying the stability of thermally induced vascular adaptations, associated with signal transduction, under conditions of multifactorial stress.

This communication discusses the effect of the combined stresses of heat and hypohydration on adaptive features associated with force generation and vascular tone in aged acclimated rats.

MATERIALS AND METHODS

Male rats (Zabar strain, albino variation) 12 months old were used. Heat acclimation was carried out by continuous exposure at $34^{\circ}C$ for one month with food and water ad libitum. Hypohydration was attained by water deprivation to achieve about 8–10% weight loss. Vascular responsiveness to thermal and catecholamine stimulation was studied in intact animals and in isolated organ preparations, respectively. This allowed evaluation of central and peripheral contributions to the adaptive response.

Experiments in intact, heat stressed animals (at 42oC) comprised measurements of portal blood flow (PBF). Since the entire splanchnic circulation is drained via this route, PBF was taken as a measure of the changes occurring in the splanchnic vascular bed during heat stress. PBF was measured using a Transonic blood flow meter (T 106) with flow probe 3S, placed around the portal vein under pentobarbital anesthesia. Blood flow measurements were carried out in conscious rats, several hours after their recovery from anesthesia, coincidentally with on–line recording of rectal temperature using a YSI #402 thermistor probe.

For measuring blood vessel responsiveness to catecholamines, cumulative dose response curves to α and β catecholamines were determined, using aortic rings (and hearts) were carried out in an organ bath.

RESULTS

The Portal blood flow of euhydrated, non–acclimated, aged rats is shown in Fig. 1. Following the thermoregulatory vasoconstriction and preceding circulatory failure, portal vasodilatation implying blood congestion in this vascular bed was observed. Heat acclimated aged rats showed similar basal PBF. However, splanchnic vasoconstriction was elicited at a significantly higher temperature threshold (T_{sh}, Table 1), was greater (not shown) than that of the controls and did not show late vasodilatation. Hypohydrated rats showed significantly lower PBF (Table 1). Portal vasoconstriction was elicited immediately on subjection to heat stress and was more pronounced compared to that observed in control euhydrated rats (Fig. 1). Basal PBFs and T_{sh} for vasoconstriction for the various groups are presented in Table 1.

Table 1. Basal portal blood flow and temperature threshold (Tsh) for vasoconstriction in control euhydrated (C), hypohydrated (CD) and acclimated euhydrated (AC) aged rats

	C	CD	AC
Basal PBF (ml/min)	23.50±4.30	11.70±3.05**	22.50±2.78
T_{sh} (°C)	39.60±0.40	37.40±0.70*	41.40±0.40**

**P<0.001 and *P<0.01 compared to C rats.

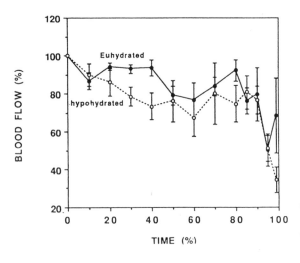

Fig.1. Portal blood flow in euhydrated and hypohydrated control aged rats. PBF (M±SE) is presented as the percentage of basal normothermic flow. Maximal resistence time for each rat was taken as 100%. No significant difference in heat endurance between the groups was observed. n for euhydrated and hypohydrated rats was 8 and 6 respectively.

Aortic rings excised from aged rats showed very poor response to isoprenaline (β agonist) and propranolol (β antagonist). In contrast, the α agonist (phenylephrine) produced a remarkably significant effect. The force developed by vessels excised from acclimated rats was significantly greater (~20%) than in those from the non-acclimated rats (Fig. 2). By contrast, in the cardiac muscle, inotropic response to both α and β adrenergic receptors (AR) was greater following acclimation, suggesting organ specificity of the heat acclimation effect. Hypohydration abolished the acclimation effect, thus no difference in the force produced by acclimated and non-acclimated aortas was observed (Fig. 2). Concomitantly, hypohydration affected sensitivity to the transmitter. Both non-acclimated and acclimated hypohydrated vessels showed a significant left shift of the dose response curve, suggesting increased sensitivity to the agonist (Fig. 3).

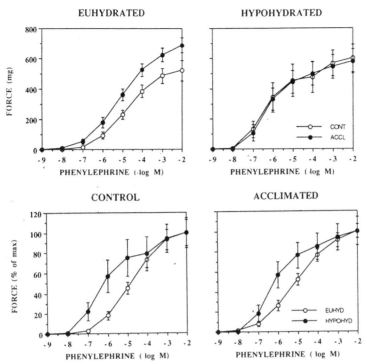

Fig. 2 (upper panel). The force (mg) developed by aortic rings excised from non–acclimated (cont) and acclimated (accl) aged rats following administration of increasing doses of phenylephrine to the organ bath. Data are represented as M±SEM, n for cont and accl: euhydrated – 28 and 21 and hypohydrated – 6 and 8, respectively.

Fig. 3 (lower panel). The force (% of maximum) developed by aortic rings excised from cont euhydrated and hypohydrated (left) and accl euhydrated and hypohydrated (right) aged rats following administration of increaseing doses of phenylephrine to the organ bath. Curves were calculated from the data presented in Fig.2.

DISCUSSION

Our data provide evidence for increased α adrenergic – induced force generation in aortas of heat acclimated aged rats. Sensitivity to the transmitter was not affected with acclimation, thus implying quantitive rather than qualitative effects. The findings may suggest that following heat acclimation peripheral changes at the effector level can modulate the intensity of the autonomic reflex. Previous findings on young acclimated rats showed similar induced–response. However, while increased α adrenergic responsiveness seems to be responsible for this effect in aged rats, a reduction in sensitivity of the β adrenergic pathway simultaneously with increased responsiveness of the α adrenergic pathway is responsible for a similar outcome in young rats (Haddad and Horowitz, in preparation). Preliminary experiments in young rats, have shown down–regulation of α AR with acclimation concomitantly with the greater force generation. This may imply increased efficiency of this transduction pathway (Haddad and Horowitz personal communication).

Hypohydration increased the sensitivity of the blood vessels to phenylephrine in both non–acclimated and acclimated old rats and blunted the ability of acclimated vessels to produce greater force than that measured in the pre–acclimation state. These findings suggest that hypohydration induces qualitative changes in the vasculature, possibly at the α adrenergic pathway, as suggested from our findings. It also suggests that these changes override modulations acquired with heat acclimation. This notion matches our previous findings on the evaporative cooling system of the rat: hypohydration almost abolished salivation cooling in short–term heat acclimated rats. Matched euhydrated acclimated rats increased salivation cooling by almost 100% (8).

Subjection of the intact animals to multifactorial stress (heat superimposed on hypohydration) evoked faster and greater thermal induced portal constriction. This fits our finding of increased α adrenergic sensitivity in isolated vessels and supports a conclusion that in intact hypohydrated aged rats a local vascular response, associated with the α adrenergic pathway, contributes to the augmented vasoconstriction. Our data are the first to show that modulations in force generation (at the membranal or cellular level) brought about by acclimation can be subjected to rapid changes under conditions of multifactorial stress.

REFERENCES

1. Horowitz M. Heat acclimation: A continuum of processes. In: Thermal Physiology. ed. J. Mercer, 1989; pp: 445–450, Elsevier.
2. Horowitz M. & Meiri U. Central and peripheral contributions to control of heart rate during heat acclimation. Pfl Arch 1993; 422: 386–392.
3. Horowitz M. Heat stress and heat acclimation: The cellular response – modifier of autonomic control. in: Integrative and cellular aspects of autonomic functions. Plescka K, Gerstenber R. and Pieraue K. Fr.eds. John Libbey Eurotext Ltd. in Press.
4. Rowell LB. Human cardiovascular adjustment to thermal stress. in Sherherd, Abboud   eds. Handbook of Physiol, The cardiovascular system, 1983; Am Physiol Soc Bethesda pp:967–1025.
5 Horowitz M. & Samueloff S. Cardiac output distribution in thermally dehydrated   rodents. Am J Physiol 1988; 254: R109–R116.
6. Horowitz M. Parnes S. Hasin Y. Mechanical and metabolic performance of the rat heat. Effects of combined stress of heat acclimation and swimming training. J.Basic Clin Physiol Pharamacol 1993; 4:139–156.
7. Meiri U. Shochina M. Horowitz M. Heat acclimated hypohydrated rats: Age dependent vasomotor and plasma volume response to heat stress. J Therm Biol 1991; 16:241–247
8. Horowitz M. Meiri U. Thermoregulatory activity in the rat: effects of hypohydration, hypovolemia and hypertonicity and their interaction with short term heat acclimation. Com Biochem Physiol 1985; 82A:577–582.

Temperature Regulation
Advances in Pharmacological Sciences
© 1994 Birkhäuser Verlag Basel

EFFECTS OF SOLAR RADIATION AND FEED QUALITY ON HEART RATE AND HEAT BALANCE PARAMETERS IN CATTLE.

Arieh Brosh[1], Stephen Fennell[2], Dick Wright[3], Graham Beneke[2] and Bruce Young[2].
[1]Department of Beef Cattle, Agricultural Research Organisation, Institute of Animal Science, Ministry of Agriculture; Israel.
[2]Department of Animal Production, The University of Queensland, Gatton College, Australia.
[3]Department of Companion Animal Medicine & Surgery, The University of Queensland, St. Lucia, Australia.

SUMMARY: Effects of shade and feed quality on feed intake, water consumption, rectal temperature, respiration rate, heart rate, plasma protein and blood hematocrit were measured in growing Hereford heifers during the summer. Heifers that were fed a high energy grain diet and exposed to the sun had higher heat loads during the hotter part of the day. Plasma volume was unaffected by the treatments. Heifers on all treatments coped well with the heat load by utilising panting and changes in body temperature to compensate for the increased heat load from solar radiation and ingestion of grain. Heart rate was mainly affected by feed quality and not by solar radiation.

INTRODUCTION: Summer heat can cause a reduction in productivity, and may cause mortality in grain fed cattle (1). According to the classic thermogenesis curve (2), when the environmental temperature is high an animal has an increase in heat production (HP) and heat loads are also higher. However, heat balance can be maintained by active evaporation. In contrast, a study (3) with cattle and sheep under a mild heat load showed that while the respiration rates (RR) were faster and rectal temperatures (Tr) increased, the animals' HP did not change. It was suggested that active respiratory cooling is an energy efficient process and the higher RR were compensated for by a reduction in the blood flow to and metabolism of other tissues. Recent work (4)has shown that in growing cattle the energy consumed by splanchnic tissues accounts for about 45% of the total body HP and when metabolizable energy (ME) intake was reduced by 42% the splanchnic HP was reduced by 38% and the total HP by 31%.

Provision of shade from solar radiation has been advocated as a method to relieve heat load on animals. However, construction of shade is expensive and beef feedlot producers consider that shade reduces the marble fat that is preferred by the Japanese market. This study was designed to determine the effects of solar radiation and feed quality on physiological parameters that relate to heat balance in feedlot beef cattle during summer in sub tropical region. MATERIALS & METHODS: Effects of shade and feed quality on feed intake, water consumption, rectal temperature, respiration rate, heart rate, plasma protein and blood hematocrit were measured in young growing female cattle during the summer (January-March) of 1993 in south-east Queensland. This region is subtropical with summer temperatures of over 30 °C being common. The animals were kept individually in open feedlot pens each of 40m^2. Shade was provided in half of the pens by galvanised sheet iron, 2.2m high, over 11.5m^2 plus 70% shade cloth, 4m high. over 12m^2.

Feed intake and water consumption were measured on a daily basis. Animal measurements and meteorological data (ambient temperature (Ta), black globe temperature (BG) and relative humidity (RH)) were collected between 0700 and 0830 before feeding, and again between 1400 and 1530. Black Globe Humidity Indices (BGHI) (5) were calculated.

Ten 12 month old Hereford heifers, 345±10.8 kg BW, were used in the cross-over designed study. Treatments were changed every 2 weeks. Heart rate radio transmitters (Telonics, Mesa, Arizona) were implanted into six of the heifers approximately 1 month prior to the commencement of measurements. A higher energy diet (H) of 80% concentrate and 20% roughage was offered ad lib to half of the animals and a low energy roughage diet (L) (sorghum hay) was offered to the other half. The estimated ME of the diets (6) were 10.8 and 8.03 MJ/kg DM, respectively.

Statistical analysis for the difference between treatments was made by using multiple analysis of variance, where $P<0.05$ was accepted as significant treatment differences. When interactions between feed and shade treatments were found, paired t tests were used to analyse the significance of differences. Significant differences between morning and afternoon measurements were also done by the paired t test. All data are presented as mean ±SE.

RESULTS: Ambient temperatures (°C) were 23.5±0.8 in the morning and 30.2±0.7 in the afternoon, while relative humidities were 65±2% and 45±3%, respectively. BG temperatures were 25.2±0.8 and 30.2±0.7 under the iron shade and 37.6±1.3 and 45.0±1.7 in the sun in the morning and afternoon, respectively. BGHI were 72.6±1.1 and 80.1±0.7 in the shade, 84.9±1.3 and 92.1± 1.6 in the sun in the morning and the afternoon, respectively. Daily DM intake was higher for diet H (24.5g/kgBW) than for diet L (14.5g/kgBW). As the ME of the H grain diet was higher than the L roughage diet the difference in the daily ME intakes by the heifers was substantially greater (265 vs 116 kJ/kgBW). Provision of shade did not affect the intake of either diet. Water consumption (mL/kgBW/day) was highest when the heifers were fed diet H, 108±10 and 127±12 in the shade and sun treatment respectively. Water consumption on the H diet was 1.5 times that

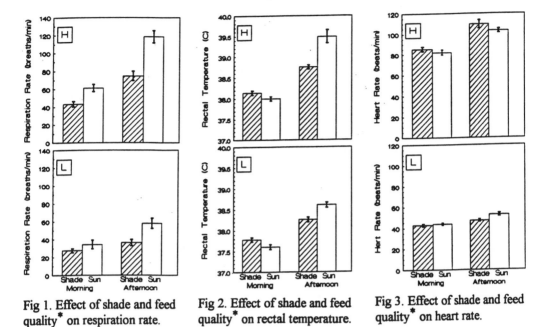

Fig 1. Effect of shade and feed quality* on respiration rate.

Fig 2. Effect of shade and feed quality* on rectal temperature.

Fig 3. Effect of shade and feed quality* on heart rate.

*H: High energy concentrate diet. L: Low energy roughage diet

Legend to Figures:
Fig 1. Effect of shade and feed quality* on respiration rate.
Fig 2. Effect of shade and feed quality* on rectal temperature.
Fig 3. Effect of shade and feed quality* on heart rate.
*H: High energy concentrate diet. L: Low energy roughage diet
Fig 1. Effect of shade and feed
quality* on respiration rate.
Fig 2. Effect of shade and feed
quality* on rectal temperature.
Fig 3. Effect of shade and feed
quality* on heart rate.
*H: High energy concentrate diet. L: Low energy roughage diet

on the L diet. Shade did not significantly affect water consumption. RR, the main active cooling channel for the heifers, was positively correlated with the energy content of the diets and with the environmental heat load (fig.1). RR was slower in the morning than in the afternoon for all treatments and also much slower when the animals were under the shade compared with those in the sun. Allowing Tr to rise can save energy and water used by reducing evaporative cooling. The Tr of the heifers (fig. 2) was lower in the morning than in the afternoon for all treatments. The greatest differences in Tr between morning and afternoon were observed in the heifers exposed to solar radiation (1.5 $^\circ$C and 1.0 $^\circ$C on diets H and L, respectively). The lowest HR was in the morning on diet L (fig. 3). It almost doubled when diet H was offered . The HR of the heifers during the morning was lower than in the afternoon by 23 beats/min on diet H but by only 7 beats/min on diet L. HR was not affected by the provision of shade. Average plasma protein concentrations ranged between 7.2 and 7.5g/100ml for all treatments. Blood hematocrits ranged between 30% and 32% for all treatments. No significant differences were observed in plasma protein concentration and hematocrits between morning and afternoon with any of the test treatments.

DISCUSSION: Heifers that were fed a high energy grain diet and exposed to the sun had higher heat loads during the hotter part of the day than did roughage fed animals or animals given access to shade. The lack of change in plasma protein concentration and hematocrit indicates that plasma volumes were kept relatively constant with all treatments. All heifers coped well with the heat load by utilising panting and changes in body temperature to compensate for the increased heat load from solar radiation and ingestion of a high energy diet.
Previous studies have shown reasonable correlations between HR and HP (7), (8). Moreover, an obvious effect of heat stress on cardiovascular function is an increase in HR (9). Even though the heat load by solar radiation was high during the present trial HR was not affected. However, the HR of the heifers was affected by feed quality. According to Conrad (10), offering feeds which are lower in fibre and higher in digestibility, will produce a lower heat increment associated with feeding and stimulate energy intake. Such an approach is often suggested as a means of lowering heat stress under hot environmental conditions. However, when the ME concentration of the diet is higher and the intake is higher, the total energy intake, heat production and heat load are also higher. Theoretically, in hot weather, we can give the animals a high energy concentrate diet in limited amounts, but this option is not practical in a commercial group feeding situation as dominant animals will continue to have a high feed intake. The present study showed that high heat loads on cattle during the hot summer can be reduced by changing from a high to a low energy diet. The action was largely through reduced feed intake and very large reduction in the

ME intake. When the hot weather passes it may be possible to rapidly transfer back to a high energy diet by adding virginiamycin to the diet (11).

In recent study (12), the effects of shade in a beef feedlot at central Queensland during summer showed that the provision of shade did not affect growth rate or feed efficiency. However, Brahman cross steers that were given access to shade had more rump and rib fat and less intramuscular fat. A possible explanation is that the heating of the skin by solar radiation causes subcutaneous fat to be mobilised and redeposited as intramuscular fat. Deposition of fat in the core of the body instead of in the periphery will give an advantage in the transfer of heat from the animals to the environment.

Heat resistance can be improved by exposing animals and allowing them to adapt to heat loads (13). Its seems that it is possible in summer, under moderate heat loads, to keep beef cattle, without detriment, in unshaded feedlots. However, by monitoring the climatic conditions, the operator must be ready to predict those days when the heat load may become too high for the animals to cope.

In conclusion: Solar radiation induced a highest heat load for all the animals, activated respiratory evaporative cooling and raised body temperature. The effect was much greater for the animals on the high energy concentrate diet. Feed quality and not solar radiation had the major effect on HR and probably HP. The delivery of new feed to the trough is the biggest stimulus to feeding (14) and HP increases during and after feeding (15). With a pending high heat load situation, reducing feed quality and/or changing the time of feeding to the late afternoon should relieve the situation (16). Feeding in the cooler hours of the day will improve passive dissipation of heat from the body to the environment.

ACKNOWLEDGMENTS: The authors wish to thank B Hall, A. Goodwin, K Rowan, J McCosker, M Josey, F Gorbacz, R Englebright, I Williams and T Schoorl for invaluable contributions.

REFERENCES: 1. Young BA, Hall AB. Heat load in cattle in the Australian environment. In: Australian Beef Coombs B, editor. Melbourne: Morescope Pty Ltd., 1993:143-8.
2. Kleiber M. The fire of life, an introduction to animal energetics. New York: Robert E, Krieger Publishing Company. 1975: 150-78.
3. Hales JRS. Physiological responses to heat. In: MTP international review of science, Physiology series 1 Vol 7. Robertshaw D. editor. London: Butterworths.1974: 107-62.
4. Reynolds CK, Tyrrell HF, Reynolds PI. Effect of diet forage-to-concentrate ratio and intake on energy metabolism in growing beef heifers: Whole body energy and nitrogen balance and visceral heat production. J Nutrition. 1991;121:994-1003.
5. Buffington DE, Collazo-Arocho A, Canton GH, Pitt D, Thatcher WW, Collier RJ. Black globe-humidity index (BGHI) as comfort equation for dairy cows. Trans ASAE. 1981; 24:711-4.

6. National Research Council, Nutrient requirement of domestic animals. National Academy of Sciences. Washington. DC.1984. No 4. Beef Cattle, 5th edn.

7. Renecker LA, Hudson R J. Telemetered heart rate as an index of energy expenditure in moose (Alces Alces). Comp. Biochem. Physiol. 1985;82A:161-5.

8. Yamamoto S. Estimation of heat production from heart rate measurement of free living farm animals. JARQ. 1989;23:134-43.

9. Rübsamen K, Hales JRS. Circulatory adjustments of heat-stressed livestock. Stress. In: Stress physiology in livestock. Vol 1: Basic principles. Yousef MK, editor. CRC Press, Inc. Boca Raton, Florida.1984: 143-54.

10. Conrad JH, Feeding of farm animals in hot and cold environments. In: Stress physiology in livestock. Vol 2: Ungulates. Yousef MK, editor. CRC Press, Inc. Boca Raton, Florida. 1985: 205-25.

11. Rowe JB, Zorrila-Rios J. Simplified systems for feeding grain to cattle in feed lots and under grazing conditions. In: Recent Advances in Animal Nutrition in Australia. Farrell DJ editor. University of New England, Armidale, N.S.W.1993: 89-96.

12. Clarke MR. Some effects of shade in cattle feedlots. In: Australian Association of Cattle Veterinarians. Gold Coast Conference, Proceedings 1993 May 16-21 QLD. 1993: 67-70

13. Nielsen M. Heat production and body temperature during rest and work. In: Hardy JD, Gagge AP, Stolwijk JAJ, editors. Physiological and behavioural temperature regulation. Springfield, Illinois: Charles C Thomas, 1970: 205-14

14. Fell LR, Clarke MR. Behaviour of lot-fed cattle. In: Recent Advances in Animal Nutrition in Australia. Farrell DJ, editor. University of New England, Armidale, N.S.W.1993: 107-16.

15. Young BA, Webster MED. A technique for the estimation of energy expenditure in sheep. Aust. J. Agric. Res. 1963;14:867-73.

16. Brosh A, Beneke G, Fennell S, Wright D and Young B. Effect of feeding on HR, HP and heat load in cattle. (In preparation).

Temperature Regulation
Advances in Pharmacological Sciences

CYTOLOGICAL CHANGES IN BROWN ADIPOSE TISSUE OF LEAN AND OBESE MICE: ACCLIMATION TO MILD COLD WITH AND WITHOUT A WARM REFUGE

Challoner, D.,[*] McBennett, S.,[+] Andrews, J.F.[+] and Jakobson, M.E.[#]

[*] Dept. of Human Genetics, 19, Claremont Place, University of Newcastle-upon-Tyne, U.K.
[+] Dept. of Physiology, Trinity College, Dublin 2, Ireland.
[#] Dept. of Life Sciences, University of East London, London E15 4LZ, U.K.

Summary

Light microscopy has revealed that brown adipose tissue of both lean and genetically obese mice shows cellular changes in response to cold. Continuous cold exposure elicits the greatest cell size reduction and increased locularity but intermittent, voluntary exposure initiates analogous changes at a lower response level. At the level of individual cells, the orders of magnitude of change in the parameters studied are similar in tissues of LN and OB mice. The difference between these tissues lies in the magnitude of recruitment into the "active" state.

Introduction

The primary function of brown adipose tissue (BAT), in small mammals is thermogenesis; either for regulation of deep body temperature (2,4,8), or as a mechanism for maintaining energy balance within the animal (10). One mechanism by which thermogenesis is controlled in the tissue is at the level of transcriptional and post-transcriptional control of "uncoupling protein" mRNA (9). This protein inserts in the inner mitochondrial membrane and allows protons to bypass the ATP synthetase complex; it is unique to BAT. The tissue receives control signals presumably from a central source mediated by sympathetic nerve fibre terminals releasing noradrenaline. Little is known of the signal pathway between receptor binding of noradrenaline at the plasma membrane and the genomic response but the correlation of triiodothyronine binding to nuclear receptors in response to cold stress and increased levels of uncoupling protein mRNA has been well recorded (9).

The effects of the biochemical and physiological controls outlined above, initiated in response to changes in the thermal environment of the animal, can be monitored via morphological changes at the cellular level (1,7). BAT of mice with a genetic lesion associated with obesity in the homozygous state has limited thermogenic capacity (11). It is not clear whether it has the potential to respond when the mouse faces a cold challenge. In this study, we have arranged for these obese mice and their lean litter mates to experience mild cold conditions either for a continuous period, or for variable, intermittent durations when they choose to leave a warm refuge to forage in the cold. Heldmaier has shown that BAT can exhibit cytological changes in response to brief cold shocks (3), and it has been demonstrated that these changes can be elicited in normal lean mice during foraging trips to a cold environment (5). Here we report cytological changes in obese mice and their lean counterparts maintained in conditions that seek to mimic the field environment; a foraging model. The parameters studied were cell size and lipid droplets per cell, changes in these predominantly reflecting changes in lipid metabolism.

Materials and Methods

Animals, environmental conditions and experimental design

For 30 days prior to study, 22 Aston mice were kept at a thermoneutral temperature of 28-30°C. at which one would expect minimal BAT lipid metabolism. Six of these mice were obese (OB), genetic constitution -/-, and 16 mice lean (LN), either +/+ or +/-. All the LN mice were male; the OB mice were of both sexes. Animals were housed either singly or in pairs; the pairings being random with respect to genetic constitution.

Some mice were exposed to temperatures of 20-22°C for either 1 day (CE1), 2 days (CE2) or 5 days (CE5), before sampling of tissue. Other pairs of mice were provided with individual cold foraging systems, of two interconnecting cages, one at the thermoneutral temperature and the other at 20-22°C, which allowed animals to select whether to remain in a warm refuge without food and water or to move, via a connecting tube, to the adjoining mild cold cage to "forage" for food and water (5). These animals were sampled at 5 days (CF5), or at 10 days (CF10). Warm acclimated mice (WA), remaining at 28-30°C were sampled too.

The light regimen were 12L:12D in the cold room and 16L:8D in the warm room.

Histology

Animals were killed by cervical dislocation. The interscapular brown adipose tissue pads were removed together with surrounding WAT and fixed overnight in formol-saline. The tissues were processed by standard histological techniques. Serial sections of 5mµ thickness were cut and treated with Shorr's trichrome stain to high-light boundaries of individual cells.

Cell area and the number of lipid droplets per cell were measured for at least a hundred cells from each tissue. The droplets per cell measurements were standardised for cell size by calculating droplets per unit area, this parameter (not reported here), was used to evaluate differences in locularity between tissues. In LN mice, peripheral cells, 4 deep from the edge of the bat pad, were excluded as being invariably unilocular even in the most cold adapted tissue. Deep to this layer, contiguous cells were measured along a vertical and horizontal line from one natural edge of the bat pad, to the other. Since "active" cells (i.e., those showing reduced cell size and increased locularity compared with WA cells), appear in OB tissues as isolated foci, no attempt was made in this study to obtain data which reflected the proportions of cells with "activation characteristics" in the BAT section as a whole, as had been attempted with the

LN mice; here, contiguous cells were counted in several of the foci (comprising between 15-50 cells), on each section.

The measurements were made manually with an optical system attached to an image analyser (Kontron Videoplan). Tests of differences between treatment groups within the lean mice and within the obese mice were made using the Kruskal-Wallis test statistic; variation between mice within treatment groups was assessed similarly.

Results and Discussion

In the warm acclimated mice the dominant cell morphology of BAT is unilocularity (Fig 1(b) & (c)); the data in Table 1 confirm this clearly for the OB mice; the modal value of the droplets/cell for the LN mice is one too (not reported). However, it may still be differentiated

	Treatment	Lean	Obese
CELL AREA (μm^2)			
	WA	497 (316-733, n=505)	1393 (965-2433, n=210)
	CE1	561 (374-740, n=151)	
	CE5	318 (260-378, n=1046)	650 (475-894, n=627)
	CF5	501 (340-644, n=333)	894 (487-1544, n=152)
	CF10	269 (205-385, n=202)	
DROPLETS/CELL			
	WA	7.0 (1-15)	1.0 (1.0-1.0)
	CE1	27.0 (15-40)	
	CE5	39 0 (30-50)	40.0 (26.0-61.0)
	CF5	17.0 (9-31)	16.0 (1-27.5)
	CF10	7.0 (1-15)	

TABLE 1   Combined Median values of cell area and the number of droplets per cell. The brackets enclose the lower and upper quartile values ; n=number of cells surveyed in the treatment group. [number of mice WA LN=4; OB=1: CE1 LN=1: CE5 LN=7; OB=4: CF5 LN=3; OB=1: CF10 LN=1].

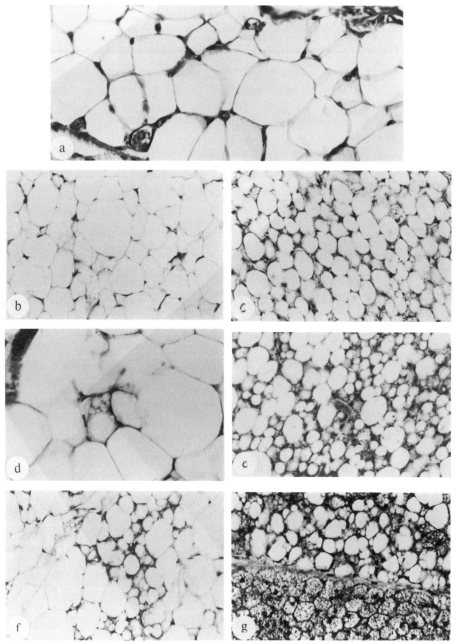

Figure 1 Specimen photomicrographs of adipose tissue from lean and obese mice: (a) WAT; (b-g) BAT. Row 2 warm acclimated: OB (b); LN (c). Row 3 intermittent cold exposure for 5 days: OB (d); LN (e). Row 4 continuous cold exposure for 5 days: OB (f); LN (g). Note the heterogeneity between adjacent cells of all BAT tissues and especially that between lobules in (g). Magnification (d) |_____| = 25μm; remainder |____| = 25 μm.

from WAT by the less angular appearance of the cells (Fig 1(a)). A most striking observation that can be drawn from Table 1 is the more than two-fold difference in cell area between LN and OB WA cells. This relationship holds approximately between the two tissues in each of the treatment groups that can be compared, suggesting that at the cellular level lipid mobilization is proceeding at least at a comparable rate in obese and lean tissues.

Comparison of the standardised median droplets/area parameter between WA, CE5 and CF5 treatment groups confirmed that cold exposure of varying duration and frequency had a significant effect on BAT tissue in increasing locularity in both LN and OB mice ($p < .001$). The depletion in lipid reflected in reduced cell size was also significant between the LN CE and WA mice and between the OB CE, CF and WA mice ($p > .001$ for both tissues). The difference between LN and OB in cell size is so clear as to make formal tests superfluous but the median droplets/cell (the non-standardised parameter), are similar in each treatment group that can be compared (Table 1).

Exposure to continuous cold (CE), caused a clear reduction in cell area in both LN and OB tissues (35% in LN and 54% in OB by day 5), and a dramatic increase in multilocularity in the cells of the areas selected for measurement; both tissues converge on a median of approximately 40 droplets/cell. These statistics do not reveal the heterogeneity which not only exists within the OB tissues (Fig 1(d)), but also between the LN CE tissues (1(e)), where differences in locularity exist between cells and between lobules.

Intermittent, voluntary exposure to cold in the CF model (10-30% of day spent in cold cage), elicited change in the same direction as that brought about by continuous cold, but the change was not so marked. A fall in cell size and increase in lipid droplets/cell were detectable in OB BAT after 5 days.

The thermoregulatory capacity of obese mice is poor (11); this is believed to be due to a defect in BAT activity or control (1). At temperatures around 10°C, commonly used experimentally to challenge the thermoregulatory response of lean mice, they develop hypothermia and die within a few hours. For this reason we chose the mild cold conditions used in the experimental design (20-22°C), to be just above the temperature at which obese mice invariably become torpid but below the critical temperature where heat exchange between the body and the external environment is balanced and the need for BAT activity minimal. However, even in these conditions, allowing body temperature to fall by 3 degrees with the assumption of a torpid state was a strategy that some OB mice adopted as a means of regulating energy expenditure in a cold challenge. Tolerating a slightly lower deep body temperature than was measured prior to cold exposure was also a strategy followed by some LN mice. These data will be reported elsewhere.

Whilst not assessed directly in this study, there was at least a 50-fold difference in the mass of tissue that assumed the appearance of cells active in lipolysis in OB mice compared with

LN. This is confirmed by a larger study which also revealed heterogeneity of tissue response between individual mice in the OB group (6). Heterogeneity of BAT response to cold was also seen in LN mice although variation within treatment groups was significantly less than between groups, as described above.

It is intriguing that individual OB cells which become "activated" appear to be capable of depleting their lipid reserves at a similar rate to stimulated LN cells. This observation raises several questions as to the nature of the phenotypic effects of the genetic lesion at the cellular level in OB mice. Is there a subset of cells which are capable of responding to stimuli at some stage of there life cycle and then loose this ability? Is there a signal communication problem either at the nerve terminal:cell membrane interface or in some paracrine interaction between stimulated cells and their less "active" neighbours. These problems remain to be investigated.

Acknowledgement

We wish to thank Professor J.V. Soames and Mr David Sales of the Department of Oral Pathology, University of Newcastle-upon-Tyne, for the use of the Videoplan and photography.

References

1. Arbuthnott, E., (1989) Brown adipose tissue: structure and function.
Proc. Nutr. Soc. **48**, 177-182.
2. Bruck, K., (1970) *In Brown Adipose Tissue,* pp. 117-254 (ed. O. Lindberg) Amsterdam: Elsevier.
3. Heldmaier, G., (1975) The effect of short daily cold exposures on development of brown adipose tissue in mice. *J. Comp. Physiol.* **98,** 281-292.
4. Himms-Hagen, J., (1986) Brown adipose tissue and cold acclimation. In: *Brown Adipose Tissue,*(ed. P. Trayhurn & D. Nicholls) p. 214 London: Edward Arnold.
5. Jakobson, M.E. & Andrews, J.F., (1989) Deep body temperature in the house mouse: foraging in the cold compared to chronic cold exposure. In: *Thermal Physiology* (ed. J.B. Mercer) p. 637, Amsterdam: Excerpta Medica.
6. McBennett, S.M., Andrews, J.F., Challoner, D. & Jakobson, M.E. (1993) Brown adipose tiss response to continuous cold exposure vs foraging in the cold in obese (ob/ob) mice. In: *Int. J. Obesity,* **17**, (in press).
7. Nechad, M., (1986) Structure and development of brown adipose tissue. In: *Brown Adipose Tissue,*(ed. P. Trayhurn & D. Nicholls) p. 1, London: Edward Arnold.
8. Nicholls, D.G. & Locke, R.M., (1984) *Physiological Reviews* **64**, 1-64.
9. Rehnmark, S., Bianco, A.C., Kieffer, & Silva, J.E (1992) Transcriptional and posttranscriptional mechanisms in uncoupling protein messenger RNA response to cold. *Amer. J. Physiol.* **262** (E), 58-67.
10. Rothwell, N.J., (1989) The role of brown adipose tissue in diet-induced thermogenesis. *Proc. Nutr. Soc.* **48**, 189-196.
11. Trayhurn, P. & James, W.P.T., (1978) Thermoregulation and nonshivering thermogenesis in the genetically obese (ob/ob) mouse. *Pflueger's Arch.* **373**, 189-193.

Temperature Regulation
Advances in Pharmacological Sciences
© 1994 Birkhäuser Verlag Basel

MANIPULATION OF BROWN ADIPOSE TISSUE DEVELOPMENT IN NEONATAL AND POSTNATAL LAMBS

M.E. Symonds, J.A. Bird, L. Clarke, C.J. Darby,
J.J. Gate and M.A. Lomax
Department of Biochemistry & Physiology, University of Reading
Whiteknights, PO Box 228, Reading RG6 2AJ, England

In precocious mammalian species such as man and sheep which are not protected from hypothermia by huddling in a nest, the ability to alter metabolic rate in response to changes in ambient temperature at birth is a prerequisite for survival. The principal tissue involved in metabolic adaptation to the extra-uterine environment is brown adipose tissue (BAT), which has the ability to rapidly generate large amounts of heat and also convert thyroxine to triiodothyronine (T_3) the dominant hormone regulating metabolic rate (1). An apparent failure to utilize BAT during neonatal and postnatal development is associated with unexpected death (1) and may be an important factor contributing to annual lamb mortality of 1-4 million in the British sheep industry. Therefore an improved understanding of the factors regulating the thermogenic function of BAT at birth and during the first month of life will aid in the manipulation of lamb thermogenic activity to improve survival rate.

The sympathetic nervous system is known to play a primary role in the initiation and regulation of non-shivering thermogenesis in BAT of altricial species such as rodents (2) and it has been proposed that rodent BAT possess a specific class of B_3 receptors (3). In lambs birth seems to be the trigger for the onset of non shivering thermogenesis which can be stimulated by noradrenaline administration and is blocked by chemical sympathectomy (1). However, the changes in BAT thermogenic activity and response to B_3 agonists have not been assessed in lambs. We have previously shown that the thermogenic activity of

ovine BAT increases during fetal life to reach a peak just after birth (4) and we now
report summaries of the results from three experiments which have examined acute and
chronic changes in BAT function during neonatal and postnatal development.

Acute response to B₃ adrenoreceptor agonist in the newborn lamb

In lambs born into a cold environment the thermogenic activity (assessed by GDP
binding to mitochondrial protein) of BAT doubles from fetal values (near to term) of 120
to 225 pmol/mg mitochondrial protein and remain at this level for the next 24 h. Over
this period, enteral administration of the B_3 adrenoreceptor agonist Zeneca D7114 results
in a further increase in BAT thermogenic activity 45 minutes after giving the drug (Figure
1). However, the effect of the B_3 agonist is only significant at 6 h after birth and is no
longer present at 24 h of life.

The smaller responses to the drug at 1.25 and 2.5 h after birth may indicate attenuation
of the response due to the release of catecholamines during birth. This proposal is
supported by the 30% decrease in BAT weight between 1 and 6 h of life, which is likely
to reflect appreciable mobilization of tissue lipid in response to catecholamine release at
birth. In addition changes in the noradrenaline content of BAT suggest a major influence
of peripheral catecholamines at birth since B_3 agonist treatment of neonatal lambs results
in a further fall in the noradrenaline content of BAT (control 3793±146; B_3 agonist
2978±40 pmoles/mg tissue (n=5) (P<0.01)) following the marked decrease in its
noradrenaline content during birth (term fetus 6660±1035 (n=4); 0.5 h old lambs
3955±986 pmoles/mg tissue (n=6) (P<0.01)). These decreases in noradrenaline content of
BAT may be due to a catecholamine induced decrease in noradrenaline synthesis and
storage in sympathetic nerve terminals. Despite the loss of BAT and altered sensitivity to
the B_3 agonist no effect on colonic temperature was apparent, which remained between
39.5-40.5°C over the first day of life in all lambs (Figure 1).

Chronic response to B₃ adrenoreceptor agonist over the first 8 days of postnatal life

Between 1 week and 1 month of postnatal life BAT cells appear to be gradually
replaced by white adipose tissue cells. This process appears to be irreversible in
ruminants but can be delayed by artificially rearing at a cool ambient temperature (1).

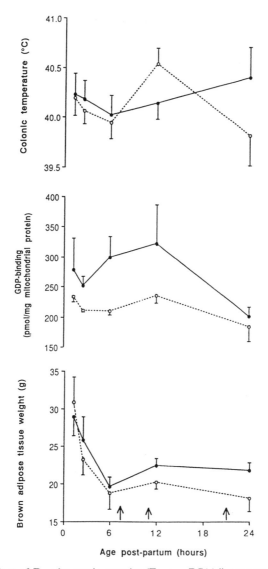

Figure 1. Effect of B$_3$ adrenergic agonist (Zeneca D7114) treatment on brown adipose tissue and colonic temperature over the first 24 h of neonatal life. Twin lambs were fed 20ml of milk replacer with B$_3$ agonist (10mg/kg body weight) (●) or with milk alone (o) 45 min prior to tissue sampling. Values are means ± S.E.M. and n = 8 at each time point. Lambs studied at 12 and 24 hours were fed 200ml milk as indicated (↑).

Since these effects of ambient temperature are likely to be the result of changes in sympathetic stimulation of BAT we have examined the effect of daily enteral B_3 agonist treatment on thermoregulation on days 7-8 of postnatal life. B_3 agonist treatment resulted in a higher thermogenic activity in BAT whilst depleting tissue lipid (Table 1). Treated lambs also produce a 2-fold greater increase in metabolic rate in response to a decrease in ambient temperature from 26 to 12°C without the occurence of shivering thermogenesis during non-rapid eye movement sleep.

Table 1. Effects of B_3 adrenergic agonist administration (Zeneca D7114; 10mg/kg body weight - fed twice daily) on the thermogenic activity of brown adipose tissue on days 7-8 of life in neonatal lambs reared artificially at a cold ambient temperature (5-15°C). Values are means S.E.M. (n=8).

	GDP binding (pmol/mg mitochondrial protein)	Lipid (mg)	VO_2 (ml/mg per kg body weight)	
			26°C	12°C
Control	129±13 +	540±120 *	15.6±0.5	18.5±1.0
B_3 agonist	180±28	1410±300	15.8±0.6	21.4±2.6

+ P=0.08, * P<0.05

Perirenal adipose tissue development over the first month of postnatal life

BAT has an appreciable iodothyronine 5'monodeiodinase (5'MDI) activity and studies in rodents have proposed that BAT makes a significant contribution to the regulation of circulating T_3 concentrations (1). Developmental changes in BAT have also been linked to alteration in 5'MDI activity so we have examined the relationships between the plasma concentrations of T_3 and the activities of Type I and II 5'MDI and GDP binding in BAT over the first month of postnatal life. In all lambs a large decrease in thermogenic activity of perirenal adipose tissue occurred between 0 and 33 days of age due to the loss of BAT from this depot. However, the activity of type I and II iodothyronine

5'monodeiodinase increased and plasma T_3 concentrations decreased over this period (Table 2). These results are in contrast to studies conducted in rodents (1) and indicate that iodothyronine 5'monodeiodinase activity may not influence thyroid status or the ability of BAT to generate heat in the postnatal lamb.

Table 2. Mean thermogenic activity and iodothyronine 5'monodeiodinase activity of perirenal adipose tissue from lambs sampled either within 2h of birth (day 0) or at 33 days of age. Values are means S.E.M. (n=6).

	GDP binding (pmol/mg mitochondrial protein)	Iodothyronine 5'monodeiodinase activity (pmol I released/mg/h) Type I	Type II	Plasma T_3 concentration (nM)
Day 0	299±50	842±171	0.65±0.06	4.50±0.32
Day 33	64±9	1230±182	0.90±0.18	2.85±0.17

Brown Adipose Tissue Development in Neonatal and Postnatal Lambs

The present studies extend our previous reports that the thermogenic activity of BAT in newborn lambs is influenced by the maternal metabolic environment during late gestation, with the period over the final two days of gestation being the most critical (4). The ability of newborn rodents to increase heat production in a cold environment appears to be dependent on a rise in BAT thermogenic activity 6-12 hours after birth which occurs following an increase in expression of uncoupling protein mRNA which is only observed 6 h after birth (5). In the newborn lamb the thermogenic activity of BAT is already at a maximum 1.25 h after birth and remains elevated for 24 h despite a loss of 39% of BAT weight during the first 6 h. This indicates that the thermogenic activity of BAT at birth is determined by regulatory factors associated with late gestation and birth rather than the immediate environment experienced by the newborn lamb. There is still potential to increase thermogenic activity of ovine BAT, since it was responsive to the B_3 agonist 6 h after birth although this stimulatory effect had disappeared after 24 h. Chronic treatment with B_3 agonist over the first week of postnatal life delayed the rate of loss of BAT by maintaining its thermogenic activity at levels 20% lower than those observed in newborn lambs. This retention of BAT appeared to benefit the developing lamb by an increased metabolic response to an acute cold challenge (12^0C) without the occurence of shivering.

It has been shown that maintained 5'MDI activity in the rat fetus is necessary for the increase in mRNA for UCP after birth (6) and for increasing UCP synthesis in cold exposed adult rats (1). A different relationship appears to exist in ovine BAT in that a clear dissociation between developmental changes in thermogenesis and 5'MDI activity in perirenal adipose tissue were observed over the first month of postnatal life. Consequently thyroid hormones may influence the appearance of white adipose tissue whilst the sympathetic nervous system acting via B_3 receptors regulates the retention of BAT.

It is concluded that ovine BAT responds to acute and chronic stimulation via B_3 agonists, indicating the tissue may possess atypical (B_3) adrenoreceptors. This could enable strategies to be developed which improve the ability of lambs to effectively thermoregulate and therefore prevent unexpected death during neonatal or postnatal life.

This work was funded by the Wellcome Trust, AFRC and studentships from MRC (L.C.), MAFF (J.J.G.) and AFRC (J.A.B.). Zeneca D7114 was a gift from Dr B R Holloway (Zeneca Pharmaceuticals, Macclesfield, Cheshire).

References

1.   Symonds M E, Lomax M A. Proc Nutr Soc 1992; **51**: 165-172.

2.   Klaus S, Casteilla L, Bouilland F, Riquier D. Int J Biochem 1991; **23**: 791-801.

3.   Champigny O, Holloway B R, Ricquier D. Mol cell Endocrinol 1992; **86**: 73-82.

4.   Clarke L, van de Waal S, Lomax M A, Symonds M E. In: Neonatal Survival and Growth. eds. Varley MA, Williams PEV, Lawrence T L J. Occ Publ No 15. Brit Soc Anim Prod 1992; 174-175.

5.   Giralt M, Maryin J, Iglesias R, Vinas O, Villarouga F, Mampel T. Eur J Biochem 1990; **193**: 297-302.

Temperature Regulation
Advances in Pharmacological Sciences
© 1994 Birkhäuser Verlag Basel

RESTING MUSCLE: A SOURCE OF THERMOGENESIS CONTROLLED BY VASOMODULATORS

M.G. Clark, E.Q. Colquhoun, K.A. Dora, S. Rattigan, T.P.D. Eldershaw,
J.L. Hall, A. Matthias and J-M. Ye

Department of Biochemistry, University of Tasmania, Hobart, Australia

Summary

Perfused, but neither incubated nor perifused, hindlimb muscle responds to a variety of vasomodulators, including noradrenaline, by rapidly altering the rate of oxygen consumption and metabolite release. The vascular tissue of muscle is identified as highly energetic and may be the major contributor to hindlimb thermogenesis. In addition, vasomodulators may control the delivery of nutrients to specialized skeletal muscle mitochondria by altering the microvascular distribution of flow. We propose that resting skeletal muscle contributes to whole body thermogenesis of endotherms and that it is controlled by total, as well as zonal (within muscle), nutrient delivery.

Introduction

Endothermic animals invoke heat producing mechanisms often referred to as facultative thermogenesis, in response to either cold or (over)eating. The mechanisms appear separate from shivering and involve, in many cases, an increase in sympathoadrenal activity. A starting point for unravelling the processes of facultative thermogenesis has been the observation that noradrenaline when injected *in vivo*, rapidly (within seconds or minutes) stimulates whole body oxygen uptake. In rats, oxygen uptake (and therefore thermogenesis) increases by up to 100 per cent when noradrenaline is injected [1]. In addition it has been assumed that the effect of noradrenaline *in vivo* can be effectively mimicked *in vitro* by exposing individual tissues to noradrenaline. For brown adipose tissue this is certainly the case and all preparations (tissue fragments, isolated cells and slices) respond markedly to the addition of catecholamine with values for oxygen uptake and heat production consistent with estimates for this tissue *in vivo* [2]. Other tissues such as liver [3] also respond positively to noradrenaline but in this case the hormone has a general effect to increase a diverse range of metabolic interconversions.

Skeletal muscle has been an enigma to researchers who study thermogenesis. It constitutes over 40 per cent of the body's mass and when working has the potential to be markedly thermogenic. Unlike brown adipose tissue, isolated muscles when incubated or perifused *in vitro* with noradrenaline do not respond by showing an increase in oxygen uptake or heat flux [4,5]. However several research groups have reported that infused sympathomimetic substances increased oxygen uptake in non-contracting skeletal muscle receiving its nutrient supply by the normal vascular route. These groups included Lundholm and Svedmar in 1965; Sutherland and Robison in 1966; and Schmitt, Meunier, Rochas, and Chatonnet in 1973 Mejsnar and Jansky in 1973; Grubb and Folk in 1977; Chapler, Stainsby, and Gladden in 1980; Richter, Ruderman, and Galbo in 1982 and Côté, Thibault and Vallières in 1985 (cited in Ref. 6, or references therein). For the perfused rat hindlimb the effects produced by noradrenaline were marked, showing rapid increases of 39 to 111% over basal oxygen uptake. Calculations from these figures indicate that oxygen uptake from the skeletal muscle would be at least 40% of the total oxygen uptake that could be contributed by brown fat. In addition data of Foster and Frydman in 1979 (cited in Ref. 6), which focused particular attention in the whole animal on BAT, also indicated a substantial role for skeletal muscle especially in warm-adapted and normal rats. These animals showed a 30% increase in skeletal muscle blood flow and a 60% increase in muscle oxygen consumption after noradrenaline administration, suggesting that muscle could produce an equal amount of heat to BAT in the whole animal, even though oxygen consumption was much less per gram of tissue.

Since the effects of noradrenaline and adrenaline on oxygen uptake by perfused rat hindlimb were mediated by α-adrenergic receptor mechanisms, and were associated with increased perfusion pressure (Grubb and Folk 1977; Richter *et al.* 1982a; Richter *et al.* 1982b; Côté *et al.* 1985, cited in Ref. 6, or references therein) it appeared possible that the increased oxygen uptake was controlled by the vascular system. Thus this communication presents our findings on the effects of various vasomodulators on perfused hindlimb, not only from the rat but from the chicken and a small Australian marsupial *Bettongia gaimardi* (bettong).

Materials and Methods

The rat hindlimb was perfused as described previously (see Ref. 6 and references therein). The lower hindlimb (16.8 ± 0.6 g) of 5-8 week old chickens (598 ± 26 g body wt) was perfused via the popliteal artery using conditions identical to that used for the rat. Similar procedures were used for the perfusion of the lower hindlimb (24.7 ± 2.8 g) of bettongs ($1,130 \pm 0.13$ g body wt). For each preparation, perfusion pressure and venous PO_2 were continuously monitored using in-line arterial pressure transducer and venous O_2 electrode, respectively. Details for stimulation of the lower calf muscles in the perfused rat hindlimb are given elsewhere [7]. Venous samples were collected for lactate and glycerol assays. Lactate was determined

Table 1. Vasomodulator effects on perfused rat hindlimb[a]

Parameter:	Vasoconstrictors	
	Type A[b]	Type B[c]
Perfusion pressure	↑	↑
Oxygen uptake	↑	↓
Lactate efflux	↑	↓
Glycerol efflux	↑	↓
Urate efflux	↑	↓
Uracil efflux	↑	↓
Insulin mediated glucose uptake	n.t.[d]	↓
Skeletal muscle contraction	n.t.[d]	↓
Perfusate distribution volume	↑	↓
Effect of the following on vasoconstriction and associated changes:		
Removal of external Ca^{2+}	B[e]	NB[e]
Replacement of O_2 by N_2	B	NB
Addition of N_3^-, CN^-	B	NB
Addition of vasodilators	B[f]	B[g]

a. Hindlimb perfused at 25°C with constant flow. Data is from references 6-10 or references therein, or is unpublished.
b. Includes α adrenergic agonists: noradrenaline, adrenaline, phenylephrine, methoxamine, amidephrine, norephedrine, ephedrine; peptides: vasopressin, angiotensins I, II, III, oxytocin, neuropeptide Y; "vanilloid" agonists: capsaicin, dihydrocapsaicin, gingerols, shogaols, piperine, resiniferatoxin. Also low frequency sympathetic nerve stimulation.
c. Includes serotonin (≥0.1 μM), noradrenaline at high doses (≥1 μM), high dose vanilloids and high frequency sympathetic nerve stimulation.
d. Not tested.
e. B = blocked; NB = not blocked.
f. Includes nitroprusside, nifedipine, isoprenaline, adenosine, AMP, ADP, ATP and UTP.
g. Includes nitroprusside, carbamyl choline and isoprenaline (partial blockade).

Figure 1. α-Adrenergic stimulation of glycerol production by the perfused rat hindlimb. Perfusions were conducted at 25°C and contained 20 μM DL-propranolol.

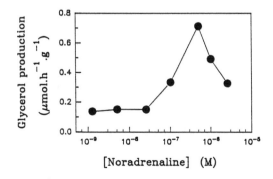

spectrophotometrically and glycerol spectrofluorometrically using standard enzymatic procedures.

Results and Discussion

Table 1 summarizes our findings and shows that the perfused rat hindlimb responds to many different vasoconstrictors that fall into two categories which we have called Type A and B. Type A vasoconstrictors stimulate a marked increase in oxygen consumption simultaneously with a rise in perfusion pressure in the rat hindlimb perfused at constant flow [6,10 and references therein]. The Type A category is associated with other changes including increased lactate [5], glycerol (Figure 1), urate [7], and uracil [7] efflux as well as increased perfusate distribution volume (unpublished). The vascular sites responsible for Type A vasoconstriction require external Ca^{2+} and oxygen; in the absence of either moiety or with respiratory poisons present, the pressor effect of Type A vasoconstrictors does not occur [9]. Table 1 also shows that vasodilators with differing modes of action block the vasoconstrictors, inhibiting the increases in perfusion pressure and oxygen uptake as well as the metabolic changes. It is important to note that the effect of the Type A vasoconstrictors to increase oxygen uptake is additive to the oxygen uptake due to skeletal muscle contraction and that the nitrovasodilators are selective, having no effect on the latter. Alpha and α_1-antagonists block sympathomimetic stimulation of oxygen and pressure but not the actions of other vasoconstrictors such as angiotensin II or capsaicin. Beta antagonists augment the action of most sympathomimetic vasoconstrictors especially those recognized as having significant beta actions. The increased glycerol release mediated by noradrenaline in the presence of propranolol (Figure 1) relates closely to the α_1-adrenergic effect of this catecholamine to cause vasoconstriction and oxygen uptake and implies that both of these may be supported in part by fatty acid oxidation.

Recently our hindlimb experiments have included species where the presence of brown adipose tissue is doubtful. Table 2 shows that constant-flow perfused hindlimbs from either the chicken or the bettong, respond positively to noradrenaline with increased pressure and oxygen uptake. This implies that resting muscle thermogenesis may be widely used amongst the endotherms regardless of the presence, or absence, of brown fat. The data of Tables 1 and 2 and Figure 1, together with rat hindlimb perfusions at varying but fixed flows and fixed pressures (Ye *et al.* 1990, cited in Ref. 6) suggests that working vascular tissue may be responsible for the Type A vasoconstrictor-induced increase in oxygen uptake. However for this to be so, the vascular smooth muscle cells responsible must be capable of high rates of oxygen consumption under load and be present in sufficient quantity in the hindlimb to account for the rates of oxygen consumption noted.

Finally Table 1 also identifies a group of vasoconstrictors (Type B) that lead to decreased oxygen consumption with increased vascular resistance in the constant-flow perfused rat hindlimb. This group includes serotonin, noradrenaline at high doses (similar to predicted concentrations at vascular smooth muscle synapses), high frequency sympathetic nerve stimulation and high dose vanilloids. In all respects the metabolic effects of the Type B vasoconstrictors are the opposite to those of Type A and are therefore potentially negatively thermogenic. We have proposed that Type B vasoconstrictors result in functional vascular shunting which coincides with reduced nutritive flow even though overall flow through the hindlimb remains unaltered [9]. The sites (vessels) controlling vascular shunting appear to be distinguishable in terms of their fuel and Ca^{2+} requirements (Table 1). Neither serotonin nor high dose noradrenaline require oxygen for vasoconstriction if glucose is present [9] and vasoconstriction is reduced but still present if Ca^{2+} is omitted from the buffer [9].

Table 2. Effects of noradrenaline on perfused hindlimbs[a]

Species	Perfusion pressure (mm Hg)		Oxygen uptake (μmol/g per h)	
	Basal	Noradrenaline[b]	Basal	Noradrenaline[b]
Rat	29±1	45±2[c]	6.4±0.2	9.6±0.3[c]
(n)	(19)	(3)	(19)	(3)
Chicken	39±4	56±3[c]	7.2±0.3	8.9±0.3[c]
(n)	(6)	(6)	(6)	(6)
Bettong	31±2	83±10[c]	4.6±0.4	9.5±0.8[c]
(n)	(5)	(5)	(5)	(5)

a. All perfusions were conducted with constant-flow of approx. 0.28 ml/g per min at 25°C with medium containing 2% serum albumin and 1.27 mM $CaCl_2$ (Ref 6. and references therein). Data for rat from Colquhoun et al. 1988 in Ref. 6.
b. Maximum oxygen uptake occurred at 50, 20 and 1000 nM noradrenaline for rat, chicken and bettong hindlimbs, respectively.
c. Significantly greater ($P<0.05$) than companion "basal" values.

We propose that resting muscle has the potential to contribute to whole body thermogenesis in endotherms and its contribution, in either a positive or negative manner, is controlled by the vascular system and cannot be observed with isolated incubated preparations. In addition to the effects of noradrenaline to increase oxygen uptake by constant-flow perfused muscle mediated by α_1-adrenergic receptors, increases in total flow to muscle in vivo e.g.

resulting from increased cardiac output (β-adrenergic receptor-mediated) have the potential to increase the thermogenic contribution by skeletal muscle.

Overall the results are consistent with our earlier proposals that work performed during constriction and resting flow [6] by the vascular smooth muscle ("hot pipes") of the hindlimb may account for the increase in oxygen consumption. The release of lactate, and to a lesser extent glycerol, during vasoconstriction have the potential to add further to thermogenesis *in vivo*. If vascular tissue does not consume sufficient oxygen, the data still suggest that it plays a key role in controlling thermogenesis by neighbouring skeletal muscle mitochondria. Thus increased oxygen uptake may be the result of vasoconstrictor-induced change in the distribution of flow so as to supply oxygen to previously unaccessed regions of muscle that contain specialized mitochondria adapted for thermogenesis. Vasoconstrictors that act to inhibit muscle oxygen uptake might do so by opening functional vascular shunts diverting flow away from thermogenic vasculature on thermogenic skeletal muscle mitochondria.

Acknowledgements: Supported in part by NH&MRC and ARC of Australia.

References
1. MacDonald IA, Siyamak AY. Plasma noradrenaline levels and thermogenic responses to injected noradrenaline in the conscious rat. Exper Physiol 1990; 75:639-648.
2. Girardier L, Stock MJ. Mammalian Thermogenesis. London: Chapman and Hall, 1983.
3. Binet A, Claret M. α-Adrenergic stimulation of respiration in isolated rat hepatocytes. Biochem J 1983; 210:867-873.
4. Dubois-Ferriere R, Chinet A.E. Contribution of skeletal muscle to the regulatory non-shivering thermogenesis in small mammals. Pflugers Arch 1981; 390:224-229.
5. Hettiarachchi M, Parsons KM, Richards SM, Dora KA, Rattigan S, Colquhoun EQ, Clark MG. Vasoconstrictor-mediated release of lactate from the perfused rat hindlimb. J Appl Physiol 1992; 73:2544-2551.
6. Colquhoun EQ, Clark MG. Open question: has thermogenesis in muscle been overlooked and misinterpreted? NIPS 1991; 6:256-259.
7. Clark MG, Richards SM, Hettiarachchi M, Ye J-M, Appleby GJ, Rattigan S, Colquhoun EQ. Release of purine and pyrimidine nucleosides and their catabolites from the perfused rat hindlimb in response to noradrenaline, vasopressin, angiotensin II and sciatic-nerve stimulation. Biochem J 1990; 266:765-770.
8. Rattigan S, Dora KA, Colquhoun EQ, Clark MG. Serotonin-induced vasoconstriction associated with marked insulin resistance in the perfused hindlimb. J Hypertens 1992; 10 (Suppl 4):S51.
9. Dora KA, Richards SM, Rattigan S, Colquhoun EQ, Clark MG. Serotonin and norepinephrine vasoconstriction in rat hindlimb have different oxygen requirements. Am J Physiol 1992; 262:H698-H703.
10. Eldershaw TPD, Colquhoun EQ, Dora KA, Peng Z-C, Clark MG. Pungent principles of ginger (*Zingiber officinale*) are thermogenic in the perfused rat hindlimb. Int J Obesity 1992; 16:755-763.

Temperature Regulation
Advances in Pharmacological Sciences
© 1994 Birkhäuser Verlag Basel

THYROID STATUS MODULATES HYPOTHALAMIC THERMOSENSITIVITY, VASOPRESSIN AND CORTICOSTEROID SECRETION IN RABBITS

R. Keil, W. Riedel and E. Simon

Max-Planck-Institut für physiologische und klinische Forschung, W.G. Kerckhoff-Institut,
D-61231 Bad Nauheim, Germany

SUMMARY

Hypothalamic thermosensitivity was estimated in eu-, hyper- and hypothyroid rabbits by correlation of respiratory rate with experimentally altered hypothalamic temperature. Hypothalamic thermosensitivity was reduced in the hypothyroid, and augmented in the hyperthyroid status. Plasma vasopressin (AVP) was elevated in the hyper-, and lowered in the hypothyroid status, with a positive relationship between AVP-immunoreactivity of hypothalamic paraventricular neurones and the status of the thyroid system. Aldosterone and corticosterone levels were elevated only in hyperthyroidism. The results suggest modulatory actions of thyrotropin-releasing hormone neurones in thermo- and osmoregulation.

INTRODUCTION

Few investigations have been performed to elucidate the role of thyroid hormones in central integration of thermoregulatory function and of their action in modulating the hormone secretion pattern of the hypothalamo-pituitary axis. Among the studies where thyroid hormone status was manipulated, thermoregulatory effector activities were found altered in a way that, at identical core temperatures, hypothyroidism enhanced cold defence, while hyperthyroidism manifested itself with augmented heat loss activities[1,2]. Besides these effects of thyroid hormones on thermoregulation, hypothyroidism has been found to reduce plasma levels of cortisol, whilst the opposite was observed in hyperthyroidism[3]. In another study[4], plasma vasopressin was diminished in hypothyroidism and elevated in hyperthyroidism, supporting the assumption of a modulatory role of thyroid hormones at some central level. The present study was performed to evaluate whether the thyroidal status alters hypothalamic thermosensitivity, using respiratory rate as a purely centrally controlled thermoregulatory effector activity, which is not directly affected by thyroidal hormones.

MATERIALS AND METHODS

The experiments were carried out in 10 conscious Chinchilla rabbits weighing between 4.0 and 5.8 kg. Eight weeks before the first experiment, each animal was chronically implanted with a hypothalamic thermode[5]. The position of the thermode in the anterior hypothalamus was verified at the end of the experiment. Perfusion of the thermodes with warm or cold water resulted in hypothalamic temperatures of 41.0 °C (HW) and 36.9 °C (HC), respectively. The thermal stimulation periods lasted two hours and were performed first in the euthyroid status. Thereafter, the animals were randomly either thyroidectomized or made hyperthyroid by receiving intraperitoneal injections of 500 μg/kg l-thyroxine within ten days. Reversal of the thyroid status was initiated six to ten weeks after recovery from hyperthyroidism, or one week after completion of the experiments in the hypothyroid status. All experiments were performed between 08.00 and 16.00 h with the rabbits sitting unrestrained in a temperature-controlled rabbit box fixed at 28 °C, with back skin and ears exposed to room temperature of 20-22°C. For collection of blood samples and for monitoring arterial blood pressure (MAP) and heart rate (HR), a central ear artery was cannulated under local anaesthesia. Ear skin and rectal temperatures (TRE) were measured continuously, using thermocouples attached to the ear or inserted 10 cm deep into the rectum, respectively. Oxygen consumption (OC) was determined using open flow respirometric techniques (Taylor Servomex, OA 184). Respiratory rate (RR) was counted visually. Blood samples (2 ml) were collected before, and in intervals of 60 min during hypothalamic thermal stimulations, centrifuged and the plasma stored at -24°C. Plasma osmolality was measured by vapour pressure osmometry. Plasma concentration of vasopressin (AVP) was determined by radioimmunoassay[6]. Aldosterone (ALD) and corticosterone (B) were assayed after dichloromethane extraction of the steroids with commercially available antisera (ICN Biomed., FRG) and ^{3}H-labelled tracers (DuPont/NEN, FRG). To define an effect of altered thyroidal status on AVP-content of hypothalamic paraventricular (PVN) neurones, hypothalami of three hypo- and three hyperthyroid animals were, in pairs, subjected to immunohistochemistry using the avidin-biotin-peroxidase technique. Statistical analysis was done using the Student's t test and a value of $P \leq 0.05$ was regarded as statistically significant.

RESULTS AND DISCUSSION

The removal of the thyroid gland elicited, within one week, significant alterations of metabolic rate and cardio-respiratory effector activities, when measured at warm ambient temperature conditions. As shown by the data in Table 1, the hypothyroid rabbits exhibited lower values of resting OC, RR and ear skin temperature (data not shown). Despite augmented autonomic heat conservation effector activities the hypothyroid rabbits maintained lower TRE. Likewise, MAP and HR were lowered. Hypothyroidism induced a significant decrease of plasma AVP, but did not significantly affect the plasma concentrations of ALD and B. In contrast, hyperthyroid rabbits showed elevated values of resting TRE, OC, RR and HR together with an elevated MAP. The plasma levels of AVP

Table 1: Effects of altered thyroidal status and of hypothalamic heating (HW) and cooling (HC) on core temperature (TRE), oxygen consumption (OC), respiratory rate (RR), mean arterial blood pressure (MAP), heart rate (HR), plasma concentrations of vasopressin (AVP), corticosterone (B) and aldosterone (ALD). Values are means±SE, n=10. [a] denotes data statistically significant from euthyroid control values; [b] denotes thermal stimulation data statistically significant from prestimulation (CTRL) values.

	HYPOTHYROID				EUTHYROID				HYPERTHYROID			
	CTRL	HC	CTRL	HW	CTRL	HC	CTRL	HW	CTRL	HC	CTRL	HW
TRE °C	39.5[a] ±0.1	39.9[b] ±0.1	39.5[a] ±0.2	39.2[b] ±0.2	39.9 ±0.2	40.3[b] ±0.1	40.0 ±0.2	39.8[b] ±0.2	40.7[a] ±0.1	41.3[b] ±0.1	40.7[a] ±0.2	40.9 ±0.2
OC ml/kg/min	6.2[a] ±0.3	6.9 ±0.3	5.9[a] ±0.6	4.9 ±0.4	7.8 ±0.5	8.0 ±0.6	7.3 ±0.3	7.0 ±0.5	11.2[a] ±0.3	11.5 ±0.4	10.9[a] ±0.6	11.0 ±0.8
RR br/min	100[a] ±21	66[b] ±14	126[a] ±20	219[b] ±46	240 ±28	132[b] ±24	243 ±26	408[b] ±21	316[a] ±28	126[b] ±23	235 ±34	376[b] ±23
MAP mm Hg	60 ±3	63 ±3	64 ±3	62 ±2	65 ±3	68 ±4	67 ±2	64 ±4	79[a] ±3	81 ±2	79[a] ±3	79 ±3
HR b/min	186[a] ±7	212[b] ±10	192[a] ±10	183 ±7	220 ±8	228 ±6	220 ±6	217 ±7	315[a] ±16	335[b] ±17	323[a] ±21	329 ±21
AVP pg/ml	7.7[a] ±1.8	7.7 ±2.4	8.9 ±1.9	8.1 ±1.8	11.2 ±1.0	10.7 ±1.3	11.2 ±1.3	13.2[b] ±1.8	18.0[a] ±2.7	14.3[b] ±1.7	15.9[a] ±1.7	19.2[b] ±3.1
B ng/ml	21.4 ±5.6	29.8 ±9.0	21.9 ±8.2	17.4 ±3.7	12.8 ±4.2	11.6 ±2.2	15.2 ±4.2	4.1[b] ±0.9	22.8[a] ±7.3	33.6 ±9.9	27.5[a] ±5.1	25.5 ±5.4
ALD pg/ml	139 ±14	179 ±24	156 ±22	134 ±15	152 ±17	138 ±13	148 ±19	129 ±20	220[a] ±40	291 ±38	242[a] ±32	267 ±42

and ALD were considerably elevated, and B to some extent. Changes in plasma osmolality were not observed, neither in the hyper- nor hypothyroid status.

As shown in Fig. 1, HC lowered at all experimental conditions RR, however, to different degrees. The most prominent depression of RR occurred in hyperthyroid animals, while in the hypothyroid status the same rabbits lowered RR only modestly. Heating the hypothalamus affected RR likewise only moderatly in the hypothyroid status, in contrast to the euthyroid status where maximum panting values were elicited. Due to higher TRE of hyperthyroid rabbits, the experimentally induced additional increase of hypothalamic temperature was small, nevertheless, RR increased with a distinctly higher rate as compared with the euthyroid status. Based on RR values obtained within the entire span of hypothalamic thermal stimulations, a displacement of 1 °C hypothalamic temperature revealed a change of 31 br/min in the hypothyroid status, 66 br/min in the euthyroid status, and 82 br/min in the hyperthyroid status.

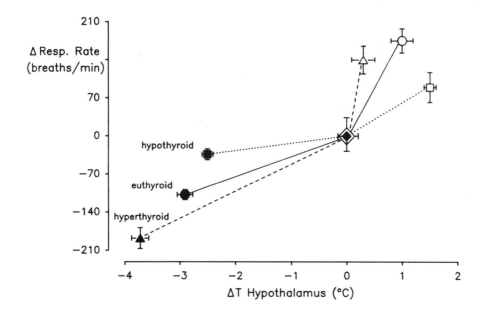

Figure 1: Mean changes ± SE in respiratory rate of 10 rabbits after heating (open symbols) or cooling (closed symbols) the hypothalamus for 20 min in the euthyroid status (circles), hypothyroid status (squares) or hyperthyroid status (triangles).

HW induced a significant systemic release of AVP in euthyroid and hyperthyroid rabbits, while HC affected AVP release only in the hyperthyroid status. Effects of hypothalamic heating and cooling in eu-, hyper- or hypothyroid rabbits on plasma concentrations of ALD or B were not observed. The influence of thyroid hormones on the reaction pattern of AVP is further substantiated by findings illustrated in Fig. 2, which shows in the hyperthyroid status augmented AVP-immunoreactivity in perikarya and fibers of the paraventricular nucleus of the hypothalamus, and nearly absent immunoreactivity in the hypothyroid status.

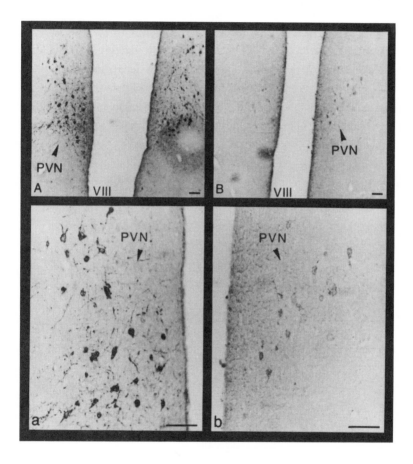

Fig. 2: AVP-immunostaining in frontal sections of hypothalamus of a hyperthyroid (A,a) and a hypothyroid (B,b) rabbit. PVN: N. paraventricularis. VIII: third ventricle. Bar represents 100 μm.

The present study shows that the thyroidal status profoundly affects, besides direct effects on metabolic rate, autonomic thermoregulatory effector activities like respiratory rate and, as shown in previous studies[2,4] visceral and cutaneous blood flow. Especially the expression of the specific autonomic pattern of cold defence with antagonistic blood flow changes between skin and viscera[7] in hypothyroidism, and its reversal in hyperthyroidism has favoured the assumption of a central, most likely hypothalamic origin of the pattern. A considerable body of evidence points to a negative feedback relationship between peripheral thyroid hormone concentrations and the activity of hypothalamic paraventricular neurones producing thyrotropin-releasing hormone (TRH)[8]. The elevated plasma concentrations of AVP, ALD and B in hyperthyroidism, a status with low TRH activity, suggest an inhibitory action of hypothalamic TRH neurones on hypothalamic control of AVP and corticotropin release. Peripheral cold signals have been found to specifically activate TRH neurones[9], which is concordant with lowered plasma AVP concentrations in cold ambient temperature. Accordingly, hypothalamic cooling in the hyperthyroid status lowers, and hypothalamic heating increases AVP secretion. Together with the presented immunohistochemical study, these results support the idea of an inhibitory action of TRH on hypothalamic AVP neurones, and might explain why hypothalamic thermal stimulations in hypothyroidism do not affect AVP secretion. By comparison with a similar experimental situation, where hypothalamic cooling and heating lowers and elevates plasma AVP also only under conditions of hyperosmotic stress[10], one could conclude that TRH neurones, thermoregulatory and osmoregulatory pathways converge on hypothalamic vasopressinergic neurones.

ACKNOWLEDGEMENTS

The authors greatly appreciate the skillful technical assistance of Mrs. Ulrike Schlapp.

REFERENCES

(1)   Andersson, B., Ekman, L., Hökfelt, B., Jobin, M., Olson, K., and Robertshaw, D., Acta Physiol. Scand. 69 (1967) 111-118
(2)   Riedel, W., Pflügers Arch. 399 (1983) 11-17
(3)   Riedel, W., and Burke, S.L., J. Auton. Nerv. Syst. 24 (1988) 157-173
(4)   Riedel, W., Städter, W.R., and Gray, D.A., J. Auton. Nerv. Syst. Suppl.(1986) 543-52
(5)   Inomoto, T., Mercer J.B., and Simon, E., J. Physiol. (London) 322 (1982) 139-150
(6)   Gray, D.A., and Simon, E., J. Comp. Physiol. 151 (1983) 241-246
(7)   Simon, E., and Riedel, W., Brain Research 87 (1975) 323-333
(8)   Taylor, T., Wondisford, F.E., Blaine, T., and Weintraub, B.D., Endocrinology 126 (1990) 317-324
(9)   Redding, T.W., and Schally, A.V., Proc. Soc. Exp. Biol. Med. 131 (1969) 420-425
(10)   Keil, R., Gerstberger, R., and Simon, E., Pflügers Arch. 420 Suppl N1 (1992) R40

Temperature Regulation
Advances in Pharmacological Sciences
© 1994 Birkhäuser Verlag Basel

ROLE OF PROLACTIN IN BROWN ADIPOSE TISSUE THERMOGENIC ACTIVITY

Takehiro Yahata and Akihiro Kuroshima
Department of Physiology, Asahikawa Medical College, Japan

Summary

Cold-exposure decreased plasma prolactin (PRL) levels in rat, but after cold-acclimation the levels returned to the control values. Saline injection elevated the plasma PRL level and this elevation was blocked by noradrenaline (NA). Haloperidol treatment suppressed, while bromocriptine treatment enhanced the *in vitro* responsiveness of brown adipose tissue (BAT) to glucagon (G), but not to NA. PRL may inhibit nonshivering thermogenesis, especially by modifying the thermogenic action of G in BAT.

Introduction

The thermogenic activity of brown adipose tissue (BAT), the major site of nonshivering thermogenesis (NST), is regulated by many hormonal factors such as thyroidal, adrenocortical and pancreatic hormones (1), in addition to the major sympathomimetic noradrenaline (NA) (2). The activity of BAT is enhanced after cold acclimation, while it decreases from parturition to lactation (3). It has been recently claimed that prolactin (PRL), which has significant roles in the regulation of such diverse physiological processes as reproduction, osmoregulation and growth, suppresses BAT thermogenic activity during lactation (4). In the present study, the possible effects of hyper- and hypo-prolactinemia on BAT thermogenesis were investigated directly in *in vitro* to evaluate a role of PRL in BAT-NST.

Materials and Methods

The following experiments were done on the 12 wks-old adult rats of Wistar strain. Exp 1: Male rats were exposed to cold (5°C) for 1 hour, 1 day or 4 wks, and the plasma PRL levels were measured by RIA (Amersham, Amersham, England). Exp 2: The effect of NA (40 μg/100g bw, ip injection) on plasma PRL level was observed in the warm controls (WC) and the cold-acclimated animals (CA). Exp 3: Haloperidol (250 μg in 30% ethanol with 0.3% tartaric acid/0.1 ml/ 100 g bw, sc, 42 and 10 hrs before killing) or bromocriptine (300 μg in the same vehicle as above/0.1 ml/100g bw, sc, 42, 29, 18 and 5 hrs before killing) was injected to female rats to see the effects of hyper- and hypo-prolactinemia (5) on BAT thermogenic dopamine (DA; 40 μg in the same vehicle as above/0.1 ml/100g bw, sc, 42, 29, 18 and 5 hrs before killing) on BAT thermogenic activity was also examined.

The thermogenic activities of interscapular BAT in these animals were estimated from NA- or glucagon-induced maximum responses in oxygen consumption of the tissue blocks in Krebs-Ringer phosphate buffer containing 4% BSA and 5 mM glucose (37°C, pH 7.4) by using of Clark type oxygen electrodes (Rank Brothers, Kambridge) (6).

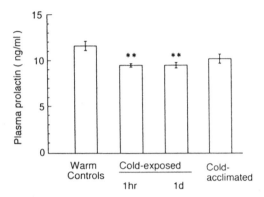

Fig. 1. Cold-induced changes in plasma prolactin level.
Each column indicates the mean of 5 samples with SE. **: Significantly different from warm controls, p<0.01.

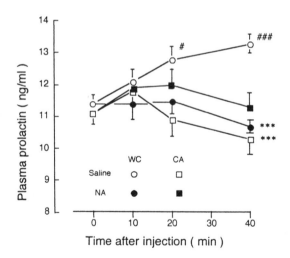

Fig 2. Changes in plasma prolactin level after saline or noradrenaline (NA) injection. Each point indicates the mean of 10 samples with SE. WC: Warm controls. CA: Cold-acclimated rats. # and ###: Significantly different from 0 min, p<0.05 and 0.001, respectively. ***: Significantly different from saline-injected WC, p<0.001.

Results were denoted as mean±SEM. Statistical significance was assessed by Student's *t*-test.

Results

Changes in the plasma PRL levels by cold and NA.

Cold exposure for 1hr or 1 day decreased significantly the plasma PRL levels from 11.7 ± 0.45 ng/ml in WC to 9.6 ± 0.22 and 9.6 ± 0.32 ng/ml in 1hr- and 1 day-cold exposure, respectively (p<0.01), but after cold acclimation the level returned to the control values (10.3 ± 0.47 ng/ml) (Fig. 1).

In WC plasma PRL increased significantly after saline ip injection, while NA did not change it. The initial plasma PRL level and that at 40 min after injection of saline or NA were 11.4 ± 0.31, 13.3 ± 0.33 (p<0.001) and 10.7 ± 0.19 ng/ml, respectively. On the other hand, either saline- or NA-injection did not affect the plasma PRL level in CA. The initial level and that at 40 min after injection of saline or NA were 11.1 ± 0.33, 10.3 ± 0.39 and 11.3 ± 0.45 ng/ml, respectively (Fig. 2).

Effects of haloperidol (Halo), bromocriptine (Bromo) or dopamine (DA) treatment on in vitro BAT thermogenic activity.

Halo treatment, which stimulated PRL secretion, increased BAT weight and suppressed the *in vitro* basal respiration (p<0.05) as well as the increase in oxygen consumption of BAT induced by glucagon (p<0.01), but not that induced by NA (Fig. 3).

Bromo treatment, which suppresses PRL secretion, decreased BAT weight and enhanced the responsiveness of the tissue to glucagon (224 ± 24.7 pmol O₂/mg BAT/min vs 124 ± 27.9 pmol O₂/mg BAT/min in the control, p<0.05), but not to NA (238 ± 31.8 pmol O₂/mg

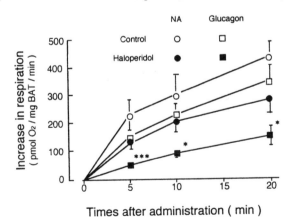

Fig. 3. Effects of haloperidol treatment on brown adipose tissue (BAT) activity. Each point indicates the mean of 6 (Control) and 7 (Haloperidol) samples with SE. * and ***: Significantly different from Control, p<0.05 and 0.001, respectively.

BAT/min vs 195 ± 39.1 pmol O_2/mg BAT/min in the control). DA treatment simulated the changes in thermogenic response caused by Bromo. The increase in oxygen consumption induced by glucagon and NA were 223 ± 23.6 and 344 ± 54.3 pmol O_2/mg BAT/min in the control, and 339 ± 30.4 (p<0.05) and 374 ± 28.8 pmol O_2/mg BAT/min, respectively.

At the time of killing plasma PRL levels did not differ from the respective control in Halo- and Bromo-treated groups.

Discussion

The present results indicate that PRL secretion induced by stress, such as saline injection, is suppressed under cold condition in which NST is enhanced. This is probably mediated by high concentration of NA, because NA blocked perfectly the saline injection-stimulated PRL secretion, though it was noted that central β2 adrenoceptor enhances stress-induced PRL secretion (7). It is, thus, surmised that PRL contributes, at least in part, to the suppression of NST as indicated in the lactating animals (4).

Halo- and Bromo-treatment affected BAT weight and the *in vitro* thermogenic response of the tissue to glucagon in opposite way, but not to NA. Although the reason why PRL suppresses the glucagon response selectively is not clear, these results suggest that PRL suppresses the thermogenic activity of BAT through, at least in part, enhancement of lipogenesis (8). It is also possible that the activation of CRH-ACTH-corticosterone axis induced by PRL (9) modify the thermogenic activity of BAT. DA is an inhibitory factor of PRL secretion, so it is reasonable that DA-treatment simulates the effect of Bromo on BAT response to glucagon.

However, it has been recently demonstrated that DA may suppress the sympathetically induced BAT function via specific DA receptor in BAT (10), and that the effect of DA on PRL secretion varies at different concentrations: suppression at high concentration and enhancement at low concentration (11). Therefore DA may affect BAT diversely via different mechanism(s) in addition to that of Bromo (12).

Failure of these drugs to influence plasma PRL levels at death may result from long duration between the last injection of the drugs and killing.

References

1.   DOI, K., KUROSHIMA, A. (1984). Economy of hormonal requirement for metabolic temperature acclimation. *J. them. Biol.* **9**, 87-91.
2.   SMITH, R.E., HORWITZ, B.A. (1969). Brown fat and thermogenesis. *Physiol. Rev.* **49**, 330-425.
3.   TRAYHURN, P., DOUGLAS, J.B., McGUCKIN, M.M. (1982). Brown adipose tissue thermogenesis is suppressed during lactation in mice. *Nature.* **298**, 56-60.

4.   CHAN, E., SWAMINATHAN, R. (1990). Role of prolactin in lactation-induced changes in brown adipose tissue. *Am. J. Physiol.* **258**, R51-R56.

5.   ARITA, J., KIMURA, F. (1986). Characterization of *in vitro* dopamine synthesis in the median eminence of rats with haloperidol-induced hyperprolactinemia and bromocriptine-induced hypoprolactinemia. *Endocrinology.* **119**, 1666-1672.

6.   YAHATA, T., KUROSHIMA, A. (in press). *In vitro* thermogenic activity of rat brown adipose tissue in neonatal period. *Biol. Neonate 1993.*

7.   HAANWINCKEL, M.A., ANTUNES-RODRIGUES, J., De CASTRO e SILVA, E. (1991). Role of central beta-adrenoceptors on stress-induced prolactin release in rats. *Horm. metab. Res.* **23**, 318-320.

8.   CINCOTTA, A.H., MEIER, A.H. (1985). Prolactin permits the expression of a circadian variation in lipogenic responsiveness to insulin in hepatocytes of the golden hamster *(Mesocricetus auratus). J. Endocrinol.* **106**, 173-176.

9.   WEBER, R.F.A., CALOGERO, A.E. (1991). Prolactin stimulates rat hypothalamic corticotrophin-releasing hormone and pituitary adrenocorticotropin secretion in vitro. *Neuroendocrinology.* **54**, 248-253.

10. NISOLI, E., TONELLO, C., MEMO, M., CARRUBA, M.O. (1992). Biochemical and functional identification of a novel dopamine receptor subtype in rat brown adipose tissue. Its role in modulating sympathetic stimulation-induced thermogenesis. *J. Pharmacol. Exp. Ther.* **263**, 823-829.

11. AREY, B.J., BURRIS, T.P., BASCO, P., FREEMAN, M.E. (1993). Infusion of dopamine at low concentrations stimulates the release of prolactin from α-methyl-*p*-tyrosine-treated rats. *Proc. Soc. Exp. Biol. Med.* **203**, 60-63.

12. HANNA, S., SHIN, S.H. (1992). Differential inhibition of dopamine and bromocriptine on induced prolactin release: multiple sites for the inhibition of dopamine. *Neuroendocrinology.* **55**, 591-599.

Temperature Regulation
Advances in Pharmacological Sciences
© 1994 Birkhäuser Verlag Basel

BIOLOGY OF ADAPTIVE HEAT PRODUCTION: STUDIES ON BROWN ADIPOSE TISSUE

Paul Trayhurn

Division of Biochemical Sciences, Rowett Research Institute, Bucksburn, Aberdeen AB2 9SB, Scotland, U.K.

Introduction

The maintenance of a relatively constant body temperature by homeotherms is dependent on mechanisms for the conservation and the generation of heat. Conservation occurs through both behavioural (e.g. huddling) and physiological (e.g. vasoconstriction) strategies. The heat that is produced as a by-product of normal metabolic processes, reflecting thermodynamic inefficiency, makes a substantial contribution to the total body heat pool. When a homeotherm is at an environmental temperature below the thermoneutral zone (lower critical temperature), additional heat is required so as to prevent hypothermia. This facultative, or adaptive, thermoregulatory heat is generated through two general mechanisms - shivering and non-shivering thermogenesis (NST). The phenomenon of shivering, involving muscle contractions, is well-documented. Progress has also been made in recent years in understanding the basis for non-shivering thermogenesis, primarily through intensive investigation of brown adipose tissue. There continues to be debate, however, about the range of mechanisms, and their tissue localization, involved in NST.

It is emphasized that the generic term "thermogenesis" is commonly used to describe the specific process of NST - or in the context of nutritional science, the heat generated in response to diet and overfeeding. Strictly, however, thermogenesis (Greek *thermos* - heat) applies to the heat resulting from all metabolic and mechanical processes in an organism.

Mechanisms for non-shivering thermogenesis

Several biochemical systems for NST have been proposed (see 1). These include the pumping of Na$^+$ across plasma membranes, substrate cycles (cycling of specific intermediates in metabolic pathways), protein turnover, and more recently "hot-pipes" in the microvasculature (Table 1). The best documented mechanism for generating heat by NST is, however, that associated with brown adipose tissue (BAT) - the sole organ in mammals specialized for heat production.

Investigation of the mechanisms of NST involves the consideration of two distinct issues - the nature of the biochemical reactions that produce heat, and the tissue(s) in which they are located. Certain criteria need to be fulfilled for a particular biochemical system to be thermogenic in the operational sense: *(i)* there should be the capacity for generating considerable amounts of heat; *(ii)* the mechanism should be capable of being rapidly turned on and off, involving changes in rate of several orders of magnitude; *(iii)* there should be long-term adaptive changes consistent with variations in the requirement for thermogenesis (e.g. during cold adaptation); *(iv)* the mechanism should be located in a tissue which physiological studies clearly indicate as being directly involved in facultative heat production (1).

Of the mechanisms for NST listed in Table 1, protein turnover can be essentially discounted on the basis of the second, third, and fourth criteria - it occurs in all tissues, and in an aggregated sense does not show the appropriate degree of flexibility. The Na$^+$ pump (Na$^+$,K$^+$-ATPase) has received considerable attention, but the difficulty in perceiving it as a major mechanism for thermogenesis lies in its quantitative importance in heat production and its presence in all tissues, where the primary role is the regulation of the ionic balance of cells. Despite early proposals that up to half of cellular energy expenditure is associated with the Na$^+$ pump (2,3), more recent work has tended to view the contribution as at most 5% (4). The correct figure is still a matter of debate, however, with work by Milligan and coworkers suggesting a contribution in tissues of farm animals closer to the initial high estimates (5). The same problems of tissue localization and quantitative significance also argue against an important role for substrate cycles. Nevertheless, it is suggested that although individual substrate cycles may produce little heat, the collective contribution to thermogenesis of a number of cycles operating in synchrony could be considerable (6).

Only the mitochondrial proton conductance pathway, which occurs exclusively in BAT, currently fulfils all the criteria for a major thermogenic mechanism. BAT, and the proton conductance pathway, therefore represent the only widely accepted thermogenic system, and on this basis have formed the main focus of recent work on NST. However, it is essential that the thesis that other tissues and other biochemical mechanisms may also contribute to the overall thermogenic capacity of an animal should continue to be explored. Indeed, there are clearly documented circumstances, such as in birds, where NST is evident in the absence of

TABLE1. Putative mechanisms for non-shivering thermogenesis

Sodium pump (Na^+,K^+-ATPase)

Protein turnover*

Mitochondrial Ca^{2+} cycle

α-glycerophosphate shuttle*

Substrate ("futile") cycles, e.g.
  fructose-6-phosphate/fructose-1,6-bisphosphate
  glucose/glucose-6-phosphate
  triglyceride/free fatty acid re-esterification cycle

Hot-pipes in the microvasculature

Proton conductance pathway in brown adipose tissue[+]

---

* Can also be viewed as substrate cycles

[+] Exclusive to brown adipose tissue; the other mechanisms are less tissue-specific

BAT (7). The recent proposal relating to hot-pipes in the microvasculature is particularly apposite in this regard (8).

Role of brown adipose tissue in non-shivering thermogenesis

BAT (or brown fat) was first described by Conrad Gesner in 1551. Until 1978 the thermogenic role of this tissue was broadly considered to be restricted to two specific situations - the arousal from hibernation and the thermoregulation of newborn mammals. Blood flow measurements with [86]Rb[+] had suggested that BAT had little direct quantitative importance in the generation of heat in adult animals (rodents) exposed to the cold (9,10). However, subsequent studies demonstrated that the [86]Rb[+] technique seriously underestimates blood flow to BAT, while overestimating the flow to skeletal muscle (11). Measurements of the distribution of the cardiac output with radioactively-labelled microspheres (15 μm diam.) have clearly demonstrated that BAT is the major site of NST in adult rats acclimated to the cold (12). In fully cold-acclimated rats, BAT accounts for approximately two-thirds of the capacity for NST, with much of the remainder being due to increases in energy expenditure in the heart and respiratory muscles, i.e. the tissues that play a supportive role in thermogenesis (12). Similar blood flow studies have been performed on partially cold-acclimated mice, and a figure of approximately 50% estimated for the contribution of BAT to the total thermogenic capacity (13).

Not only have blood flow measurements with microspheres demonstrated that BAT is the main site of adaptive thermogenesis in small laboratory rodents, but as a corollary they have also indicated that other tissues do not play a major role. This is particularly significant with respect to skeletal muscle, which has long been considered as a potential site of non-shivering - as well as shivering - thermogenesis. Whether skeletal muscle does have some direct thermogenic function, is not, however, a closed question, with recent work suggesting that the tissue may indeed have a role in NST (7,8,14,15). Since skeletal muscle accounts for approximately 40% of body mass in a normal animal, a relatively minor thermogenic mechanism at the cellular level could in principle make a significant contribution to NST on a whole-body basis.

BAT is extensively vascularized (16), which is critical to the ability to receive a high blood flow during peak thermogenesis. The high blood flow enables heat to be rapidly distributed to the body core, and also ensures that the substrates (including oxygen) required to fuel thermogenesis are provided at rates appropriate to the exceptional demands of the tissue.

Mechanism of heat production in brown adipose tissue

The unifying property of the adipose tissues is their ability to store large quantities of lipid in the form of one or more triacylglycerol droplets. Brown and white adipose tissues have, however, quite different roles in energy metabolism. White adipose tissue is the main long-term energy store in the body, providing substrates in the form of fatty acids for utilization in other tissues. In contrast, the central function of BAT is the generation of heat, and heat is the *primary* product of metabolism in the tissue (16,17).

Heat is generated in BAT through a unique proton translocation mechanism operating in the mitochondrial inner membrane (16,18). This acts as a proton "short-circuit" across the membrane such that the proton gradient which is normally generated during respiration is dissipated as heat rather than being linked to the synthesis of ATP. The proton translocation of BAT mitochondria is regulated by a specific "uncoupling" protein (also termed "thermogenin") in the inner membrane (16-18). Uncoupling protein (UCP), M_r 32-33000, is subject to acute regulation such that the proton translocation of the mitochondria, and thus the heat produced, varies according to the physiological requirements for thermogenesis (19).

Free fatty acids (or possibly acyl CoA's) provide the intracellular signal for the activation of the proton conductance pathway, through an interaction with UCP (18). Thus fatty acids play a dual role, as both primary fuel and a signal for the acute activation of thermogenesis. The initial event in the stimulation of heat production is the release of noradrenaline from the sympathetic nerves which extensively innervate BAT. The tissue contains a novel ß3-adrenoceptor (20), to which noradrenaline binds, and this leads, through a cascade of events, to the activation of a hormone-sensitive lipase with the consequent stimulation of lipolysis.

Both noradrenaline and insulin play an important role in the regulation of the level of UCP, the latter through an interaction with the sympathetic nervous system (see below). Other hormones, including corticosteroids, oestrogens, and neuropeptide Y also modulate the activity (or capacity) of BAT, either directly or via the sympathetic system.

Molecular biology of uncoupling protein

UCP is a member of the family of mitochondrial carriers, and there is considerable homology between it and both the ADP/ATP translocase and the mitochondrial phosphate carrier (21). UCP consists of a single polypeptide chain with 306 amino acids. Several models have been proposed for the organization of the protein within the inner mitochondrial membrane. The various models are similar in suggesting that there are six hydrophobic transmembrane α-helices, although the most recent has proposed that the hydrophobic segment nearest to the C-terminal end of the molecule does not completely span the membrane (22). There appear to be two hydrophilic loops on the matrix side of the membrane, with three hydrophilic loops and the C-terminal region of the molecule on the cytosolic side (22). The region of the protein responsible for the translocation of protons may be located close to the N-terminal end.

The most important recent developments in the study of BAT have come from the application of molecular biology. The gene coding for UCP has been cloned and sequenced in several species - rats, mice, rabbits, Syrian hamsters, cattle and humans (see 22,23). Heterogeneity between species in the sequence of the UCP gene is evident. Comparison of the derived amino acid sequences suggests, however, that there is considerable species homology in the primary structure of the protein itself (22). A central issue is the basis for the tissue-specific expression of the UCP gene. This question is under active investigation, and the regulatory elements (promoter sequences) that allow hormonal responsiveness are being explored.

The availability of cDNA's for UCP has provided a tool for the detection and measurement of the mRNA for the protein, so that factors (environmental, nutritional, hormonal) which affect the transcription of the UCP gene can be examined. Acute cold exposure has been shown, for example, to rapidly induce a striking increase in the level of UCP mRNA (24), as the initial event underlying the production of more UCP itself. UCP mRNA is also increased during overfeeding with a cafeteria diet, and in response to ß-agonists (22-24). On the other hand, fasting (at least in the short-term) results in a fall in the level of UCP mRNA (22,23). It is apparent that physiologically the UCP gene is transcriptionally regulated by noradrenaline from the sympathetic nervous system, and that T_3 also plays a role (25).

Large, or full length, rat cDNA's for UCP do not appear to hybridize readily with UCP mRNA from widely divergent species (e.g. cattle, humans), reflecting the heterogeneity between species in the sequence of the gene. One substantive region of 32 bases is, however, identical in rats and cattle, and shows a difference of only a single base in mice, rabbits, Syrian hamsters and humans (Table 2). The 32-base segment is an extension of the 27-base fragment that we have previously

TABLE 2.   Comparison of a highly conserved 32-base sequence of the uncoupling protein gene, or cDNA, from rats, cattle, mice, rabbits, Syrian hamsters and humans

Species		Sequence	
Rat	88	*AT*CACCTTCCCGCTGGACACCGCCAAAGTCC*CG*	3'
Bovine	44	*AT*CACCTTCCCGCTGGACACCGCCAAAGTCC*CG*	3'
Mouse	321*	*AT*CACCTTCCCGCTGGACAC_T_GCCAAAGTCC*CG*	3'
Hamster		*AT*CACCTTCCCGCTGGACAC_A_GCCAAAGTCC*CG*	3'
Human	2324	*AT*CACCTTCCCGCTGGACAC_G_GCCAAAGTCC*CG*	3'
Rabbit	88	*AT*CACCTT_T_CCGCTGGACACCGCCAAAGTCC*CG*	3'
32-mer oligonucleotide		*TA*GTGGAAGGGCGACCTGTGGCGGTTTCAGG*C*	5'

The sequences were obtained from published work, as described previously (26), and are given with the original positioning. The hamster sequence is from the deposition in the EMBL database by Raimbault *et al.* (accession no. X73138). *Relates to exon 1. Differences between species are underlined. The bases added at both the 3' and 5' ends to extend the highly conserved 27-base region reported previously are shown in bold italics.

described (26), following a re-analysis of the published sequences. The original 27-base region has been extended by three bases at the 5' end, and by two bases at the 3' end, these being identical in the six species for which sequences are now available (Table 2). The highly conserved region is rich (62.5%) in guanine and cytosine residues (i.e. the bases which form triple hydrogen bonds), and we have recently synthesized the complementary sequence as a 32-mer oligonucleotide (3'-TAGTGGAAGGGCGACCTGTGGCGGTTTCAGGC-5'), to provide a probe for studying the expression of the UCP gene across a wide range of species (Table 2); this replaces the 27-mer designed earlier for a similar purpose (26).

   Using the 27-mer oligonucleotide we have been able to detect the mRNA for UCP, and thus the expression of the UCP gene, in species as diverse as laboratory rodents, pipistrelle bats, and newborn ruminants such as goats and red deer (26). The extension to a 32-mer should, in principle, further increase the specificity of a simple cross-species probe for UCP mRNA; the T_m of the new probe is some 5°C higher than that of the 27-mer oligonucleotide.

Adaptive changes in brown adipose tissue

   Much is known of the responses in BAT of exposing animals to the cold (see 27,28). Acutely, the proton translocation of the mitochondrial inner membrane is increased, through the activation of pre-existing UCP (18,19). This acute activation is often reflected in what is described as an "unmasking of GDP binding sites" on the protein, mitochondrial GDP binding being a widely used *in vitro* assay for assessing the thermogenic activity of BAT (19). As noted above, exposure to cold leads to a rapid increase in the level of the mRNA for UCP, implying that there is a cold-induced stimulation of the transcription of the UCP gene. UCP is not the only gene whose expression in BAT appears to be increased by cold; the level of the mRNAs for lipoprotein lipase (29), glyceraldehyde phosphate dehydrogenase, and the insulin-responsive glucose transporter GLUT-4 (30) have each been shown to increase on cold-exposure. The expression of other genes in BAT may also prove to be cold-inducible.

Chronic exposure to cold produces a constellation of changes in BAT. These include an increase in cellularity, and an increase in the mitochondrial content of each cell (mitochondriogenesis). The concentration of UCP in the mitochondria also changes in response to prolonged cold-exposure (see 27,31). The net effect of these changes is to greatly augment the *capacity* for thermogenesis in BAT - which is dependent on the total tissue content of UCP - and this is driven principally by increased stimulation from the sympathetic system (27,28). In general, parallel changes occur in the mitochondrial content of BAT and in the specific mitochondrial concentration of UCP (19,31).

The effects on BAT of overfeeding with a variable and palatable "cafeteria" diet consisting of human food items has also been extensively studied (see 28,32). Such a dietary regimen leads to a stimulation of "diet-induced thermogenesis", a process that provides a counter-regulatory mechanism in the control of whole-body energy balance such that energy expenditure is increased on overfeeding and energy storage (as lipid) reduced (32). Both the thermogenic activity and capacity of BAT are increased in young animals (rats, mice) fed a cafeteria diet (32). Thus, as with cold exposure - with which there are many parallels - mitochondrial GDP binding, mitochondrial content, and the specific mitochondrial concentration of UCP are each augmented (together with other measures linked to the activity of the tissue). Again, such changes are the result of an increase in the sympathetic stimulation to BAT (28,32). Additional situations in which the thermogenic activity of BAT is augmented include the arousal from hibernation (33), fever, and the cachetic state associated with cancer (see 28).

There are several conditions which lead to a functional atrophy of BAT, in addition to long-term maintenance in warm environments. The most widely studied is that of obesity, and the activity of the tissue has been investigated in a variety of obese animals (see 28). In general, it is evident that there is a relative atrophy of the tissue in obesity, with a reduction in mitochondrial GDP binding, mitochondrial content, and the concentration of UCP (28). The decrease in GDP binding is an early event, the other changes following later. Decreased thermogenic activity and capacity of BAT in obese animals contributes substantially to the positive energy balance which leads to the obese state.

Fasting also results in a major atrophy of BAT, with a substantial fall in both the specific mitochondrial concentration and total tissue content of UCP. Such changes are reversed by refeeding, but this needs to be prolonged in order for there to be a full restoration of thermogenic capacity (34). Lactation leads to a similar functional atrophy of BAT as that in fasting, and the lactation-induced changes are largely reversed following weaning of the pups (see 28). The decline in thermogenesis occurring in lactation is considered to reduce the energy costs of milk production, through a reduction in the non-lactational component of maternal energy expenditure. Equally, energy expenditure is reduced by the atrophy of BAT occurring during fasting.

In the same way that the activation of thermogenesis results from sympathetic stimulation, functional atrophy of BAT is attributable to a fall in sympathetic tone (see 28). Regulation of the tissue is, however, also dependent on other factors, such as insulin and glucocorticoids. This is clearly illustrated in relation to the effects of experimentally-induced diabetes. A major atrophy of BAT occurs in animals made diabetic with streptozotocin, there being a substantial fall in the amount of UCP (35). The protein, and therefore thermogenic capacity, is restored in a dose-dependent manner by the long-term infusion of insulin from implanted osmotic minipumps (35). Studies in which the sympathetic innervation was surgically cut suggest that insulin regulates the amount of UCP in BAT through an interaction with the sympathetic system (35).

Insulin may also regulate the expression of the gene coding for UCP. In preliminary studies, the level of UCP mRNA was found to be greatly reduced in BAT of streptozotocin-diabetic animals (26).

Species distribution of brown adipose tissue

Although the presence of BAT has long been recognized in small rodents, the extent to which it occurs in other - especially larger - species has been unclear. This uncertainty is principally a consequence of the absence, until recently, of a satisfactory basis for the identification of BAT, and specifically its differentiation from white fat (see 18,36). The problem is compounded by the fact that in some species BAT is apparent only at restricted stages of development. From a general viewpoint it is, of course, essential to establish the species distribution of BAT, and thus the extent to which the tissue represents a common system for NST. The issue is also of importance from an evolutionary perspective, particularly with respect to the origin and development of UCP. Systems specialized for the generation of heat are, of course, not generally required by poikilotherms. There are, nonetheless, special circumstances in which an ability to generate heat for a thermal function is evident in poikilotherms; an example is in the warming of

TABLE 3. Species distribution of brown adipose tissue in mammals based on the immunological identification of the tissue-specific uncoupling protein

Rodents:

    Laboratory - Mouse (*Mus musculus*); Rat (*Rattus norvegicus*); Guinea pig *(Cavia porcellus)*

    Hibernators - Syrian hamster (*Mesocricetus auratus*); European hamster (*Cricetus cricetus*); Turkish hamster (*Mesocricetus brandti*); Richardson's ground squirrel (*Spermophilus richardsonii*); Columbian ground squirrel (*Spermophilus columbianus*); 13-lined ground squirrel (*Spermophilus tridecemlineatus*); Edible dormouse (*Eliomys quercinus*)

    Other rodents - Djungarian hamster (*Phodopus sungorus*); Wood lemming (*Myopus schisticolor*); Wood mouse (*Apodemus sylvaticus*); Orkney vole (*Microtus arvalis orcadensis*); Short-tailed vole (*Microtus agrestis*)

Chiroptera:

    Pipistrelle bat (*Pipistrellus pipistrellus*)

Lagomorphs:

    Rabbit (*Oryctolagus cuniculus*)

Carnivores:

    Dog (*Canis canis*)

Artiodactyla:

    Sheep (*Ovis aries*); Cattle (*Bos taurus*); Goat (*Capra hircus*); Reindeer/Caribou (*Rangifer tarandus*); Red deer (*Cervus elaphus*)

Primates:

    Rhesus monkey (*Macaca mulatta*); Cynomolgus monkey (*Macaca fascicularis*); Humans - newborn and adult (*Homo sapiens*)

Adapted from Trayhurn (36); this article contains detailed references to the original observations

the flight muscle of bumble bees by substrate cycling, enabling flight to occur at low ambient temperatures (37).

Until recently, BAT was diagnosed in various species, and distinguished from white adipose tissue, on the basis of histological appearance (36). However, since this is now recognised to be an unsatisfactory and potentially misleading approach, recent studies have focussed on the immunological detection of UCP as providing the critical identifier for brown fat. The most effective way of detecting UCP is by Western blotting (immunoblotting), as this ensures that immunoreactivity occurs at the molecular weight characteristic of the protein. Western blotting for UCP has been applied to a wide variety of mammals, to establish the species distribution of

BAT. The animals studied range from rodents through to primates; Table 3 shows the species in which UCP has now been identified. In some cases, such as ruminants (lambs, cattle, goats, reindeer) UCP, and hence BAT, is present only for a restricted period over the first days or weeks of postnatal life - presumably because the tissue is required for only a limited period following birth, to meet the initial thermal needs.

As more animals are examined, and UCP identified, species and situations where the tissue is *absent* take on growing significance. Among mammals, the domesticated pig is the sole species where UCP has not been detected (see 36). We have also been unable to identify the protein in several avian species adapted to the sub-arctic winter (see 36); nor has UCP been detected in common poorwills (*Phalaenoptilus nuttallii*), the sole species of bird for which there is strong evidence of true hibernation (Brigham and Trayhurn, unpublished observations). Current indications would suggest that BAT is restricted to mammals.

Coda

Much is now known of the biology of BAT, and the contribution of the tissue to NST. BAT continues to be the site of the only extensively defined biochemical system for the generation of heat - with firmly established physiological significance. It is important to emphasize, however, that there are mechanisms for NST which do not involve brown fat. Elucidation of these mechanisms, together with their regulation and tissue localization, should be regarded as a high priority for future studies.

Acknowledgements

I am grateful to my colleagues at the Rowett Research Institute for their support. The financial contribution of the Scottish Office Agriculture and Fisheries Department is acknowledged.

References
(1)   Trayhurn P, Milner REM. Mechanisms of thermogenesis: brown adipose tissue. Obesity Weight Regulat. 1987; 6: 147-161.
(2)   Whittam R, Willis JS. Ion movements and oxygen consumption in kidney cortex slices. J. Physiol. (Lond.) 1963; 168: 158-177.
(3)   Whittam R, Blond DM. Respiratory control by an adenosine triphosphatase involved in active transport in brain cortex. Biochem. J. 1964; 92: 147-158.
(4)   Chinet A, Clausen T, Girardier L. Microcalorimetric determination of energy expenditure due to active sodium-potassium transport in the soleus muscle and brown adipose tissue of the rat. J. Physiol (Lond.) 1977; 265: 43-61.
(5)   Milligan LP, McBride BW. Energy costs of ion pumping by animal tissues. J. Nutr. 1985; 115: 1374-1382.

(6)   Newsholme EA. A possible metabolic basis for the control of body weight. New Engl. J. Med. 1980; 302: 400-405.

(7)   Duchamp C, Cohen-Adad F, Rouanet JL, Barré H. Histochemical arguments for muscular non-shivering thermogenesis in muscovy ducklings. J. Physiol. (Lond.) 1992; 457: 27-45.

(8)   Colquhoun EQ, Clark MG. Open question - Has thermogenesis in muscle been overlooked and misinterpreted. News Physiol. Sci. 1991; 6: 256-259.

(9)   Kuroshima A, Konno N, Itoh S. Increase in the blood flow through brown adipose tissue in response to cold exposure and norepinephrine in the rat. Jap. J. Physiol. 1967; 17: 523-537.

(10) Jansky L, Hart JS. Cardiac output and organ blood flow in warm-and cold-acclimated rats exposed to cold. Can. J. Physiol. Pharmacol. 1968; 46: 653-659.

(11) Foster DO, Frydman ML. Comparison of microspheres and ^{86}Rb$^+$ as tracers of the distribution of cardiac output in rats indicates invalidity of ^{86}Rb$^+$-based measurements. Can. J. Physiol. Pharmacol. 1978; 56: 97-110.

(12) Foster DO, Frydman ML. Nonshivering thermogenesis in the rat: II. Measurements of blood flow with microspheres point to brown adipose tissue as the dominant site of the calorigenesis induced by noradrenaline. Can. J. Physiol. Pharmacol. 1978; 56: 110-122.

(13) Thurlby PL, Trayhurn P. Regional blood flow in genetically obese (ob/ob) mice. The importance of brown adipose tissue to the reduced energy expenditure on non-shivering thermogenesis. Pflügers Archiv 1980; 385:193-201.

(14) Dubois-Ferriere R, Chinet AE. Contribution of skeletal muscle to the regulatory non-shivering thermogenesis in small mammals. Pflügers Archiv 1981; 390: 224-229.

(15) Challis RAJ, Arch JRS, Newsholme EA. The rate of substrate cycling between fructose 6-phosphate and fructose 1,6-bisphosphate in skeletal muscle from cold-exposed, hyperthyroid or acutely exercised rats. Biochem. J. 1985; 231: 217-220.

(16) Nicholls DG, Locke RM. Thermogenic mechanisms in brown fat. Physiol. Rev. 1984; 64: 1-64.

(17) Cannon B, Nedergaard J. The biochemistry of an inefficient tissue: Brown adipose tissue. Essays Biochem. 1985; 20: 110-164.

(18) Nicholls DG, Cunningham SA, Rial E. The bioenergetic mechanisms of brown adipose tissue thermogenesis. In: Trayhurn P, Nicholls DG, editors. Brown adipose tissue. London: Edward Arnold, 1986: 52-85.

(19) Trayhurn P, Milner RE. A commentary on the interpretation of in vitro biochemical measures of brown adipose tissue thermogenesis. Can J. Physiol. Pharmacol. 1989; 67: 811-819.

(20) Arch JRS. The brown adipocyte ß-adrenoceptor. Proc. Nutr. Soc. 1989; 48: 215-223.

(21) Klingenberg M. Mechanism and evolution of the uncoupling protein of brown adipose tissue. Trends Biochem. Sci. 1990; 15: 108-112.

(22) Klaus S, Casteilla L, Bouillaud F, Ricquier D. The uncoupling protein UCP - a membraneous mitochondrial ion carrier exclusively expressed in brown adipose tissue. Int. J. Biochem. 1991; 23: 791-801.

(23) Ricquier D, Casteilla L, Bouillaud F. Molecular studies of the uncoupling protein. FASEB J. 1991; 5: 2237-2242.

(24) Bouillaud F, Ricquier D, Mory G, Thibault J. Increased level of mRNA for the uncoupling protein in brown adipose tissue of rats during thermogenesis induced by cold exposure or norepinephrine infusion. J. Biol. Chem. 1984; 259: 11583-11586.

(25) Bianco AC, Sheng X, Silva JE. Triodothyronine amplifies norepinephrine stimulation of uncoupling protein gene transcription by a mechanism not requiring protein synthesis. J. Biol. Chem. 1988; 263: 18168-18175.

(26) Brander F, Keith JS, Trayhurn P. A 27-mer oligonucleotide probe for the detection and measurement of the mRNA for uncoupling protein in brown adipose tissue of different species. Comp. Biochem. Physiol. 1993; 104B: 125-131.

(27) Himms-Hagen J. Brown adipose tissue and cold-acclimation. In: Trayhurn P, Nicholls DG, editors. Brown adipose tissue. London: Edward Arnold, 1986: 214-268.

(28) Himms-Hagen J. Brown adipose tissue thermogenesis and obesity. Prog. Lipid Res. 1989; 28: 67-115.

(29) Mitchell JRD, Jacobsson A, Kirchgessner TG, Schotz MC, Cannon B, Nedergaard J. Regulation of expression of the lipoprotein lipase gene in brown adipose tissue. Am. J. Physiol. 1992; 263: E500-E506.

(30) Olichon-Berthe C, Van Obberghen E, Le Marchand-Brustel Y. Effect of cold acclimation on the expression of glucose transporter GLUT 4. Mol. Cell Endocrinol. 1992; 89: 11-18.

(31) Trayhurn P, Ashwell M, Jennings G, Richard D, Stirling DM. Effect of warm or cold exposure on GDP binding and uncoupling protein in rat brown fat. Am. J. Physiol. 1987; 252: E237-E243.

(32) Rothwell NJ, Stock MJ. Brown adipose tissue and diet-induced thermogenesis. In: Trayhurn P, Nicholls DG, editors. Brown adipose tissue. London: Edward Arnold, 1986: 269-298.

(33) Milner RE, Wang L, Trayhurn P. Brown fat thermogenesis during hibernation and arousal in Richardson's ground squirrel. Am. J. Physiol. 1989; 256: R42-R48.

(34) Trayhurn P, Jennings G. Nonshivering thermogenesis and the thermogenic capacity of brown fat in fasted and/or refed mice. Am. J. Physiol. 1988; 254: R11-R16.

(35) Géloën A, Trayhurn P. Regulation of the level of uncoupling protein in brown adipose tissue by insulin requires the mediation of the sympathetic nervous system. FEBS Lett. 1990; 267: 265-267.

(36) Trayhurn P. Species distribution of brown adipose tissue: Characterization of adipose tissues from uncoupling protein and its mRNA. In: Carey C, Florant GL, Wunder BA, Horwitz B, editors. Life in the cold III: Ecological, physiological and molecular mechanisms. Colorado: Westview Press, 1993. In press.

(37) Newsholme EA, Crabtree B. Substrate cycles in metabolic regulation and in heat generation. Biochem. Soc. Sym. 1976; 41: 61-109.

Temperature Regulation
Advances in Pharmacological Sciences
© 1994 Birkhäuser Verlag Basel

BROWN ADIPOSE TISSUE:
RECEPTORS AND RECRUITMENT

Jan Nedergaard and Barbara Cannon
The Wenner-Gren Institute, The Arrhenius Laboratories F3,
Stockholm University, S-106 91 Stockholm, Sweden

SUMMARY: The thermogenic capacity of brown adipose tissue can be augmented in response to the physiological needs of an animal (i.e. it can be recruited). This happens e.g. perinatally, although at different time points for precocial, altricial or immature newborns. At least in the case of the altricial and the immature newborns, the causative agent for the recruitment process is probably norepinephrine. This is e.g. verified by the ability of norepinephrine to induce both cell proliferation (via β_1 receptors) and an enhanced degree of differentiation, including enhanced expression of the genes coding for lipoprotein lipase and the uncoupling protein thermogenin (via β_3 receptors), in a cell culture system.

Different patterns of perinatal recruitment

When discussing perinatal recruitment of brown adipose tissue, it is important to realize that newborn mammals of different species are born in different degrees of development, which we may refer to as precocial (e.g. lamb), altricial (e.g. mice) or immature (e.g. hamster) (for general review see [1]) (Fig. 1). Also their thermoregulatory properties differ in parallel with this general level of maturity at birth, with the precocial newborns showing a high capacity for nonshivering thermogenesis at birth (but this capacity normally declines rapidly after birth), with the altricials showing a rapid postnatal increase in capacity, and the immature newborns not being able to initiate such an increase until a lag phase after birth has passed.

The function of brown adipose tissue

There is today no doubt that the anatomical site of thermoregulatory nonshivering thermogenesis is the brown adipose tissue of the newborns. In the altricial newborn, this tissue is thus recruited *after* birth, i.e. there is a concerted activation of cell proliferation (although this has not so far been directly

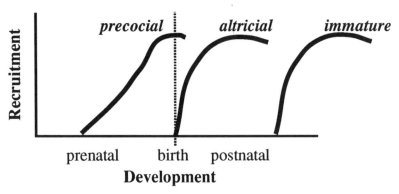

Fig. 1. The three typical developmental patterns of perinatal recruitment.

shown), mitochondriogenesis and expression of certain genes, the products of which are involved in the heat production process. The most notable of these is, of course, the uncoupling protein thermogenin (UCP). Thermogenin is found in the inner membrane of the mitochondria (Fig. 2), and although quite a lot is known today about its structure, the exact mechanism for its function as a protonophore, as well as for the regulation of its protonophoric capacity, is still not understood [2]. It is, however, clear that the postnatal changes in capacity for nonshivering thermogenesis are fully explainable as reflecting the changes of thermogenin content in the tissue. The amount of thermogenin is apparently determined simply by the amount of thermogenin mRNA [3], and e.g. in the altricial newborns, the level of thermogenin mRNA increases immediately from birth. This increase is a response to the – relative – cold normally experienced by the newborn after parturition; if a high environmental temperature is kept, an increased expression of the thermogenin gene *ucp* is not observed [4,12]. Thus, this recruitment process of brown adipose tissue in the newborn is parallel to that observed in adult, cold-exposed small mammals, and is therefore probably induced by the chronic sympathetic stimulation of the tissue, i.e. by norepinephrine stimulation.

The study of brown adipose tissue recruitment in-vitro

The effect of norepinephrine as an inducer of the recruitment process is, however, preferentially studied in-vitro, in cell cultures starting with undifferentiated brown-fat cell precursors isolated from the tissue. If these cells are allowed to proliferate and differentiate in culture, they gain the ability

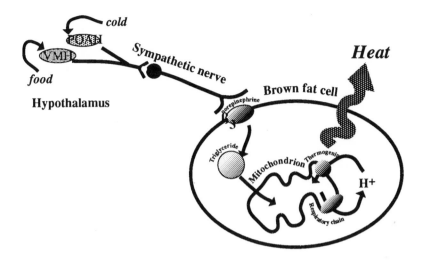

Fig. 2. Scheme of stimulation of brown adipose tissue thermogenesis.

to respond to norepinephrine addition by e.g. inducing thermogenin gene expression [5]. This is a very rapid process, which with time also results in an increase in the amount of thermogenin itself [6].

The adrenergic receptors involved in this process may be studied by the use of selective adrenergic agents or by the addition to the culture of agents that may mimic the early cellular effects of adrenergic stimulation. Such experiments have demonstrated that agents that lead to an increased level of cAMP in the cells, such as β-adrenergic agonists or forskolin, as well as cAMP analogues, are able to induce *ucp* expression, but it would seem to a somewhat smaller extent than norepinephrine; a further analysis seems to indicate that it is the α_1-part of the adrenergic stimulation which is missing. The β-response can be analyzed in order to determine whether it shows β_1, β_2 or β_3 characteristics; since both CGP-12177 and BRL-37344 are good inducers, the response is mediated via β_3 receptors [5].

It has also recently been demonstrated that the gene for lipoprotein lipase can be stimulated in cell culture by norepinephrine, probably also through β_3-receptors [7]. Thus, in these differentiated, mature cells in culture, β_3-receptors are activated by norepinephrine and this stimulates expression of structural genes, via increases in cyclic-AMP.

The expression of structural genes is controlled by a multitude of DNA-binding transcription factors, some of which have been suggested to be related to adipocyte differentiation, e.g. FOS and members of the C/EBP family. We could show that although *c-fos* expression was not spontaneously

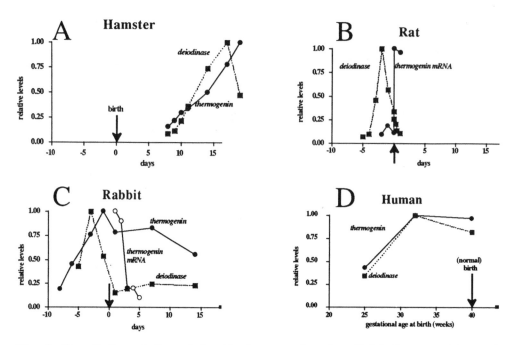

Fig. 3. Compilation of the relationship between thyroxine 5'-deiodinase activity and thermogenin emergence in brown adipose tissue of different species.
Data from A: [13, 14]; B: [15, 16, 17]; C: [18, 19, 20]; D: [21]. The data have been normalised to 1.0 as the maximal value in all cases.

induced during differentiation, the expression could be stimulated by norepinephrine [8]. The transcription factor C/EBPβ was constitutively expressed in the cells and its expression further enhanced by norepinephrine [9]. In contrast, C/EBPα, although also constitutively expressed, responded very differently. In the differentiated cells, it was stimulated, as the the other factors. However, in cells still in a proliferative phase, the expression was markedly decreased by norepinephrine [9].

It would appear, from the above results, that only β_3-receptors are coupled to adenylyl cyclase in the differentiated cells. However, also during the proliferative phase of culture, the cells are responsive to norepinephrine, and DNA synthesis is positively stimulated by all agents which elevate cyclic-AMP levels; remarkably, at this stage, it is the β_1-receptor which is coupled to adenylyl cyclase [10, 11].

These results have led to the conclusion that during development – at least in culture – there is a sudden switch in expression of β-adrenergic receptor subtype and also in the genomic interpretation of increases in cyclic-AMP level.

The influence of thyroid hormone

For the ability of norepinephrine to induce thermogenin gene expression to be present, the cells in the cell culture, as well as the the cells in the brown adipose tissue of the newborn, have to be in an euthyroid state, and the increase in thermogenin mRNA levels is always parallelled or perhaps slightly preceded by an increase in the activity of the enzyme thyroxine deiodinase which converts T_4 to T_3. As seen in Fig. 3 this is equally true for the newborns of the precocial, altricial or immature type. The activity of thyroxine deiodinase is apparently necessary for the ability of the newborn pups to induce enhanced thermogenin expression postnatally, as treatment of the dams with the thyroxine deiodinase inhibitor iopanoic acid abolishes the postnatal increase in thermogenin mRNA [12].

The enigma of brown adipose tissue recruitment in the precocial neonates

As can be realized from the arguments above, it is possible to understand the postnatal recruitment of brown adipose tissue in the altricial and the immature newborns as being a response to the cold experienced postnatally, and it is likely that the recruitment processes in these newborns are identical to those found in adults mammals undergoing cold acclimation and that they therefore represent responses to chronic norepinephrine stimulation, similar to those which can be induced in the cell culture system. However, the induction of recruitment in the precocial group of newborns is not readily understood in such terms, as uterine norepinephrine stimulation is unlikely. The mechanisms allowing for this recruitment to take place within the uterine environment are still not understood.

References

1. Nedergaard, J., Connolly, E. & Cannon, B. (1986) in Brown Adipose Tissue (Trayhurn, P. & Nicholls, D. G., eds.), pp 152-213, Edward Arnold Ltd., London
2. Nedergaard, J. & Cannon, B. (1992) in New Comprehensive Biochemistry vol. 23: Molecular Mechanisms in Bioenergetics (Ernster, L., eds.), pp 385-420, Elsevier, Amsterdam
3. Jacobsson, A., Mühleisen, M., Cannon, B. & Nedergaard, J. (1993) Subm. for publ.
4. Obregon, M. J., Jacobsson, A., Kirchgessner, T., Schotz, M. C., Cannon, B. & Nedergaard, J. (1989) Biochem. J. **259**, 341-346
5. Rehnmark, S., Néchad, M., Herron, D., Cannon, B. & Nedergaard, J. (1990) J. Biol. Chem. **265**, 16464-16471
6. Puigserver, P., Herron, D., Gianotti, M., Palou, A., Cannon, B. & Nedergaard, J. (1992) Biochem. J. **284**, 393-398
7. Kuusela, P., Rehnmark, S., Cannon, B. & Nedergaard, J. (1993) In preparation
8. Thonberg, H., Zhang, S.-J., Tvrdik, P., Jacobsson, A., Cannon, B. & Nedergaard, J. (1993) In preparation.
9. Rehnmark, S., Antonson, P., Xanthopoulos, K. G. & Jacobsson, A. (1993) FEBS Lett. **318**, 235-241

10. Bronnikov, G., Houstek, J. & Nedergaard, J. (1992) J. Biol. Chem. **267**, 2006-2013
11. Bronnikov, G., Bengtsson, T., Golozoubova, V., Cannon, B. & Nedergaard, J. (1993) In preparation.
12. Giralt, M., Martin, I., Iglesias, R., Vinas, O., Villarroya, F. & Mampel, T. (1990) Eur. J. Biochem. **193**, 297-302
13. Houstek, J., Janíková, D., Bednár, J., Kopecky, J., Sebastián, J. & Soukup, T. (1990) Biochim. Biophys. Acta **1015**, 441-449
14. Houstek, J., Kopecky, J., Baudysová, M., Janikova, D., Pavelka, S. & Klement, P. (1990) Biochim. Biophys. Acta **1018**, 243-247
15. Obregon, M. J., Pitamber, R., Jacobsson, A., Nedergaard, J. & Cannon, B. (1987) Biochem. Biophys. Res. Commun. **148**, 9-14
16. Obregon, M. J., Ruiz de Ona, C., Hernandez, A., Calvo, R., Escobar de Rey, F. & Morreale de Escobar, G. (1989) Am. J. Physiol. **257**, E625-E631
17. Giralt, M., Martin, I., Mampel, T., Villarroya, F., Iglesias, R. & Vinas, O. (1988) Biochem. Biophys. Res. Commun. **156**, 493-499
18. Freeman, K. B. & Patel, H. V. (1984) Can. J. Biochem. **62**, 479-485
19. Rozon, D. K., Harris, W. H. & Verrinder-Gibbins, A. M. (1989) Can. J. Physiol. Pharmacol. **67**, 54-58
20. Brzezinska-Slebodzinska, E. & Kapluk, J. (1990) Biol. Neonate. **57**, 358-366
21. Houstek, J., Kopecky, J., Pavelka, S., Tvrdik, P., Baudysova, M., Vizek, K., Hermanska, J. & Janikova, D. (1992) in Adenine Nucleotides in Cellular Energy Transfer and Signal Transduction (Papa, S., Azzi, A. & Tager, J. M., eds.), pp 447-458, Birkhäuser Verlag, Basel

HIGH-ENERGY FOOD SUPPLEMENT, ENERGY SUBSTRATE MOBILIZATION AND HEAT BALANCE IN COLD-EXPOSED HUMANS

André L. Vallerand & Ira Jacobs. Defence & Civil Institute of Environmental Medicine
P.O. Box 2000, 1133 Sheppard Ave. W., North York, Ontario, M3M 3B9 CANADA

SUMMARY

It is hypothesized that our inability to confirm that energy substrate mobilization is a limiting factor for cold-induced thermogenesis (M) in humans (1, 2) is due to a sub-optimal dose of ingested substrates. This hypothesis was tested in healthy males exposed twice to the cold (3h at 7°C, 1 m.s^{-1} wind, nude, fasting) following the ingestion of either a placebo or a high-energy food supplement (710 kcal or 2,970 kJ). The supplement did not influence body temperatures, M or heat debt, even though it enhanced carbohydrate mobilization and oxidation ($P<0.05$), albeit at the expense of lipid mobilization and oxidation ($P<0.05$). The results suggest that, under normal conditions, energy substrate mobilization per se is not a limiting factor for M in humans.

INTRODUCTION

Several years ago, it was suggested that one of the main physiological factors limiting maximal cold-induced thermogenesis (M) was neither the saturation of the cellular oxidative capacity nor the cardiorespiratory or cardiovascular functions for gas transport, but the timely supply of metabolic fuels (3). The supply of metabolic fuels can be modulated in several ways. Ingesting energy substrates increases exogenous substrate mobilization whereas pharmacological agents such as methylxanthines (caffeine-like substances), enhance endogenous substrate mobilization. The ingestion of substrates and/or methylxanthines in the cold has been used to delay the onset of hypothermia, apparently via an energy substrate mobilization effect (1, 2, 4). Although this theory has been inferred in several studies, we were not able to confirm it in a recent series of human studies (5-7). One possible explanation is that the usual dose of energy substrates (approximately 300 kcal or 1,255 kJ) may not have been optimal in our lab and that a higher dose was required to observe a beneficial effect. The goal of this study was therefore to determine whether a much higher dose of energy substrates (710 kcal or 2,970 kJ) that improves substrate mobilization can enhance M, reduce the heat debt and ensure warmer body temperatures.

METHODS

Seven healthy male volunteers participated in the present study. They were thoroughly familiarized with both the cold air and the protocol. Each fasting subject was then exposed to two cold tests (1 wk apart), each for 3 h at 7°C, 1 m.s^{-1} wind speed, sitting at rest and wearing jogging shorts only. One test was performed after the blind ingestion of a high-energy supplement (710 kcal or 2,970 kJ; Ensure Plus™; Abbott Pharmaceutical, Montréal P. Qué.; 53%, 32% and 15% of energy derived from carbohydrate, fat and protein, respectively) and another following the ingestion of a placebo drink (artificially-sweetened/flavoured water with dietary fibres). Subjects were instrumented with a rectal temperature (T_{re}) probe (Sherigan, Argyle NY), 12 calibrated heat flux transducers (Thermonetics Co., San Diego CA) for the measurement of mean skin temperature ($\overline{T}_{sk}$) and mean dry heat loss as well as a forearm intravenous catheter for the determination of plasma substrates and hormones as before (9). Oxygen consumption and carbon dioxide production (in L.min^{-1} STPD) were measured for 20 min of every half hour (Beckman Horizon Metabolic Cart, Anaheim CA).

M, non-protein respiratory exchange ratio (NPRER), rates of carbohydrate and lipid utilization were calculated exactly as previously reported (8, 9). The *rate of heat debt* ($\dot{S}$ in W.m^{-2}) was determined minute by minute as the balance of M minus all heat losses (dry, respiratory and evaporative heat losses). The *heat debt* in kJ.kg^{-1} (S) was then obtained by the integration of $\dot{S}$, after taking into account the body surface area and the body mass (5, 10). Main effects of time, treatments (placebo vs food supplement) as well as time vs treatment interactions were tested by repeated measures ANOVA (Biomedical Computer Programs, BMDP-90, Los Angeles CA). ANOVA were corrected by the Huynh-Feldt epsilon adjusted degrees of freedom when sphericity was significant (BMDP-90). When interactions were significant, paired t-tests (adjusted for multiple comparisons) were used to locate significant differences. Results are expressed as mean ± standard error of the mean (SEM).

RESULTS

All subjects completed the 3 h test in both the placebo and high-energy food supplement conditions. A repeated measures ANOVA revealed that the oxygen consumption was affected by a significant main effect of time ($P<0.01$) but not by a main effect of treatment. In contrast, the NPRER was significantly altered by a main effect of treatment ($P<0.05$) as NPRER values during the high-energy supplement were generally higher than those of the placebo. Lipid oxidation, another important metabolic parameter, was affected by an interaction of effects ($P<0.01$; Fig. 1A).

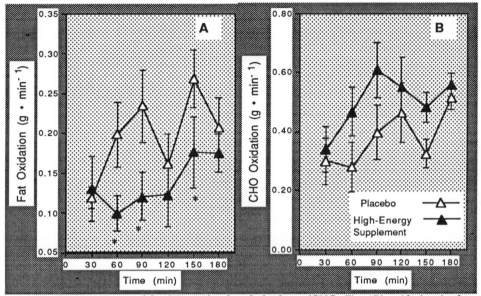

Figure 1: Rates of fat (Fig. 1A)and carbohydrate (CHO; Fig. 1B) oxidation in the cold following the ingestion of a placebo or high-energy food supplement. Significant differences from the placebo condition at the same point in time are indicated by "*" (at least $P<0.05$).

Overall, the supplement reduced the rates of lipid oxidation by 32% compare to the placebo ($P<0.05$). In contrast, the food supplement increased rates of carbohydrate oxidation by approximately 32% in comparison to the placebo ($P<0.05$).

Cold exposure increased M ($P<0.01$), and by min 90 of the test, M was maintained at about 140 $W.m^{-2}$ (Fig. 2A). This represents an increase of about 3.5 times resting values, assuming normal resting values of about 38 $W.m^{-2}$ (6). Nevertheless, M remained unaffected by the food supplement (Fig. 2A). Therefore, since there was no influence of the supplement on the rates of dry, evaporative and respiratory heat loss (data not shown), this resulted in no effect as well on the heat debt profiles shown in Fig. 2B. It is thus not surprising that absolute and relative values of T_{re} and $\overline{T}_{sk}$ were not significantly altered by the ingestion of the food supplement in the cold.

Plasma insulin levels were significantly increased by the food supplement, as levels peaked at min 30 with a 6.4-fold rise ($P<0.05$). This is likely the result of the corresponding increment of plasma glucose levels that occurred at the same time (22%; $P<0.01$). Plasma glycerol and free fatty acids (FFA) levels were greatly increased throughout the 3 h of the placebo test ($P<0.01$), whereas this effect was largely prevented by the high-energy supplement ($P<0.05$).

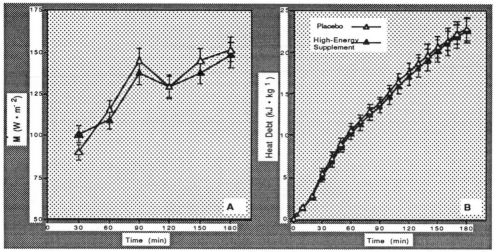

<u>Figure 2</u>: Metabolic rate (M; W.m-2; Fig. 2A) and heat debt (kJ.kg-1; Fig. 2B) in the cold following the ingestion of a placebo or high-energy food supplement.

DISCUSSION

The present results confirm previous studies where it was shown that the ingestion of energy substrates does not alter cold-induced thermogenesis and cold resistance of subjects that are exposed to either a mild or a more severe cold test, while at rest or while performing intermittent exercise (5-7). The present study also extends these results to the ingestion of a higher dose of energy substrates (710 kcal or 2,970 kJ). The results of these studies therefore contrast with the theory formulated more than a decade ago (3) that food ingestion and/or the administration of substrate-mobilizing drugs, enhances thermogenesis and delays the onset of hypothermia via a substrate mobilization effect, a factor thought to be a limiting for cold-induced thermogenesis (1-4). The present and our previous results conflict with those supporting the energy substrate theory and the present study demonstrates that such conflict is probably not related to an insufficient amount of substrate being used.

Before reviewing and criticizing the above theory, it is important to assess whether energy substrate mobilization was affected *at all* by the present protocol. Changes in some plasma substrates and hormones implicated in energy metabolism were used as indices of substrate mobilization, as before (6, 9) The notion that the present ingestion of a high-energy food supplement under resting conditions in the cold accelerated carbohydrate mobilization is supported by the increments in both plasma glucose and insulin levels. The data also suggest that the supplement reduced the mobilization of lipids, and this is supported by a reduction in plasma FFA and a smaller rise in plasma glycerol levels, an index of lipolysis. Further to these important changes in mobilization, it is essential to realize that they were without any impact whatsoever on the rate of metabolic heat production (Fig. 2A), since carbohydrate oxidation was increased but at the expense of lipid oxidation (Fig. 1). In addition to M, all avenues of heat loss were not altered by the supplement (see Results), resulting in no change in the heat debt (Fig. 2B) or body temperatures (see Results). Surely, if energy substrate mobilization was limiting for the cold-induced increase in thermogenesis, the ingestion of almost 3,000 kJ should have affected those variables.

It is quite interesting to note that food supplements have also been used to enhance exercise performance. Several studies have demonstrated the beneficial effects of substrate ingestion (mainly carbohydrates) during prolonged exercise. It enhances carbohydrate mobilization, increases exogenous carbohydrate oxidation, preserves glycogen reserves and enhances exercise performance (11). But the key to this concept seems to be whether hypoglycemia is observed (<2.5 mM; 11). Maintenance of plasma glucose thus appears to be a limiting factor during prolonged exercise. But in the cold however, levels are hardly in the hypoglycemic range. The lowest plasma glucose levels observed in this study were 4.4 mmol.L^{-1} (at min 120) and about 4.8 mmol.L^{-1} in another study (2). Similarly, there is no reason to believe that, under normal conditions, the availability of intramuscular reserves of carbohydrates, or muscle glycogen availability, is limiting in the cold. It is known that glycogen utilization goes up during shivering in humans although it does not induce a marked depletion. However, it has been demonstrated that, when glycogen reserves, including blood glucose, are depleted prior to cold exposure, their availability could become limiting for thermogenesis and cold tolerance (12).

In conclusion, under our experimental conditions, the ingestion of high-energy food supplement does not offer any advantage on cold-induced thermogenesis, heat debt or body temperatures in comparison to the ingestion of a placebo during a semi-nude cold test. Further, our data do not support the theory that energy mobilization is a limiting factor for cold-induced thermogenesis either at rest (5, 6), during intermittent exercise (7) or even with a high dose of substrates (Figs. 1, 2). We suggest that this theory may be more applicable to animals or to humans exposed to the cold under conditions of energy deficiency. It is also proposed that other factors, such as an increase in sympathoadrenal system activity, are required to enhance cold-induced thermogenesis, which in turn can effectively reduce the heat debt and thus ameliorate cold resistance. Further studies are required to explore these concepts.

ACKNOWLEDGEMENTS

The authors would like to acknowledge the expert technical assistance of Mr. Carl Bowen, Ms. Debbie Kerrigan-Brown, Mr. Robert Limmer, Mr. Jan Pope and in particular, Ms. Ingrid Schmegner. Medical coverage was kindly provided by Maj Hugh O'Neill.

REFERENCES

1. Wang LCH, Man SFP &, Belcastro A. Improving cold tolerance in men: Effects of substrates and aminophylline. In: Cooper, Lomax, Schonbaum & Veale editors. Homeostasis and thermal stress.Karger: Basel CH, 1986: 22-26.

2. Wang LCH, Man SFP & Belcastro AN. Metabolic and hormonal responses in theophylline-increased cold resistance in males. J Appl Physiol 1987; 63: 589-596.

3. Wang LCH. Factors limiting maximum cold-induced heat production. Life Sci 1978; 23: 2089-2098.

4. Wang LCH, Man SFP, Belcastro AN & Westly JC. Single, optimal oral dosage of theophylline for improving cold resistance in man. In: P. Lomax & E. Schonbaum editors. Thermoregulation: Research and clinical applications, Karger: Basel CH, 1989: 54-58.

5. Vallerand AL, Tikuisis P, Ducharme MB & Jacobs I. Is energy substrate mobilization a limiting factor for cold thermogenesis? Eur. J. Appl. Physiol, In Press.

6. Vallerand AL, Schmegner IF & Jacobs I. Influence of the Cold Buster™ Sports bar on heat debt, mobilization and oxidation of energy substrates. Defence & Civil Institute of Environmental Medicine, North York Ont., 1992 DCIEM Report No.: 92-60.

7. Vallerand A.L. & Jacobs I. Interaction of a food supplement, intermittent exercise and cold exposure on heat balance. Defence & Civil Institute of Environmental Medicine, North York Ont., 1993 DCIEM Report No.: 93-19.

8. Vallerand AL & Jacobs I. Rates of energy substrate utilization during human cold exposure. Eur J Appl. Physiol 1989; 58: 873-878.

9. Vallerand AL, Jacobs I & Kavanagh MF. Mechanism of enhanced cold tolerance by an ephedrine/caffeine mixture in humans. J Appl Physiol 1989; 67(1): 438-444.

10. Vallerand AL, Savourey G & Bittel JHM. Determination of heat debt in the cold: partitional calorimetry vs conventional methods. J Appl Physiol 1992; 72(4): 1380-1385.

11. Coyle EF. Carbohydrate supplementation during exercise. J Nutr 1992; 122: 788-795.

12. Martineau L & Jacobs I. Muscle glycogen availability and temperature regulation in humans. J Appl Physiol 1989; 66: 72-78.

EMERGING THEMES IN THERMOREGULATION AND FEVER.

K. E. Cooper & W. L. Veale. Department of Medical Physiology, University of Calgary, Health Sciences Centre, 3330 Hospital Drive N.W., Calgary, Alberta Canada. T2N-4N1.

Summary. This paper is partly a report of recent findings and partly speculative. It deals with the problem of identification of neural circuits involved in thermoregulation and fever, the use of the anaesthetized animal to study those circuits, the significance of so-called endogenous antipyresis, the interesting finding of suppression of fever by cortical spreading depression.

New approaches and extremely important new technologies are coming into use in physiology, but prediction of those most likely to succeed and of what new advances are probable is risky. The unpredictable factor is serendipity, the chance observation which gives new insights into the secrets of nature. But serendipity occurs usually in the course of patient and painstaking research so that if we outline our choices for avenues of new investigation we may lay the foundation for the occurrence of some important new accidental discovery.

1. Neuronal circuitry.

a. The problem.

The outstanding problem in thermoregulation still remains the unravelling of the neuronal circuitry involved. With the involvement of the preoptic area and the anterior and lateral hypothalamic areas together with the cortical and perhaps other subcortical regions in behavioural thermoregulation, there are many interwoven pathways as yet unknown. There are many hypotheses dependant on assumptions about the function of thermosensitive neurones in the hypothalamus and elsewhere the characteristics of which have been studied mostly in the anaesthetized animal, in the absence of observations on the accompanying thermoregulatory activities. Indeed apart from a few observations (1,2,3) there are few observations relating firing rates of thermosensitive hypothalamic neurones with thermoregulatory behaviour in the awake animal. While theoretically sensible, the link between neuronal behaviour and thermoregulatory responses is still weak. Indeed, the correlation between the high Q_{10} thermosensitive neurones in the frontal cortex, which were described by Barker & Carpenter (4) and thermoregulatory behaviour is unknown. It is reasonable to suggest that high thermosensitivity may be a characteristic of many neurones but that some of them have been incorporated, during evolution, into thermoregulating systems.

b. *Tools for identifying active brain regions,application and collection of putative neurotransmitters.*

At least four methods are currently available for the detection of synapses,cells and brain regions which become active in response to various types of stimuli. They are the uptake of radioactively labelled 2-deoxyglucose, the expression of proto-oncogenes such as c-fos, the use of fast magnetic resonance functional imaging (MRI) and PET scans. The use of these techniques can demonstrate regions of the brain in which metabolic and other activities increase. However, though these increases could indicate regions primarily involved in the thermoregulatory control or fever inducing mechanism, they could indicate regions involved secondarily in the activation of the efferent mechanisms necessary for thermoregulation or fever induction; or they could represent non-specific activation of the regions if, for example, the thermoregulatory stimuli or fever induction induced a "stress" response. There is need therefore for very careful control experiments to be built into the use of these activity demonstrating mechanisms in order to get a valid interpretation of the results as relating to primary or secondary mechanisms or to non-specific stimuli.

Sokoloff et al. (5) pioneered the method for using 2-deoxy-D-[^{14}C]glucose ([^{14}]DG), which is taken up into active neurones and which can be detected within the neurones by autoradiography. The activity demonstrated with this method is mainly in the synaptic regions of the neurones. Technical details were more fully elaborated by Hand (6). This method has been used by Cooper et al., (7) to show reduced activity in the bed nucleus of the stria terminalis, the ventral diagonal bands of Broca and the lateral septal area during prostaglandin evoked hyperthermia.

Morimoto et al., (8) found uptakes indicating increased metabolic activity in the lateral pre-optic area, the posterior hypothalamus, the dorsomedial thalamus and the bed nucleus during endogenous pyrogen induced fever.

The activation of the nuclear proto-oncogene c-fos can be judged by immunostaining for the c-fos protein (9). The protein can be detected in many cell bodies as they are activated. Not all active neurones will express c-fos, so that a positive result has more meaning than a negative. C-fos expression in the central nervous system can occur in the anaesthetized animal as well in the conscious animal, and this could be an advantage in some studies since "stress" responses could be partly or wholly eliminated. This technique is still in its infancy in thermoregulatory studies but we believe that, carefully used, it will yield important results in localizing areas of neuronal activity in relation to cold and heat exposure and in fever, particularly when used in collaboration with 2-deoxyglucose uptakes. MRI will, we think, soon become available for use with small animals as well as humans. It measures the patterns of blood flow by detecting the oxygenated and reduced haemoglobins and thus detecting where oxygen rich blood gives up its oxygen (10). It uses magnetic fields and thus is free of the problems of X-ray or radioactive exposure. It has been used in the human subject to visualize occipital cortex activity during pulsed photic stimulation of the retina and in other areas during somatosensory stimulation as well as during mental activity. The use of this technique to study human brain areas activated during thermoregulation and fever could be very exciting.

The PET-scan is expensive but again adds another possible dimension to showing up active brain areas, particularly in animals, during fever or extremes of thermal exposure.

c. Tracing Connexions.

Once regions are shown to be active by these techniques then the effects of electrical stimulation can be observed in order to see what thermoregulatory responses are mimicked. In addition, the micro-injection of identifiable compounds which are transported along axons both in anterograde or retrograde directions can be used to suggest connectivities between active brain regions, and these suggested connectivities must then be tested by the standard electrophysiological methods of stimulation, recording and collision methods. In addition the effects of lesioning can be used to correlate the identified regions with their natural responses. The heat coagulation and electrolytic lesions are often rather gross, but finer lesions can be produced by microinjecton of minute quantities of kainic acid or of ibotenic acid. It is also possible to use temporary and reversible lesions. These can be achieved by the use of local anaesthetics and colchicine (11) which blocks axonal transport.

A very important recent study (12) has drawn attention to the differences in thermoregulatory responses to administration of norepinephrine into the pre-optic region by micro-injection as compared to micro-dialysis. The latter seems to give a truer response since the micro-injection may release PGE locally. The microdialysis probes can be made cheaply in the laboratory (13,14). They can also be made very small.

2. *The use of the anaesthetized animal.*

One of the more exciting pieces of work in our laboratory in recent years has been the characterizing of the fever occurring in the urethane anaesthetized rat and rabbit (Malkinson et al., (15). The urethane anaesthetized rat is unable to maintain its body temperature when exposed to heat or cold. The anaesthetized rats' core temperature can be maintained at a steady level by placing it on an electrically heated pad and adjusting the electrical energy supplied to the pad. When such thermal stability has been maintained for one hour, ICV PGE_2, ICV IL-1, or IV endotoxin causes a characteristic rise in temperature accompanied by shivering and a rise in oxygen consumption. It is likely that the increase in oxygen uptake is mostly due to brown adipose tissue (BAT) stimulation (16). The rise in core temperature occurred at all baseline temperatures studied from 33°C to 39°C, but the increases were greater at the lower resting core temperatures. Possibly the fever producing agents can act on the initial processes of fever, e.g., release of pyrogenic prostaglandin, in this unconscious model, and that the PGE_2 acts outside the mechanisms for normal thermoregulation. It may be that the final mediator(s) of fever provide a potent stimulus to the efferent pathways involved in heat production and heat conservation, or it may be that the fever mediators are able to stimulate the central connexions normally stimulated by peripheral and central cold receptors but the connexions to which are suppressed by the urethane. The urethane anaesthetized rat could,then, provide a useful experimental approach to further study of the regions of the brain involved in cold defence and in fever. Using the methods described above for demonstrating activity in discrete brain loci comparison can be made between the awake and the anaesthetized animal both at neutral and low environmental temperatures. The normally active regions which are blocked by the anaesthetic should be made evident.

3. Endogenous Antipyresis.

Over the last decade our group and that of Kasting in Vancouver have been working on the action of arginine vasopressin (AVP) within the ventral septum as an endogenous antipyretic, (17). Intensive study of this has also been undertaken by Zeisberger and his colleagues (18). The evidence strongly supports the concept that AVP released into the ventral septum, close to the diagonal band of Broca, acts via AVP-V_1 receptors to limit the extent of fever. In some species to source of the release of AVP appears to be neurones derived from the bed nucleus of the stria terminalis, and in others it seems to be mainly from neurones from the paraventricular nucleus. The released AVP can inhibit the firing rate of neurones in the ventral septum stimulated by glutamate (19). The mechanism of the suppression of fever by septally released AVP is however not fully understood. Whether the reduction of fever is entirely by neuronal circuits acting within the brain on thermoregulatory, or special fever pathways, or whether the septal and other intracerebral neurones involved act to release extracranial antipyretic substances is not known. For example, the anterior pituitary gland could be stimulated to release hormones which in turn lead to release of antipyretic steroids and to stimulate immunoactive substance release. The AVP antipyretic system seems also not to be confined to the ventral septum since Federico et al., (20) have demonstrated an antipyretic action of AVP within the medial amygdala, an effect which is clear but less than that seen in the ventral septum.

Recently Wilkinson and Kasting, (21), have found that in rats made "tolerant" to endotoxin, by repeated daily injections of the endotoxin, application of the AVP-V_1 blocker can reverse the "tolerance". While there is much evidence for some peripheral mechanisms of tolerance, e.g., more rapid clearance of endotoxin from the circulation (22) this is the first implication of a possible intracerebral mechanism of tolerance using a natural endogenous antipyresic substance.

The initial implication of AVP as an endogenous antipyretic was derived from the studies of Kasting on the absence of fever in the sheep given intravenous gram-negative endotoxin close to term of pregnancy. The evidence supported the notion that septally released AVP was responsible for the fever suppression. Goelst et al. (23) were able to confirm the suppression of fever at term in Dorper ewes. However, they added a most interesting and important new observation. The ewes in which endotoxin fever was suppressed had 'normal' fevers in response to intravenous staphylococcal wall pyrogen. This suggests that either the fever mechanism itself is not suppressed in near term ewes and that some possible peripheral antipyretic substance which acts on one fever pathway is released by AVP; or that some peripheral or central cytokine necessary for the release intracerebrally of AVP is released in response to endotoxin but not to staphylococcal wall pyrogen. It is probable that the two pyrogens use different pathways, in part, to cause fever. The role of the other endogenous antipyretic presently known, namely α-MSH (24), which acts on the lateral septum and probably peripherally as well deserves a great deal of further study.

4. Cortical Spreading Depression and fever.

A recent and most interesting observation was made by Monda & Pittman (25). Cortical spreading depression (CSD) was induced in rats by the method of Bures & Buresova (26). The KCl causes CSD and this can be done in the urethane anaesthetized rat and, without apparent major disturbance of the animal, in the awake chronic preparation. The CSD so caused greatly attenuates or abolishes the fever due to pyrogens or PGE_1 intraventricularly.

The increase in rat body temperature induced by corticotrophin releasing factor (CRF) was not altered by CSD nor was the normal body temperature. Further investigations by Komaromi et al. (27) demonstrated that the fever depression was pyrogen dose dependant and that, in the rat, the attenuation of fever was accompanied by a large reduction in oxygen uptake. It does not apparently seem that release of AVP into the septum is a major part of the mechanism since CSD is effective in suppressing fever in long term castrated rats in which the brain AVP levels are greatly reduced.

We would suggest that further exploration of the CSD action on fever and decortication would be fruitful lines of study.

5. Sites of action of circulating pyrogens.

The loci of action of interleukin-1 (IL-1) in the brain as a mediator of fever have been the subject of most important research recently by Blatteis et al., (28) and Stitt (29). Earlier work based on microinjection studies had shown that the anterior hypothalamus/pre-optic area (AH/POA) was very sensitive to endogenous pyrogen and also to bacterial pyrogen (30,31,32,33). Unfortunately these loci soon became accepted as the loci of action of endogenous pyrogen in fever due to intravenous pyrogen administration despite the total lack of evidence for penetration of the blood-brain barrier by IL-1. The evidence adduced by Blatteis et al., (28) and Stitt (29) that one of the circumventricular organs, the organum vasculosum laminae terminalis (OVLT), a region of specialized blood brain barrier characteristics, has provided us with a probable locus of action of IL-1 not needing penetration of the blood brain barrier to the hypothalamus. The question then arises as to whether other circumventricular organs could act as loci or portals of entry for IL-1. Since one theory of the action of IL-1 requires the release of prostaglandin E_2 Sirko et al. (34) examined the release of PGE_2 within the hypothalamus (the AH/POA) and the tuberal-posterior hypothalamus (PH-Tu) and found that its release followed intravenous injection of endotoxin or IL-1 at the same level in both sites. They suggested that blood borne pyrogens might act at several brain loci, particularly the circumventricular organs. More recently, Takahashi et al., (35) have adduced evidence that the subfornical organ (SFO) is not a locus of access of circulating pyrogens into the brain. This conclusion was based on the lack of difference in febrile responses between SFO lesioned and sham operated rats. Multiple loci of action of circulating pyrogens in the central nervous system could provide for multiple effects of pyrogens, determined within the central nervous system, without the need for complex intracerebral circuitry, while if it could be shown that the true action of pyrogens is at only one site then the multiple pyrogen actions would require a more complex intracerebral network of communications. Thus we suggest that a more comprehensive study of all potential primary loci of action of circulating pyrogens would be of great value.

It is interesting that IL-1 can be demonstrated in the human hypothalamus and in various hypothalamic nuclei, as well as in the hippocampus and other brain regions, in the rat (36,37). Prostaglandin also can be released in the hypothalamus by applied IL-1, and in this locus causes fever. If such IL-1 containing neurons can be excited to cause fever in response to infection then either circulating pyrogens could cause the IL-1 release by a secondary cascade after initial action at the OVLT or one could speculate that such release could result from intracerebral infection. If we accept that the febrile component of the acute phase response is of survival value it would make sense to have a mechanism to cause fever in response to infection, e.g., in encephalitis, within the blood brain barrier, and another to deal with infection outside the blood brain barrier. Such an idea is teleological and purely speculative but possible.

A further hypothesis suggested by Milton (38), that the production of the PGE_2 which acts to mediate the fever response may be peripheral and that this PGE acts at the appropriate locus such as the OVLT to cause fever deserves a vigorous follow up. If true it could have profound effects on the range of antipyretic drugs becoming available to name but one possibility.

The authors wishes to thank their colleagues Drs Pittman, Kasting, Naylor, Mathison, Davison, Fyda, Komaromi, Mr. Terrance J. Malkinson and many others whose results have been reported in this review and with whom it has been such a pleasure to work on so many of these projects.

References.

1.   Hellon, R. F. (1967).   Thermal stimulation of hypothalamic neurones in unanaesthetized rabbits. J. Physiol. (Lond). 193. 381-395.

2.   Reaves, T. A, & Heath, J. E. (1975). Interval cooling of temperature by CNS neurones in thermoregulation. Nature (Lond) 257. 688-690.

3.   Mercer, J. B, Jessen, C, & Pierau, Fr-K. (1978). Thermal stimulation of neurons in the rostral brain stem of conscious goats. J. Thermal Biol. 3. 5-10.

4.   Barker, J. L, & Carpenter, D. O. (1970). Thermosensivity of neurons in the sensorimotor cortex of the cat. Science. 169. 597-598.

5.   Sokoloff, L, Reivich, M, Kennedy, C, Des Rosiers, M. H, Patlak, C. S, Pettigrew. K. D, Sakurada, O, & Shinohara, M. The [^{14}C]deoxyglucose method for the local cerebral glucose utilization: theory, procedure, and normal values in the conscious and anesthetized albino rat. J. Neurochem. 28. 897-916.

6.   Hand, P. J. (1980). The 2-deoxyglucose method. In "Methods in contemporary neuroanatomy: The tracing of central pathways." Ed. L. Heimer & L. Robarts. Plenum Press, New York, U. S. A. pp 511-538.

7.   Cooper, K. E, Hickie, J, Malkinson, T. J, Davison, J, & Veale, W. L. (1989). ^{14}C labelled deoxyglucose uptake into rat septal regions during PGE$_1$ induced hyperthermia.Proc. XXXI International Congress of Physiological Sciences, Thermal Physiology Satellite Symposium, Tromsø, Norway, pg.8.

8.   Morimoto, A, Ono, T, Watanabe,, T, & Murakami, N. (1986). Activation of brain regions of rats during fever. Brain Res. 381. 100-105.

9.   Dragunow, M, & Faull, R. (1989). The use of c-fos as a metabolic marker in neuronal pathway tracing. J. Neuroscience Methods. 29. 261-265.

10. Ogawa, S, Menon, R. S, Tank, D. W, Kim, S. G, Merkle, H, Ellermann, J. M, & Ugurbil, K. Functional brain maping by blood oxygenation level-dependant contrast magnetic resonance imaging. Biophys. J. 64. 803-812.

11. Thornton, S, N, Sirinathsinghji, D. J, & Delaney, C. E. (1987). The effects of reversible colchicine-induced lesion of the anterior ventral region of the third cerebral ventricle in rats. Brain Res. 437. 339-344.

12. Quan, N, & Blatteis, C. M. (1989). Intrapreoptically microdialyzed and microinjected norepinephrine evokes different thermal responses. Am. J. Physiol. (Regulatory Integrative Comp. Physiol.26): R816-R821.

13. Landgraf, R, & Ludwig, M. (1991). Vasopressin release within the supraoptic and paraventricular nuclei of the rat brain: osmotic stimulation via microdialysis. Brain Res. 558. 191-196.

14. Neumann, I, Russell, J. A, & Landgraf, R. (1993). Oxytocin and vasopressin release within the supraoptic and paraventricular nuclei of pregnant, parturient and lactating rats: a micro-dialysis study. Neuroscience. 53. 65-75.

15. Malkinson, T. J, Cooper, K. E, & Veale, W. L. (1988). Physiological changes during thermoregulation and fever in urethan-anesthetized rats. Am. J. Physiol. 255 (Regulatory Integrative Comp. Physiol.24): R73-R81.

16. Fyda, D. M, Cooper, K. E, & Veale, W. L. (1991a). Contribution of brown adipose tissue to central PGE$_1$-evoked hyperthermia in rats. Am. J, Physiol.260 (Regulatory, Integrative Comp. Physiol.29): R59-R66.

17.   Cooper, K. E, Naylor, A. M, & Veale, W. L. (1987). Evidence supporting a role for endogenous vasopressin in fever suppression in the rat. J. Physiol. (Lond). 387. 163-172.

18.   Zeisberger, E, Merker, G, & Blähser, S. (1980). Fever response in the guinea pig before and after parturition and its relationship to the antipyretic reaction of the pregnant sheep. Brain Res. 212. 379-392.

19.   Disturnal, J. E, Veale, W. L, & Pittman, Q. J. (1987). Modulation by AVP of glutamate excitation in the ventral septal area of the rat brain. Can. J. Physiol. Pharm. 65. 30-35.

20.   Federico, P, Malkinson, T. J, Cooper, K. E, Pittman, Q. J, & Veale, W. L. (1992). Vasopressin perfusion within the medial amygdaloid nucleus attenuates prostaglandin fever in the urethane-anesthaesthetized rat. Brain Res. 587. 319-326.

21.   Wilkinson, M. F, & Kasting, N. W. (1990). Centrally acting vasopressin contributes to endotoxin tolerance. Am. J. Physiol. 258. R443-R449.

22.   Cooper, K. E, & Cranston, W. I. (1963). Clearance of radioactive pyrogen from the circulation. J. Physiol. (Lond). 41-42P.

23.   Goelst, K, Mitchell, D, Macphail, A. P, Cooper, K. E, & Laburn, H.(1992) Fever response of the sheep in the peripartum period to gram-negative and gram-positive pyrogens. Pflügers Arch. 420. 259-263.

24.   Lipton, J. M. (1985). Antagonism of IL-1 fever by the neuropeptide α-MSH. In "The Physiologic, Metabolic and Immunologic Actions of Interleukin-1. Alan Liss Inc. pp 121-132.

25.   Monda, M, & Pittman, Q. J. (1993). Cortical spreading depression blocks prostaglandin E_1 and endotoxin fever in rats. Am. J. Physiol. 264. R456-R459.

26.   Bures, J, & Buresova, O. (1969). Inducing cortical spreading depression. In: Methods in Psychobiology, R. D, Myers (Ed), Academic Press, New York, pp. 319-343.

27.   Komaromi, I, Malkinson, T.J, Veale, W. L, Pittman, Q. J, Rosenbaum, G, & Cooper K. E. (1993). The effect of cortical spreading depression on fever in rats. Proc. XXXII Congress of Physiological Sciences, Glasgow, Scotland. (In press).

28.   Blatteis,C. M, Bealer, S. L, Hunter, W. S, Llanos-Q. J, Ahokas, R. A, & Mashburn, T. A. (1983). Suppression of fever after lesions of the anteroventral third ventricle in guinea pigs. Brain Res. Bull. 11. 519-526.

29.   Stitt, J. T. (1985). Evidence for the involvement of the organum vasculosum laminae terminalis in the febrile response of rabbits and rats. J. Physiol. (Lond). 368. 501-511.

30.   Cooper, K. E. Cranston, W. I, & Honour, J. A. (1967). Observations on the site and mode of action of pyrogens in the rabbit brain. J. Physiol. (Lond). 191. 325-338.

31.   Jackson, D. I. (1967). A hypothalamic region responsive to localized injections of pyrogens. J. Neurophysiol. 30. 586-602.

32.   Repin, I, S, & Kratskin, I. L. (1967). Hypothalamic mechanisms of fever. Neurosci. Transl. 3. 336-340. (Translated from Fiziol. Zh. SSSR I.M. Sechenova 53 1206-1211.

33.   Villablanca, J, & Myers, R. D. (1965). Fever produced by microinjection of typhoid vaccine into the hypothalamus of cats. Am. J. Physiol. 208. 703-707.

34.   Sirko, S, Bishai, I, & Coceani, F. (1989). Prostaglandin formation in the hypothalamus in vivo: effect of pyrogens. Am. J. Physiol. 256. (Regulatory Integrative Comp. Physiol. 25): R616-624.

35.   Takahashi, Y, Smith, P, Wilkinson, M, J, & Cooper. K. E. (1993) Pyrogen access into the brain. Proc. Soc. Neurosci. (In press).

36.   Breder, C.D, Dinarello, C.A, & Saper, C.B. (1988). Interleukin-1 immunoreactive innervation of the human hypothalamus. Science, 240. 321-324.

37.   Lechan, R. M, Toni, R, Clark, B. D, Cannon, J. G, Shaw, A. R, Dinarello, C. A, & Reichlin, S. (1990). Immunoreactive interleukin-1ß localization in the rat forebrain. Brain Res. 514. 135-140.

38.   Milton, A. S. (1989). Endogenous pyrogen initiates fever by a peripheral and not a central action. Proc. XXXI International Congress of Physiological Sciences, Satellite thermal physiology symposium, Tromsø, Norway. Pg.47.

Author Index

Adler, M.W.	115	Jacobs, I.	351
Andrews, J.F.	303	Jadeszko, M.	139
Barriga-Briceno, J.A.	81	Jakobson, M.E.	303
Beneke, G.	297	Jessen, C.	145
Bird, J.A.	309	Kandori, Y.	127
Blatteis, C.M.	41, 65,	Keil, R.	321
	71, 75,	Kenney. W.L.	151
	81	Kizaki, T.	241
Bock, M.	11	Kluger, M.J.	11, 17
Boulant, J.A.	93	Kolka, M.A.	165
Brinck, H.	201	Kosaka, M.	35, 87
Bristow, G.K.	213	Kuroshima, A.	327
Brosh, A.	297	Laburn, H.P.	229
Cade, J.R.	261	Langer, T.	145
Cannon, B.	345	Lee, J.M.	35
Carlisle, H.J.	247	Lewko, J.	139
Cervos-Navarro, J.	267	Lin, M.T.	103
Challoner, D.	303	Liu, H.J.	103
Chen, F.	219	Lomax, M.A.	309
Clark, M.G.	315	Mariak, Z.	141
Clarke, L.	309	Matsumoto, T.	35, 87
Colquhoun, E.Q.	315	Matthias, A.	315
Cooper, K.E.	17, 357	McBennett, S.	303
Darby, C.J.	309	McClellan, J.L.	11, 17
Dascombe, M.J.	47	McConaghy, F.F.	189
Davidson, J.	29	Meiri, U.	291
Dey, P.K.	267	Menon, V.	65
Dora, K.A.	315	Milton, N.G.N.	01, 59
Ducharme, M.B.	223	Mitchell, D.	229
Dudek, H.	139	Morimoto, T.	279
Eldershaw, T.P.D.	315	Morita, T.	285
Fennell, S.	297	Murakami, M.	285
Fregly, M.J.	261	Murzenok, P.P.	23
Gate, J.J,	309	Nagasaka, T.	207
Geller, E.B.	115	Natsume, K.	183
Ghosh (nee Biswas), S.	121	Nedergaard, J.	345
Giesbrecht, G.G.	213	Nielsen, B.	145, 195
Goelst, K.	229	Nilsson, H.	219
Gonzalez, R.R.	165	Nishida, M.	127
Gourine V.N.	23	Nose, H.	279
Griefahn, B.	159	Nyberg, F.	267
Gurevitch, V.S.	133	Ogawa, T.	127, 183
Haddad, W.	291	Ohira, Y.	241
Hales, J.R.S.	189, 195	Ohnishi, N.	127, 183
Hall, J.L.	315	Ohno, H.	241
Handler, C.M.	115	Ohwatari, N.	35, 87
Heath, M.E.	171	Oladehin, A.	81
Hillhouse, E.W.	59	Ookawara, T.	241
Hodgson, D.R.	189	Pehl, U.	109
Holmer, I.	219	Pertwee, R.G.	177
Horowitz, M.	291	Piliero, T.C.	115
Howell, R.B.	71	Pleschka, K.	87
Hunter, W.S.	75	Poddar, M.K.	121
Imai, K.	127, 183	Puerta, M.	253
Imamura, R.	183	Rattigan, S.	315
Ishizuka, A.	127	Ricklefs, R.E.	273

Riedel, W.	321	Tikuisis, P.	223
Romanovksy, A.A.	41	Tokura, H.	285
Roth, J.	11, 17	Trayhurn, P.	333
Rothberg, S.	247	Tsuchiya, K.	35
Rotondo, D.	53	Umeyama, T.	127
Rowland, N.E.	261	Ungar, A.L.	71
Sakurada, S.	207	Vallerand, A.L.	351
Sato, Y.	241	Veale, W.L.	357
Schmid, H.A.	109	Visser, G.H.	273
Schroeder, M.	213	Wakatsuki, T.	241
Schwarzenau, P.	159	Werner, J.	201
Sehic, E.	65, 75	Westman, J.	267
Sharma, H.S.	267	Wright, D.	297
Shido, O.	207	Yahata, T.	327
Shimazu, M.	35, 87	Yaitchnikov, I.K.	133
Shochina, M.	291	Yamamoto, M.	241
Sidara, J.Y.	47	Yamashita, H.	241
Simon, E.	109, 321	Yamashita, Y.	183
Stephenson, L.A.	165	Yamauchi, M.	87
Stevenson, L.A.	177	Yanase, M.	195
Stock, M.J.	247	Yang, G.J.	35
Sugenoya, J.	127, 183	Ye, J-M	315
Swanton, E.	59	Young, B.	297
Symonds, M.E.	309	Yutani, M.	285
Székely, M.	65	Zeisberger, E.	11, 17
Takamata, A.	279	Zimmer, C.	267
Thomas, J.R.	171		

KEYWORD INDEX

acclimation 159, 303

acoustically evoked brainstem potentials 145

ACTH, see adrenocorticotrophin

active vasodilation 151

acute cold exposure 87

acute phase 53

acute phase reactants 53

acute-phase reaction 81

adenyl cyclase 345

ADH, see antidiuretic hormone

adrenaline 345

adrenergic transduction responsiveness 291

adrenoceptors β 247

adrenoceptors β1 241,345

adrenoceptors β2 345

adrenoceptors β3 309, 345

adrenocorticotrophin (ACTH) 01

adrenocorticotrophin release 59

aging 291

aldosterone 321

aluminium 219

ambient temperature 121

γ-aminobutyric acid (GABA) 121

amygdala 133, 357

analysis of body temperature 87

anandamide 177

angiotensin II 261

angiotensin II receptors 261

anterior hypothalamus 93

anteroventral third ventricle (AV3V) 75

antibodies 01

anticholinergic 165

antidiuretic hormone (ADH) 01

antipyresis 17, 65

antipyretic 01, 29

arginine vasopressin (AVP) 01, 357

arterial occlusion 223

arteriovenous anastoma 171

atropine 121, 165

attenuation 23

axonal transport 357

AVP, see arginine vasopressin

basic fibroblast growth factor (bFGF) 241

BAT, see brown adipose tissue

bettong hindlimb 315

bicuculline 121

birth 229

blood flow 201, 223

blood temperature 201

blood vessels 291

blood-brain barrier 75, 267

body temperature 17, 121, 229, 351

body temperature: neural control 93

brain 53

brain damage 127

brain lesions 75

brain oedema 267

brain regions, activation by LPS 81

brain temperature 127, 133, 139, 145, 189

bright light 285

BRL 35135 247

bromocriptine 327

brown adipose tissue 115, 229, 241, 247, 253, 303, 309, 327, 333, 345, 357

c-fos 81, 357

CA-4 267

cannabinoid 177

carbohydrates 351

cardiac strain 159

cardiopulmonary baroreflex 279

carotid blood temperature 139

catecholamines 351

cattle 297

central venous pressure 279

cerebral blood flow 127

chick 273

chicken hindlimb 315

circadian rhythm 93, 207, 285

circumventricular organ 75, 81

colchicine 357

cold environment 201

cold exposure 327, 351

cold sensitive neurone 93

cold tolerance 87

cold-adaptation 333

cold-induced thermogenesis 253

colonic temperature 17, 261

combined stressors 71

competing regulatory systems 71

computed tomography (SPECT) 127

contact cold 219

convective heat transfer 223

core temperature 65, 71, 75, 159, 183, 213, 285, 351

cortical spreading depression 357

corticosterone 321

corticotrophin-releasing-factor-41 (CRF-41) 01, 59, 357

cortisol 01

CP 55,940 177

CRF, see corticotrophin-releasing-factor

cross-tolerance 177

curarization 183

cutaneous blood flow 151

cyclic AMP 345

cyclooxygenase inhibitors 65

cyclophosphamide 23

cytokines 11, 17, 29

delta-9-tetrahydrocannabinol 177

2-deoxyglucose 357

development of homeothermy 273

dexamethasone 29, 53

diabetes 333

diet-induced thermogenesis 333

dim light 285

direct calorimeter 207

dopamine 327

DPDPE 115

duck 273

dynorphin A1-17 115

ear skin temperature 41, 71

electroencephalogram 133

endogenous antipyresis 357

endogenous cannabinoid 177

β–endorphin 01

endothelial cells 53

endotoxin 17, 53, 357

endotoxin tolerance 357

energy substrates 351

environment, cold 309

epinephrine, see adrenaline

equine 189

equivalent climates 159

escherichia coli 17

evaporative heat loss 165

exercise 145, 189, 198

exposure, cold 219, 241

exposure, heat 219

face fanning 145, 189

fasting 333

febrile patterns 65

febrile response 01, 17, 53

feed quality 297

feeding 207

feotus 229

fever 01, 05, 11, 17, 23, 29, 35, 47, 53, 59, 71, 75, 81

fever index 17

fever, neuronal circuitry 357

fever, phases of 41

finger 219, 223

food intake 253, 351

forced air warming 213

forskolin 345

functional vascular shunts 315

GABA, see g-aminobutyric acid

gestation 229

glial fibrilliary acidic protein (GFAP) 267

gliosis 65

glucagon 327

glucocorticoid antagonist 29

glucose 93

glutamate 109, 121

guinea pig 17, 41, 75

haemoconcentration 351

haloperidol 327

hamster, syrian 87

heart rate 297

heat acclimation 207, 291

heat balance 351

heat exchange (countercurrent) 201

heat exchange 121

heat exposure 207

heat gain 223

heat load 297

heat loss 223, 351

heat loss, evaporative, 207

heat loss index 41

heat loss, non-evaporative 207

heat production 207, 229, 297, 351

heat shock protein (HSP) 35

heat stress 189, 195, 267, 291

heat transfer 201

hibernation 87

hippocampus 267, 257

hot environment 198, 201

"hot pipes" 315, 333

HPA-axis, see hypothalamo-pituitary adrenocortical axis

human thermoregulation 151,139, 145, 159, 165, 183, 207, 219, 285

5-hydroxytryptamine receptor agonists 103

5-hydroxytryptamine (5-HT) 47, 103

hyperthermia 41, 145, 267, 279

"hyperthermia-induced hypothermia" 41

hypervolemia 279

hyperthyroidism 321

hypohydration 291

hypothalamic temperature 41, 189

hypothalamic tissue slice; in vitro 93

hypothalamo-pituitary-adrenocortical axis (HPA axis) 01

hypothalamus 01,11, 93, 109, 121, 133, 321, 357

hypothermia 47, 87, 177, 213, 261

hypovolemia 279

hypothyroidism 321

ibotenic acid 357

IL-1, see interleukin-1

immune response 53

immunocompetent cells 23

immunocytochemistry 81

immunomodulators 23

immunostimulation 53

implantation trauma 65

indomethacin 65

insensible heat loss 165

insulin 241, 333, 351

interferons 01, 59

interleukin-1 (IL-1) 01, 11, 23, 29, 35, 53, 59, 357

interleukin-6 (IL-6) 11, 17

intracellular recording 93

intraperitoneal heating 41

irradiation of central motor command 183

isolated brown adipocytes 253

isoproterenol 261

kainic acid 357

ketaniserine 103

ketoprofen 53

lactation 333

lamb 229

laser Doppler blood flowmetry 171

laterality 127

lesion 75, 357

leukotriene B$_4$ 53

lipids 351

lipocortin-1 29

lipopolysaccharide (LPS) 01, 11, 17, 53, 59, 75

losartan potassium 261

LPS, see lipopolysaccharide

macrophage 35, 53

malaria 47

meclofenamate 65

α-melanocyte stimulating hormone (α-MSH) 01

melatonin, 285

metabolic heat production 223

mice (genetically obese) 303

mice (lean) 303

microdialysis 357

microdialysis, intracerebral 65

microvascular blood flow 171

mini-osmotic pump 17

minimal thermal conductance 273

monocytes 53

mouse 177

MRI 357

α-MSH, see α-melanocyte stimulating hormone

muramyl dipeptide 59

muscimol 121

negative feedback 29

neonate 229, 309, 345

neuronal thermosensitivity 93

neurones, temperature sensitive 109

neuropeptide-Y (NPY) 171

newborn 229

nonshivering thermogenesis 229, 315, 333

noradrenaline 151, 241, 253, 315, 327, 333, 345

noradrenaline turnover rate 253

norepinephrine, see noradrenaline

normothermia 139

normovolemia 279

NPY, see neuropeptide Y

obesity 333

oestradiol 93,253

oestrogen 93

oligonucleotides 333

ontogeny 229

opioid agonists 115

opioid antagonists 121

opioid peptides 115

opioid receptors δ 133

opioid receptors κ 133

opioid receptors μ 133

opioids 115

organum vasculosum laminae terminalis (OVLT) 75, 357

osmotic pressure; osmosensitive 93

oxygen consumption 115, 121, 327

p-chlorophenylalanine (p-CPA) 267

parturition 229

peak metabolic rate 273

peripheral actions 01

peripheral blood flow 171

PET scan 357

PGE$_2$, see prostaglandin-E$_2$

$PGF_{2\alpha}$, see prostaglandin-$F_{2\alpha}$

physostigmine 121

pika rabbit 35

piperazine 103

pituitary 59

PL-017 115

placenta 229

plasma volume 351

plasmodium berghei 47

polyadenylic:polyuridylic acid 59

Poly I:C, see polyinosinic: polycytidylic acid

polyinosinic: polycytidylic acid (poly I:C) 01, 29, 53, 59

polytidylic: polyguanuylic acid 59

portal blood flow 291

postsynaptic potentials; IPSP; EPSP 93

posture 127

potentiation 23

precocial 273

preference 247

pregnancy 229

pregnant animals 17

preoptic area 65

preoptic region 109

preoptic region; PO/AH 93

prepotential; pacemaker potential 93

pro-opiomelanocortin hormones 01, 59

pro-pyretic 01, 59

progesterone 253

prolactin 327

propanolol 103

prostaglandin E_2 (PGE$_2$)01, 11, 29, 53, 59, 65, 357

prostaglandin-$F_{2\alpha}$ (PGF$_{2\alpha}$)01, 53

prostaglandins 53, 229

protein 351

protein turnover 333

pyrogens 01, 23, 59

rabbit 01

radiant temperatures 159

radiotelemetry 17, 229

rat 01, 59, 109, 171, 247

rat hindlimb 315

rat, wistar 87

receptor antagonist 01

receptors 151

rectal temperature 189, 297

respiration rate 297

resting muscle thermogenesis 315

restraint stress 81

RU 38486 29

selective brain cooling 145, 189

selective opioid agonists 121

sensible heat loss 165

serotonin, see 5-hydroxytryptamine

sheep 01, 229

shivering 213, 351

shorebird 273

single photon emission 127

skin blood flow 151, 165, 198

skin temperature 213, 219

skin wettedness 165

skinfold thickness 223

sodium pump 333

sodium-potassium pump; Na-K pump 93

solar radiation 297

spinal cord 109

splanchnic circulation 291

staphylococcal wall pyrogen 357

stellate ganglion block 127

steroid 29

stress, thermal 279

subcutaneous fat 223

subfornical organ 357

substance P 109

substrate cycles 333

suprachiasmatic nucleus 93

sweat gland 165

sweat rates 159

sweating 165, 279

sweating regulation during exercise 183

sympathetic activity 333

sympathetic nervous system 151

sympathetic stimulation 253

synaptic blockade 93

synpatic network 93

tail blood flow 171

tail blood volume 171

tail skin temperature 261

temperature insensitive neurone 93

temperature regulation 109

testosterone 93

THC, see tetrahydrocannabinol

thermal environments 133

thermal stress 171

thermally-induced vasomotor response 291

thermode 93

thermogenesis 229, 333, 351

thermogenesis, cold induced 253

thermogenesis, heat induced 253

thermogenic activity 327

thermogenin 345

thermoregulatory capacity 303

thermosensitive neurones 357

threshold dissociation 41

thromboxane A_2 (TXA_2) 53

thymic cells 59

thymic corticotrophin releasing factor 59

thyroid hormones 321

thyrotropin-release hormone (TRH) 321

thyroxine 309, 345

tissue blocks 327

TNF, see tumour necrosis factor

tolerance 17, 177

total vascular conductance 279

tumor necrosis factor (TNF) 01, 11, 17

tympanic temperature 139, 127

uncoupling protein 303, 333

uncoupling protein gene 333

uncoupling protein mRNA 333

urethane 267

urethane anaesthetized rat 357

vascular control of thermogenesis 315

vascular thermogenesis 315

vasoactive agents 171

vasoconstriction 151

vasoconstrictors 315

vasodilators 315

vasopressin (AVP) 01, 321

venous occlusion plethysmography 171

warm sensitive neurone 93

water immersion 223

wet bulb globe temperature 159

white adipose tissue 333

widedead band temperature regulation 41

WIN 55,212-2

zonal nutrient delivery 315

zonal nutrient efflux 315

B I R K H Ä U S E R

LIFE SCIENCES

Experientia Supplementum

Comparative Molecular Neurobiology

Edited by
Y. Pichon, *Université de Rennes, France*

1993. 434 pages. Hardcover. ISBN 3-7643-2785-5 (EXS 63)

Most comparative studies of the physiological and pharmacological properties of the receptors and ionic channels of various animal species have so far stressed the differences. More recent studies based on the knowledge of the primary structure of these proteins as obtained using molecular cloning techniques emphasize the common features and have led to the concept of superfamilies. These superfamilies are believed to be derived from common ancestors through evolution. To understand how this happened, it is necessary to compare the sequences and the properties of the receptors in species sufficiently distant in the evolutionary tree. Until recently, this kind of information was lacking. In the present volume, specialists in the field of comparative molecular neurobiology, most of them working on both vertebrate and invertebrate species, report their recent findings concerning the three most important superfamilies: the Ligand-Gated Ion Channels superfamily (n-ACh, $GABA_A$, glycine), the Second-Messenger Linked receptor superfamily (m-ACh, catecholamines, peptides) and the Voltage-Gated Ion Channels (Na^+, K^+ and Ca^{2+}) superfamily.

Please order through your bookseller or directly from:
Birkhäuser Verlag AG, P.O. Box 133,
CH-4010 Basel / Switzerland (Fax ++41 / 61 / 721 7950)
Orders from the USA or Canada should be sent to:
Birkhäuser Boston
44 Hartz Way, Secaucus, NJ 07096-2491 / USA
Call Toll-Free 1-800-777-4643

Birkhäuser

Birkhäuser Verlag AG
Basel · Boston · Berlin

Prices are subject to change without notice. 9/93

BIRKHÄUSER
LIFE SCIENCES

Adenine Nucleotides in Cellular Energy Transfer and Signal Transduction

Edited by

S. Papa, *Bari, Italy*
A. Azzi, *Berne, Switzerland*
J.M. Tager, *Amsterdam, The Netherlands*

1992. 488 pages. Hardcover. ISBN 3-7643-2673-5 (MCBU)

Control of cell growth, differentiation and oncogenesis, and cellular energy metabolism and its adaptation to physiopathological states, in particular ageing, are major topical areas in biomedical research.

Adenine nucleotides play a major role in cellular metabolism and functions serving as high-potential phosphate transfer compounds in energy metabolism and as substrates and co-factors for proteins involved in signal transduction.

During the last few years, definite advancement has been made in elucidating the molecular and genetic aspects of *ATP synthase*. Non-invasive NMR technologies have been developed to monitor in vivo the energy level of tissues, based on determination of the concentrations of adenine nucleotides, phosphate and phosphate esters.

Enormous progress has been made in defining the role played by *protein phosphorylation in cellular signal transduction* and control of cell growth, differentiation and oncogenesis. A further topic of growing interest concerns the discovery of the *ATP- binding* cassette *(ABC)* superfamily of transport proteins which includes systems of primary importance in medicine such as the multi-drug resistance P glycoprotein, the cystic fibrosis transmembrane conductance regulator (CFTR) and the 70 kD peroxisomal membrane protein.

These topics are dealt with in the present book by leading experts. It should be of immediate interest to investigators in basic medical sciences as well as to clinicians.

Please order through your bookseller or directly from:
Birkhäuser Verlag AG, P.O. Box 133,
CH-4010 Basel / Switzerland (Fax ++41 / 61 721 79 50)
Orders from the USA or Canada should be sent to:
Birkhäuser Boston
44 Hartz Way, Secaucus, NJ 07096-2491 / USA
Call Toll-Free 1-800-777-4643

Birkhäuser

Birkhäuser Verlag AG
Basel · Boston · Berlin

N